ENVIRONMENTAL PROBLEMS AND CHALLENGES
Recent Issues and Opportunities

A FESTSCHRIFT FOR
PROFESSOR RAJ KUMAR SEN

ENVIRONMENTAL PROBLEMS AND CHALLENGES

Recent Issues and Opportunities

A FESTSCHRIFT FOR PROFESSOR RAJ KUMAR SEN

Edited by

SOMNATH HAZRA

REGAL PUBLICATIONS

New Delhi - 110 027

ENVIRONMENTAL PROBLEMS AND CHALLENGES
Recent Issues and Opportunities

A Festschrift for
Professor Raj Kumar Sen

ISBN 978-81-8484-250-0

Typeset by
RAHUL COMPOSERS
New Highway Apartments, Lakshmi Niwas
760, Pocket-D, Lok Nayak Puram, New Delhi - 110 041

Printed in India at
MAYUR ENTERPRISES
WZ Plot No. 3, Gujjar Market, Tihar Village, New Delhi - 110 018

Published by
REGAL PUBLICATIONS
F-159, Rajouri Garden, New Delhi - 110 027
Phone : 45546396, 25435369
E-mail : regalbookspub@yahoo.com, regaldeepbooks@yahoo.com

Contents

Message

Physical environment is the most important feature of the human habitat in this world. Till recently the western world treated them as free goods and exploited it without knowing its crucial importance in human life. Though the oriental wisdom treated environment in a holistic manner as an inseparable part of life, this concept was not present in the western thinking which always treated human beings as economic agents concentrating their activities in the areas like consumption, production, exchange, etc. Naturally we are now more interested in the discussion of environmental issues to expand our frontiers of knowledge as far as possible. So far as India is concerned, this discussion is more important as India is not only the habitat of one-sixth of humanity, it is also regarded as one of the 17 mega bio-diversity countries of the world. This present book is one such humble attempt to enrich our knowledge in this very important area in the study of which I am engaged since the 1990s.

My student and scholar Dr. Somnath Hazra has done a yeoman service by preparing and editing this volume highlighting the challenges and opportunities of the environmental issues with emphasis on India and its constituent states. I am personally thankful to Somnath as he has dedicated this very important volume in my honour. I also consider myself fortunate as this is the third festschrift prepared for me and another one is forthcoming. I wish Somnath and all others every success and peace in life.

In brief, this book consists of as many as 32 research articles authored by distinguished scholars from all parts of India and even abroad from countries like Australia. The articles are distributed over areas like water, air, land, forest, bio-diversity, energy, climate change, environmental pollution and waste management besides a few articles on general issues but relevant for India. The readers will obtain from this book a near exhaustive view about the theme very correctly described in its well chosen title. And I also hope that this book will not only be a light-bearing one but also become fruit-bearing too.

Kolkata

DR. RAJ KUMAR SEN
Former Professor and Head, Dept. of Economics
Former Director, Centre for Studies on Environment and Sustainable Development (CSESD) and
Director, Tagore-Ikeda International Centre for Global Citizenship (TICGC)
Rabindra Bharati University, Kolkata and
Chairman, International Institute of Development Studies (IIDS)

About Professor RAJ KUMAR SEN and this Festschrift

I had the privilege of being both PG student and PhD research scholar of Professor Raj Kumar Sen during the periods 1999-2001 and 2004-09 and planned the preparation of this book when in 2008 the first festschrift for Professor Sen was published being authored by his research scholars only and also being edited by one of them. As my thesis is related to environmental economics and as the current research interest of Professor Sen turned around this fast growing area of economics among others and in my professional and teaching career, also I am connected with this subject, I have decided to concentrate on it as the theme of this book. I am thankful to the authors of the 32 contributed articles from not only all parts of India but also from outside the country who responded to my appeal inspite of their busy schedule as a mark of their admiration and respect for Professor Sen. I am also grateful to my teacher for his kind permission to go ahead with this project and his thoughtful message for it.

Professor Sen has successfully travelled to different frontiers of economics during his long career of teaching and research of nearly five decades since 1964. His first book on environmental economics (*Economic Development and Environment: A Case Study of India*) was published by Oxford in 1992 and as it was a grand success, its paperback edition appeared in 1995. His other books on this subject are:

Sustainable Economic Development and Environment: India and Other Developing Countries (1996), *Environment and Sustainable Agricultural Development* (1996), *Economic Globalisation: Social Conflicts, Labour and Environmental Issues* (2004) {published by Edward Elgar jointly with Clem Tisdell}, *Environment and Sustainable Development in India* (2010), *Environment Ecology and Sustainable Economic Development* (2011), *Environment and Sustainable Development: Some Emerging Issues* (2011), *Forest Management and Sustainable Development* (2012), and *Environmental Ideas of Tagore* (forthcoming). He has also authored a large number of articles on this subject, presented research papers in many international and national seminars and conferences, supervised four Ph.D. theses and conducted at least two UGC major research projects. He was also the Director of Centre for Studies on Environment and Sustainable Development (CSESD), Rabindra Bharati University, Kolkata and organised one International and three national seminars and published at least 5 books from it. Currently, he is the Director, Tagore Ikeda International Centre for Global Citizenship (TIICGC), Rabindra Bharati University. He is also the Chairman, International Institute of Development Studies (IIDS). Considering his interest in this subject and widespread contribution to it, I sincerely hope that he will be quite happy and pleased for this festschrift. I am also sure that appreciating the valuable contributions of the subject experts to this volume, there will be adequate demand and publicity of it among the researchers, advanced students and interested public in general and teachers and policy-makers in particular.

Kolkata

SOMNATH HAZRA

Preface

The environmental issues in India are now becoming serious day-by-day and turning into a mess. The industrialization and the urbanization are both growing rapidly in an uncontrolled manner with lack of proper knowledge of environmental education. As a consequence the massive deforestation is going on. The industrial boom has created the waste disposal problem. We have 7516 km of long coastline, but due to lack of legislation this long coastline is also under the attack. We are losing agricultural output day-by-day due to soil degradation, improper land management, lack of irrigation facility and improper water management. Increasing competition for water among various sectors, including agriculture, industry, domestic use, drinking, energy generation and others, is causing this valuable natural resource to dry up. Increasing water pollution is also leading to the destruction of the wildlife habitat living in water bodies. The eco-system is also adversely affected by the negative effects of the environment.

As a developing country, India has some strategic plans regarding economic development, reducing poverty and to stop environmental degradation. The massive population growth leads to slowdown of per capita income due to resource constraint, and degrading environment due to habitation, industrialization and expansion of agricultural land through deforestation. In this book we are trying to find out the real environmental issues in India and to establish the right path of sustainable development for India. The 1992 United Nations Conference on environment and development held in Rio-de-Janeiro for the discussion of global conflict of development and environment protection with the new concept of sustainable development. Sustainable development refers to creating a style of economic development which is sustainable within the context of plants ecosystem and human ecology; it is an approach that will permit continuing improvements in the present quality of life at a lower density of resource use thereby leaving behind the future generations undiminished or even enhanced stock of natural resources and other assets. Sustainable development was neglected and even degraded in India. For sustainable development of India it is useful to distinguish between necessary and sufficient conditions for it. For development to be sustainable it is necessary to enhance the capital stock on which the development depends, while capital stock includes human made capital and natural capital. In developing economy the amount of human made capital is not sufficient for development but the natural capital occupies a very strong position. Poverty is the main reason for the degradation of sustainable development in India. Poor people of India like other developing countries are mainly dependent on natural resources and primary production because of the lack of resources and economic opportunities in other sectors of the poor economies and these are important determinants of their inability to innovate sufficiently to achieve higher long-term growth rates. Controlling environmental degradation in India is difficult because of the poverty-environment trap. If the poverty and environmental degradation exist at a high level then it can be said that the environmental degradation is the main cause of poverty. Poverty and environmental degradation are positively related.

In line with this thought the present book is conceived. The articles are distributed into three parts; Part one is on general issues on environment (Tisdell, Ghosh), Part two is on natural resources consisting of areas of Water, Land, Forest (Sahu and Panda, Chatterjee and Maiti, Nampoothiri and Hrideek, Sadasivam Jhunjhunwala, Basavaraddi *et al.*, Jana, Chowdhury, Puttaswamaiah, Patnaik, Mitra and Nath, Nanjundaiah, Mitra and Singh, Basu, Ojha, Sarangi) and the Part three is on recent environmental problems in India like Climate Change, Environmental Pollution, Waste Management and Energy (Sirajuddin *et al.*, Naveen *et al.*, Anathanag *et al.*, Kanan and Varadarajan, Pandey and Mahapatra, Goswami, Pan, Umesh Babu and Puttaiah, Karmakar, Varadarajan and Sadasivam, Bhowmik, Mahanty).

It is recognized that poverty alleviation, pollution policy and environment protection were mutually reinforcing and supporting economic growth. There was recognition that environmental risk factors were major source of health problems in India. Therefore, environmental degradation produced unsustainable development and alleviation of poverty became a non-separable element of sustainable development. Poverty alleviation could no longer be kept for trickle-down effect of growth process.

From the environmental point of view India is also a unique country as it belongs to one of the 17 mega bio-diversity countries of the world. In India, there exists a vicious circle (poverty-environment degradation—more poverty) and this linkage between poverty and environment is a greater impediment for sustained development than commonly understood. As a developing country the major targets of all public policies of India since her independence is the reduction of poverty as early as possible through various ways, for example, trickle-down benefits to the poor, state intervention for the building of capability of the poor and the weak and their empowerment, etc. The scope of the environmental problem is unimaginable and now we have to think about environment protection parallely with employment and income generation.

Kolkata SOMNATH HAZRA

List of Contributors

A. Ibotombi Singh, Associate Professor of Economics, Dera Natung Govt. College, Itanagar (Arunachal Pradesh).

A. Kannan, Assistant Professor, Department of Environmental Economics, School of Economics, Madurai Kamaraj University, Madurai (T.N.).

A. Mitra, Professor of Economics, Rajiv Gandhi University, Itanagar (Arunachal Pradesh).

A.K. Maiti, Professor and Head, Department of Agricultural Economics-B.C.K.V., Nadia (W.B.).

Ananthanag B., Department of Environmental Science, Gulbarga University, Gulbarga (Karnataka).

Asim K. Karmakar, Assistant Professor, Jadavpur University, Kolkata (W.B.).

Atin Tyagi, Biodiversity and Climate Change Division, Indian Council of Forestry Research & Education, Dehradun (Uttarakhand).

B. Chatterjee, Assistant Professor; International Business and Economics, A.S.B.M., Bhubaneswar (Orissa).

B.E. Basavarajappa, Department of ESTSC/Chemistry, BIET, Davanagere (Karnataka).

B.N. Ghosh, Director, Kerala Institute of Making the Best & Visiting Professor, Institute of Management and Technology (Kerala).

Basavaraj, S., Department of Environmental Science, Gulbarga University, Gulbarga (Karnataka).

Bharat Jhunjhunwala, Journalist, Delhi.

Bimal K. Mohanty, Former Professor of Economics, Gangadhar Meher (Autonomous) College, Sambalpur (Orissa).

C. Nanjundaiah, Associate Professor, Centre for Economic Studies and Policy (CESP), Institute for Social and Economic Change (ISEC), Bangalore (Karnataka).

Chandra Dhwaj Panda, Gandhi Institute of Engineering and Technology (GIET) (Orissa).

Clement A. Tisdell, Professor Emeritus, School of Economics, The University of Queensland, Brisbane (Australia).

Debesh Bhowmik, Research Fellow, IIDS (Kolkata), Kolkata (W.B.).

Dhulasi Birundha Varadharajan, Professor and Head, Department of Environmental Economics, School of Economics, Madurai Kamaraj University, Madurai (T.N.).

E.T. Puttaiah, Vice-Chancellor, Gulbarga University, Gulbarga (Karnataka).

Hina Kousar, Department of Environmental Management, School of Environmental Sciences, Bharathidasan University, Tiruchirappalli (T.N.).

Jyotish Prakash Basu, Associate Professor & Head, Department of Economics, West Bengal State University, Kolkata (W.B.).

K. Sadasivam, Assistant Professor, Department of Environmental Economics, Madurai Kamaraj University, Madurai (T.N.).

K.G. Girish, Department of Environmental Management, School of Environmental Sciences, Bharathidasan University, Tiruchirappalli (T.N.).

K.U.K. Nampoothiri, Director, Biju Patnaik Medicinal Plants Garden and Research Centre, M.S. Swaminathan Research Foundation, (Orissa).

Kaju Nath, Research Scholar, Department of Economics, Rajiv Gandhi University, Rono Hills, Itanagar (Arunachal Pradesh).

M. Balasubramanian, Research Scholar, Department of Environmental Economics, School of Economics, Madurai Kamaraj University, Madurai (T.N.).

M. Ravichandran, Department of Environmental Management, School of Environmental Sciences, Bharathidasan University, Tiruchirappalli (T.N.).

M.S. Umesh Babu, Doctoral Fellow, Department of Environmental Sciences, Kuvempu University (Karnataka).

Nabaghana Ojha, Consultant, DFID (Orissa).

Naveen, D., Department of Environmental Science, Gulbarga University, Gulbarga (Karnataka).

Nirmal Chandra Sahu, Department of Economics, Berhampur University (Orissa).

Parth Sarathi Mahapatra, IIMT, Bhubaneswar (Orissa).

Rajiv Pandey, Scientist, Biodiversity and Climate Change Division, Indian Council of Forestry Research & Education, Dehradun (Uttarakhand).

S. Puttaswamaiah, Lecturer, Department of Economics, Bangalore University, Bangalore (Karnataka).

S.B. Basavaraddi, Department of Environmental Science, Kuvempu University (Karnataka).

S.J. Veeresh, Department of Environmental Management, School of Environmental Sciences, Bharathidasan University, Tiruchirappalli (T.N.).

S.R. Pravitha, Department of Environmental Management, School of Environmental Sciences, Bharathidasan University, Tiruchirappalli (T.N.).

Sanjoy Patnaik, State Director, Rural Development Institute (RDI), Orissa Chapter (Orissa).

Sebak Jana, HoD, Economics and Rural Development, Vidyasagar University, Medinipur (W.B.).

Sirajuddin M. Horaginamani, Department of Environmental Management, School of Environmental Sciences, Bharathidasan University, Tiruchirappalli (T.N.).

Srijit Choudhury, Assistant Professor of Economics, Vidyasagar College for Women, Kolkata (W.B.).

Subrata Goswami, Bhangar Mahavidyalaya, South 24 Parganas (W.B.).

Suvranshu Pan, Assistant Professor in Economics, Kashipur Michael Madhusudan Mahavidyalaya, Purulia (W.B.).

T.K. Hrideek, Scientist, Biju Patnaik Medicinal Plants Garden and Research Centre, M.S. Swaminathan Research Foundation (Orissa).

Tapas Kumar Sarangi, ICSSR Doctoral Fellow, P.G. Department of Economics, Sambalpur University (Orissa).

Part I
General Issues

Chapter 1

Sustainable Development and Intergenerational Equity

Issues Relevant to India and Globally

CLEMENT A. TISDELL

ABSTRACT

As outlined, recurring concerns have surfaced since the 1700s that economic growth may prove to be unsustainable. These concerns have been expressed again and have intensified in recent decades but their foundation differs from that of Malthus. The rapid economic growth of China and India has added to these worries. Recent discussions by economists of the desirability of achieving sustainable economic development have mainly focused on measures to attain intergenerational equity in resource-use and the dominant view is that each succeeding generation should be at least as well-off as its predecessor. While this is said to be an implication of Rawls' principle of justice, this dominant rule does not fully reflect Rawls' principle and it also can violate the Paretian improvement criterion. However, the full application of Rawls' principle leads to questionable results. For example, it assumes a greater degree of risk-aversion than seems likely in practice and it ignores the importance of intergenerational altruism, for example, the sacrifices that parents willingly make for their children. Rawls' principle also displays cosmological bias which results in it being at odds with the teachings of Hinduism and Buddhism. The mainstream stance on sustainable economic development does not oppose economic growth. However, another neo-Malthusian point of view, expressed for example by Daly and Georgescu-Roegen, does. It is opposed to an increase in levels of global material production, that is, increased throughput of natural resources for economic production. These views are given some attention. Even if there is agreement about what constitutes a desirable path for economic development, uncertainty limits the scope for identifying measures that will achieve it. That raises the question of how far into the future should existing generations attempt to sustain economic development. This is discussed. In conclusion, it is pointed out that the nature of market systems and international relations make it very difficult to implement policies that can significantly reduce global economic growth and foster sustainable economic development. These problems are global problems and no country, India and China included, can afford to ignore them.

Keywords: China, economic growth, India, intergenerational equity, Rawls' principle of justice, sustainable economic development.

JEL Classifications: O13, O44, Q01.

1. INTRODUCTION

The Problem and its Changing Nature

The question of how sustainable economic development is has a history of recurrence in economic thought. In the late 1700s, when the Industrial Revolution was under way and into the early 1800s when Britain was experiencing a 'take-off' in its economic growth, doubts were being raised about how sustainable this economic growth would be. Malthus (1798) argued that rising levels of per capita income associated with economic growth might fail to be maintained because, in the long-run, there was a tendency for human population to increase in response to higher levels of incomes and eventually wipe out the benefits of economic growth. He envisaged that in the long-term, income levels above the subsistence level would not be sustained because of population growth and the operation of the law of diminishing returns in food production (Tisdell, 2005, pp. 12-14). In the absence of measures to counter population growth, Malthus believed that the welfare benefits of economic growth would prove to be transitory and in the long-run, as a result of population pressures, economic systems would tend to a stationary state with the population living at a subsistence level.

Ricardo (1817) modified this theory by pointing out that continuing technological progress could stave-off a return to the stationary state following the occurrence of economic growth. Marxists, such as Engels (1959), flatly rejected the Malthusian theory arguing that advances in science and technology would enable growth to proceed at a much faster rate than population growth. For these technological optimists, there are no limits to economic growth. This view seems to have been widely accepted by both Marxists and non-Marxists in the latter part of the 19th century and during most of the 20th century. However, in the last part of the 20th century, renewed doubts emerged about the sustainability of economic growth and debates occurred about the desirability of such growth.

Concerns were expressed about the following:

(1) The depletion of non-renewable natural resources, such as fossil fuels (consider the views expressed by the Club of Rome as outlined by Meadows *et al.* (1972);
(2) the loss of renewable but extinguishable resources such as occurs with biodiversity loss;
(3) the increasing scarcity of recirculating resources, such as water;
(4) the pollution of natural resources such as air and water, thereby reducing their utility; and
(5) alterations in the composition of some natural resources, such as air, due to human activity, for example, the elevation of the levels of greenhouse gases in the atmosphere with consequential impacts on global warming (a subject reviewed in Tisdell, 2009a, Ch. 11).

All of these developments have rekindled the view that unlimited economic growth is impossible and that continuing economic growth based on the principle of business as usual is unachievable.

The modern neo-Malthusian view is based on wider considerations than those of Malthus and rejects the view that scientific and technological developments will be able to more than adequately overcome any limits to economic growth. Proponents of neo-Malthusian limits to economic growth accept that scientific and technological advances can help to sustain economic growth but that these advances have limited ability to do so.

Neo-Malthusian concerns are succinctly summarised in a formula suggested by Ehrlich (1989). He argues that the adverse environmental impact of human activity (which threatens sustainable economic development) depends on three factors P, the level of human population; A, consumption (or GDP) per head; and T, the extent to which environmentally damaging technology is used. He believes that these factors interact in a multiplicative way and suggests that adverse environmental impacts of human activity, I, can roughly be typified by the equation:

$$I = PAT \tag{1}$$

While this is not a precise formula, most neo-Malthusians accept the view that major influences on the sustainability of economic development are the level of human population; its per capita level of economic production, and the state of technology and its application. Other things remaining equal, the threat to sustainable economic development is believed to be greater the higher is the level of human population and the greater is its economic production per capita because this adds to strains on the natural resource base of the Earth.

A consequence of this view is that economic growth in Asia, particularly in India and China will add substantially to the drain on the globe's natural resources and hasten deterioration of natural environments. Higher income countries have already contributed to those changes and continue to do so. Consequently, some limits to economic growth are likely to be approached at an increasing pace globally as economic growth in Asia continues at a high rate.

Increasing consumption (economic production) per capita has become a more significant sustainability problem than population growth because demographic transition has eased the population expansion problem in many countries, and China has continued to maintain its one child policy in recent times. Therefore, the issues of the sustainability of economic growth is becoming a major policy consideration, namely (but not exclusively) because of the desire for ever increasing levels of per capita economic production.

In order to address this matter, it is necessary to consider to what extent sustainable economic development is desirable and how it can be achieved, if it can be achieved at all. However, the desirability of sustainable economic development hinges on ethical and normative views about the desirability of sustaining particular attributes. Although there is a dominant view in the contemporary economic literature about the desirable characteristics of sustainable development, more than one view exists. In this article, the dominant view is outlined first, its ethical foundations are outlined and assessed. An alternative approach to sustainable economic development as outlined by Daly (1980) is then considered. Then before concluding, some of the practical difficulties involved in formulating and implementing policies for sustainable development are outlined. These difficulties include a great deal of uncertainty about the conditions that will face generations in the distant future.

2. THE DESIRABLE PATH FOR SUSTAINABLE ECONOMIC DEVELOPMENT (SED): THE DOMINANT VIEWPOINT

According to Pearce (1998, p. 70), most economists addressing issues involving sustainable economic development (SED) define SED as 'non-declining consumption per capita, or GNP, or whatever the indicator of development is'. He also states that most economists would prefer to define it in terms of a non-declining level per capita "utility" or wellbeing (Pearce, 1998, p. 70). He refers to himself and co-authors as adopting this point of view (Pearce *et al.*, 1990) as well as to Maler (1990) as a proponent of it. It is also the definition that is adopted in many textbooks (for example, Tietenberg, 2003, p. 94).

Note that when real consumption per capita or GDP per head, or similar indicators are used, as the dependent variable in determining the occurrence of SED, it is usually assumed that human well-being per head increases with the magnitude of these variables. This, as discussed later, is a contentious assumption.

Given the mainstream definition of SED, it can be seen that economic development path (ACE) in Figure 1 is compatible with SED but path ABCD is not. Path ACE results in every succeeding generation being better-off than its predecessor but path ABCD results in some generations being worse-off than their preceding generation. In this figure, *U* represents utility or well-being per capita, *t* indicates time and t_n is the horizon for this problem. As time passes, and new generations emerge.

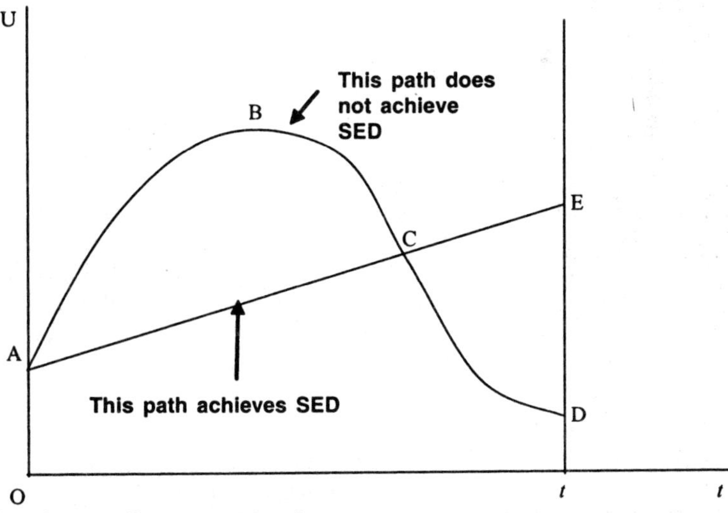

Figure 1 : An example of a development path meeting the mainstream criterion for SED and one that does not.

Rawls' Principle of Justice as the Ethical Basis for the Mainstream Criterion for SED

Frequently, Rawls' principle of justice is seen as providing the ethical basis for the above SED rule (see Tietenberg, 2003, pp. 93-94). Rawls (1971) espouses the principle that the income of everyone should be equal unless inequality is to the advantage of all. It is claimed that if all could confer before being born, they would opt for this SED rule given that they do not know in advance when they will be born.

However, Rawls' principle does not always support the rule that the per capita income, *y*, of each successive generation should not be less than that of its predecessor. This is illustrated for the alternative economic development paths down in Figure 2. There the growth path ABC satisfies the mainstream SED rule but ADE does not. Yet path ADE is clearly Paretian superior to path ABC because every generation, except initially, is better-off if it is followed (see Tisdell, 1999a).

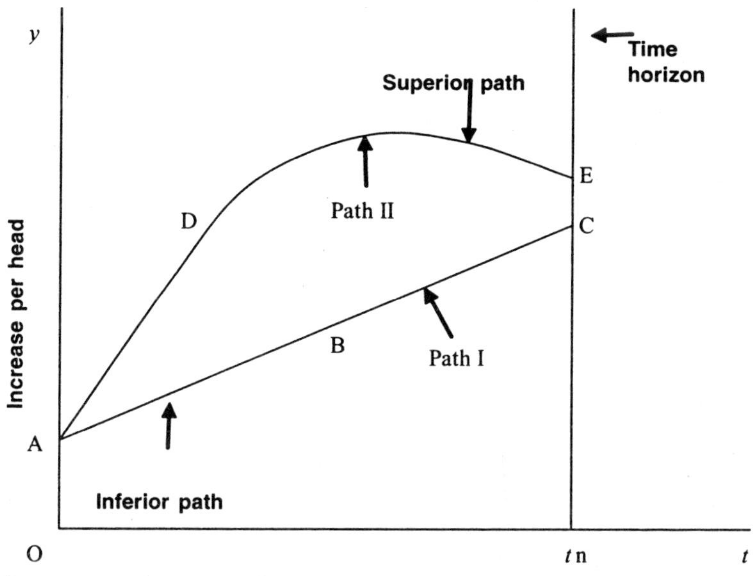

Figure 2 : The mainstream SED criterion would reject path ADE in favour of path ABC even though the former is clearly superior from a Paretian economic point of view.

Although the mainstream SED criterion would reject path ABC, the application of Rawls' principle (considered in its entirety) does not.

This is because path ADE is advantageous to all. The problem arises because economists claiming support for the mainstream SED criterion only make use of part of his principle. Nevertheless, Rawls' principle does not seem to be realistic. It assumes an extreme form of risk-aversion by individuals. I have argued elsewhere that subject to some safety constraints being met, individuals are likely to be prepared to take greater risk. They might, for example, opt for maximizing their expected level of well-being subject to being assured of its not falling below a minimum acceptable level (Tisdell, 1999a, 2011).

An Alternative to Rawls' Principle as a Foundation for the SED Criterion

Observation indicates that most parents want their offspring to be at least as well-off as they are and they may have similar feelings in relation to their grandchildren. If each successive generation has this attitude, it provides a strong social basis for the view that each successive generation should not be worse-off than its predecessor.

Of course, parents differ in their attitude towards the future of their children. Most strive to make their children at least as well-off as they are. It is only in dysfunctional families that this is not an objective. One might also expect it on biological grounds. Human beings that take good care of their children have a higher probability of passing on their genes than those that do not. Given the views expressed by Dawkins (1976), this may contribute dominance of the attitude that parents should take good care of their children and if possible, ensure that they are better-off than their parents. However, this motive is likely to result in parents forgoing consumption in order to assist their children. This is not adequately addressed by the mainstream SED principle and by Rawls.

Explicit Attention to Altruism is Lacking in the Mainstream SED Criterion and Rawls' Principle

Whether or not the sacrifices parents make for their children constitute altruism or not can be debated. Dawkins (1976) might argue that it does not constitute altruism but rather shows the self-interest of parents in ensuring the continuation of their genes. However, the above discussion highlights the possibility that altruism could be displayed by existing generations.

Existing generations have no uncertainty about when they exist (unlike in Rawls' model) and they are in a position to influence the welfare of future generations but not to do so precisely. For simplicity, however, let us suppose that the level of the current generation's per capita income does determine exactly the level of the per capita income of the next generation, that only two generations are relevant and that the problem of overlapping generations can be ignored. Then the trade-off function between the level of income per capita of the current generation, y_1, and that of the next generation, y_2, might be as shown in Figure 3 by the relationship ABCD. Given this relationship, Rawls' principle implies that point C is optimal because at this point, the income per capita of each generation is exactly equal. The income per capita of the future generation could be higher (have a value, for instance corresponding to point B) but this would require the current generation to forgo some income and have a lower per capita income than the next generation. Rawls would not consider this sacrifice to be fair. On the other hand, it would be unfair for this generation to bring about a situation corresponding to point D given Rawls' criterion.

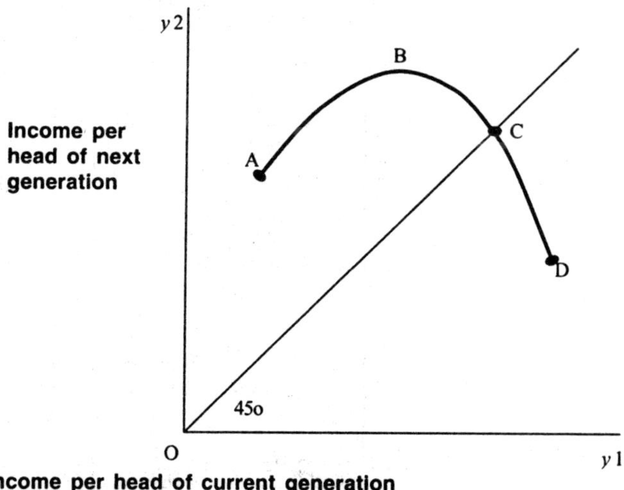

Figure 3 : Rawls' principle does not imply that the current generation should forgo income per head to ensure that the next generation (or future generations) will have levels of income higher than theirs. It ignores altruism.

However, apart from parental indulgence towards their children, the current generation may show some altruism towards succeeding generations. For example, if in Figure 3, the trade-off curve were to decline very sharply after reaching point B, this would mean that for little sacrifice of income by the current generation, the next generation could be made much better off. It would seem appealing to make this small sacrifice although it would not accord with Rawls' principle.

Again, this can be illustrated in a slightly different way. For example, suppose that a nation has the two alternative development paths shown in Figure 4. Path I, ABC, is available or Path II, DBE. Those who live in the time interval $0 < t < t_1$ will be best off if Path I is followed but those living in the time interval $t_1 < t < t_2$ will be best off if Path II is followed. Neither path is advantageous to all. The following question arises: Because the gains to those who live after t_1 are very large and the losses to those who live before t_1 are very small, it is not reasonable that the earlier generations should sacrifice some income for the benefit of future generations? Might not this choice be sometimes made? Did not some earlier generations of Indians and Chinese make sacrifices in their level of consumption to ensure rising per capita incomes for current generations? Was this justified? Such issues do not seem to be adequately addressed by Rawls' principle of justice nor by the dominant criterion for determining the occurrence of SED.

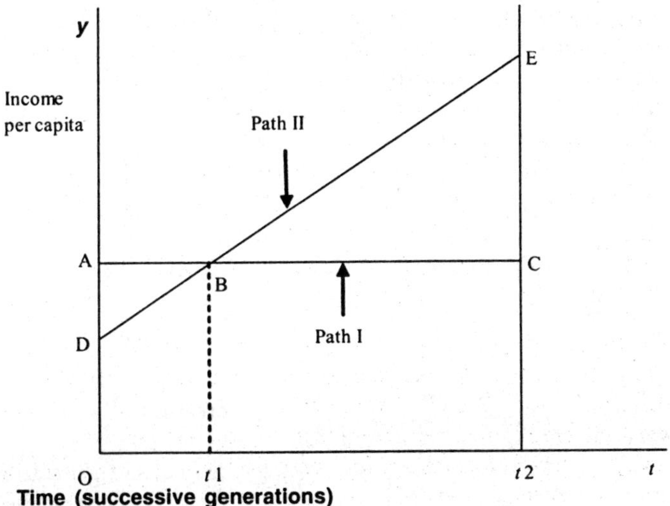

Figure 4 : Given the two alternative development paths shown, Rawls' principle would reject Path II as socially optimal. However, from some point of view, it is socially preferable to Path I. It is altruistically preferable to Path I.

Ethical and Cosmological Limitations of Rawls' Principle of Justice

Rawls' principle of justice and the mainstream SED criterion are entirely anthropocentric. It is only what humans want that counts. There is no mention of the duty of mankind to conserve nature nor show reverence towards it, for example, as highlighted by Hinduism and Buddhism. Rawls' principle presumes an outlook that predominates in Christian, Judaic and Islamic thought.

In addition, it supposes that individuals will live only once. Therefore, it does not allow for rebirth of individuals (unlike the situation in Hinduism and Buddhism) nor does it (admit) of the possibility that individuals might be reborn in another form, for example, as an animal. Thus, it is heavily influenced by Western cosmological views. This is not to say that these differing Oriental views are correct, but to highlight the cultural influences on Rawls and on the mainstream economic approach to SED.

3. ANOTHER POINT OF VIEW ABOUT SED—DALY'S PERSPECTIVE

The exact perspective of Herman Daly on sustainable development is a little difficult to pin down exactly. Basically however, he espouses the view that a steady state economy in which there is zero population growth (ZPG) and in which aggregate physical flows of production and consumption do not increase is desirable. He expects that this will enable the human race to survive for as long as is otherwise possible. However, Georgescu-Roegen (1971) thought it possible that current levels of production would need to be reduced to achieve this goal.

Daly (1980, pp. 8-9) adopts the view that economic activity should be so organized that it serves an ultimate end. This is a deontological approach and the ultimate end is something that is intrinsically good according to Daly. This might be, for example, the goal of enabling members of the human race to enjoy a satisfying life for as long as possible (as long as God wills) while at the same time limiting the adverse impacts of human activity on other creatures. Daly does not state precisely what he believes to be the Ultimate End but he clearly believes that this end is incompatible with continuing rapid economic growth involving perpetual rises in the level of physical production. He sates: "the physical flows of production and consumption must be *minimized not maximized* subject to some desirable population [level] and standard of living" (Daly, 1980, p. 21). Thus, in Figure 2, Path I would be preferred to Path II providing it gave every generation a desirable minimum standard of living. One reason why Path I might be more desirable than Path II is that it could give greater scope for the continuing existence of other species.

It seems clear that human welfare does not depend solely on per capita levels of physical consumption. While increases in physical consumption can increase welfare up to a point, very little increase in welfare may be obtained beyond that point, and the marginal environmental costs of consumption beyond that point may be high. Moderation in material consumption seems necessary in order to ensure social harmony and harmony of humankind with nature. At least in principle, Daly's approach to development opened the way to adopting a less anthropocentric stance than that taken by Rawls (1971) and is espoused in most economic growth theories, including the mainstream theory of sustainable economic development. A message that might be drawn from the writings of neo-Malthusians, such as Daly and Boulding (1980), is that those who could live an indulgent life involving high levels of material consumption should not do so. If they do, economic growth is likely to be unsustainable, serious environmental and ecological deterioration will occur and social strife will follow. Furthermore, it is unlikely that such an indulgent life will bring them happiness or contentment. This approach seems consistent with teachings that can be found in many religious works, especially in Buddhistic writings.

4. UNCERTAINTY AND SED

Theories of sustainable development focus on the long term. As a rule, we become more uncertain about future possibilities the further off they are. Although we may be able to predict the state of the world with reasonable accuracy 20 years hence, predictions for 1000 years hence, border on being fantasies. This raises the legitimate question of how far into the future should current generations endeavour to influence the sustainability of economic development.

There are actually two matters involved here. The first has to do with uncertainty about current human actions and their consequences many centuries hence. High levels of uncertainty make it difficult (impossible) to allow adequately for the well-being of individuals who might live several centuries hence when devising current economic policies. Secondly, existing generations are unlikely to show any practical concern about generations that could be alive far into the future. In this respect, Pearce (1998, pp. 70-71) observes: "The context of sustainable development has always been that of intergenerational equity. Then the time horizon must be a few generations at least but it will not be infinity. We might appeal to some 'coefficient of concern' to set some pragmatic limit on how far into the future we should look. Casual observation indicates that people care for at least their children and grandchildren. Few probably look ahead much further than that."

The concerns of politicians about sustaining the per capita issues (consumption) are likely to be a decreasing function of the amount of

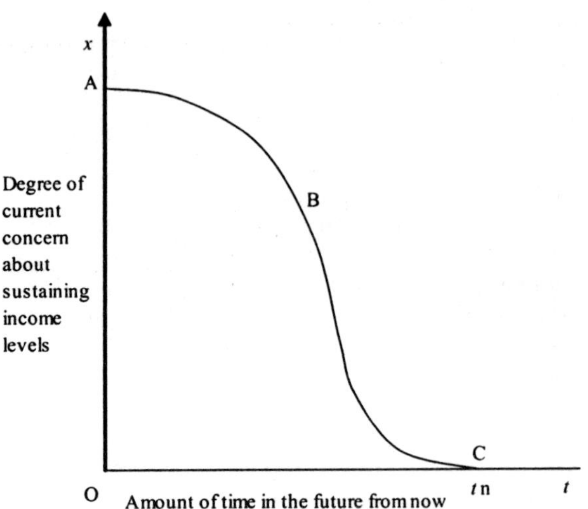

Figure 5 : Current social and political concerns for maintaining income levels are likely to be truncated in the way illustrated. Various reasons for this are outlined in the text.

time into the future that this income is to be received. Note, that this is not purely an intergenerational issue because most individuals want to have a non-declining source of income during their life-time. The politicians' concerns are merely a reflection of the electorate. Thus concern for achieving development that ensures a non-declining level of income or consumption might be of the form shown in Figure 5 by curve ABC. The greatest concern is for the near future and this concern tapers off for the more distant future, and may even fall to zero for the very distant future. In the case illustrated, there is no practical concern now about income levels from t_n onwards.

5. CONCLUDING COMMENTS

The question of whether continued economic growth and development is sustainable given the already high level of global economic production and population has received considerable attention in the last few decades. The mainstream economic view is that SED is achieved if the well-being of each future successive generation is not less than that of its predecessor. Furthermore, most mainstream economists assume that well-being is positively associated with high levels of real consumption or income per head. Therefore, their goal is to search for the strategy that ensures the highest level of non-declining real consumption per head. But some neo-Malthusians believe that this is an inappropriate goal. They are in favour of a social system in which those on higher incomes (the wealthy), whether in high income or low income countries, live frugally. They believe that humankind should endeavour to minimize its ecological and environmental footprint. However, even if the desirability of this is agreed to in principle, it is difficult to put into practice.

This is so for the following reasons. The market system relies on continuing economic growth to maintain employment. Individuals are socially trapped in this system (Tisdell, 1999b). In the absence of institutional change, they depend on employment in the market system and this requires its continuing growth. Furthermore, material wealth is a sign of social status and a source of social power. Therefore, in most societies, it is difficult to control the display of wealth. Again, if only a few follow the path of frugality, it will make little difference to the rate of natural resource extraction and depletion. Therefore, a free-rider and a type of prisoners' dilemma problem exists. Furthermore, the international economic and political power of nations appears to be related to their material wealth. For example, China's international influence has grown along with its economic growth (Tisdell, 2009b). Many nations like to obtain and retain global international influence. For example, to some extent, the United States feels challenged by the rising international influence of China. Nations still spend heavily on armaments and military hardware to maintain their international influence and this adds to levels of physical throughput of natural resources.

Despite these difficulties and the complexity of theoretical issues involved in analysing SED, we still need to confront the issues involved. It is still worthwhile for us to strive to give future generations scope to earn an adequate and secure level of income and bequeath to them conditions that will help them to live a satisfying life in harmony with nature and one another. We should ourselves try to avoid excessive levels of physical consumption and should encourage our successors to do likewise. It is not only desirable to do away with poverty but as well, excessive waste and consumption (which can accompany wealth) are worthwhile avoiding. This is desirable for environmental and ecological reasons and can result in a more satisfying life, as many religious teachings indicate.

In conclusion, it is worth noting an observation made by Daly (1999, p. 52). He states:

> "The welfare of future generations is beyond our control and fundamentally none of our business. As any parent knows, you cannot bequeath welfare. You can only pass on physical

requirements for welfare. Nowadays natural capital is the critical requirement. A bequest of a fishing fleet with no fish left is worthless. But even the bequest of a world full of both fish and fishing boats does not guarantee welfare. Future generations are always free to make themselves miserable with whatever we leave to them. Our obligation therefore is not to guarantee their welfare, but their capacity to produce in the form of a minimum level of natural capital, the limiting factor."

Nevertheless, it is not only natural capital that provides future generations with economic opportunities. For example, useful knowledge passed on to future generations is also of value to them as is to some extent, man-made physical capital. Getting the balance right between different types of resources to be left for future generations is the real issue (Tisdell, 2005, pp. 248-51). Clearly, neoclassical economic growth theorists, such as Solow (1956), were wanting in that regard because natural resources play no role in their growth models, a deficiency severely criticized by Daly (1999, Chs. 8 and 9).

In conclusion, one may ask why current debates about SED are relevant to most Indians. Reasons could include the ones that follow:

(1) SED in every country depends on global economic growth and today, natural systems and economic systems are interdependent.

(2) The rapid economic growth of India in recent times is adding to strains on the Earth's natural resources and environments thereby accelerating the speed at which global limits to economic growth are being approached. Those living in Western countries may increasingly view Asia's economic growth as an escalating threat to the maintenance of their well-being because of its consequences for the environment and the availability of natural resources.

(3) The dominant Western perspective or the desirability of SED is anthropocentric and based on Western cosmological beliefs, as is especially evident when it is grounded on Rawls' principle of justice. These views are inconsistent with the teachings of Hinduism and Buddhism because they do not consider, for example, the possibility of rebirth of individuals, including the possibility of rebirth in non-human form.

(4) There is also a major social problem that cannot be ignored. Equity in the distribution of resources between generations and their distribution within generations should be

simultaneously considered. Justice may require redistributing income and resources from the rich to the poor within generations. This would require not only a redistribution of income internationally but a redistribution within India of income from those who are well-off to those in poverty. Politically, such a redistribution is difficult to achieve even if it is felt to be justified. Therefore, the easy way out politically is to promote material economic growth on the basis that all, including the poor in society will be better-off as a result. This, together with other institutional mechanisms locks modern economic societies into striving for continual economic growth despite its potentially fateful consequences.

References

Boulding, K.E. (1980). The Economics of the Coming Spaceship Earth. pp. 253-63 in H.E. Daly (Ed.), *Economics, Ecology and Ethics*, Freeman, San Fransisco.

Daly, H.E. (1980). Introduction to the Steady-state Economy, pp. 1-31 in H.E. Daly (Ed.), *Economics, Ecology and Ethics: Essays Toward a Steady-State Economy*, Freeman and Company, San Fransisco.

Daly, H.E. (1999). *Ecological Economics and the Ecology of Economics: Essays in Criticism,* Edward Elgar, Cheltenham, UK and Northampton, MA, USA.

Dawkins, R. (1976). *The Selfish Gene,* Oxford University Press, Oxford.

Ehrlich, P.R. (1989). Facing the Habitability Crisis. *BioScience,* 39, 480-82.

Engels, F. (1959). Outlines of a Critique of Political Economy in K. Marx (Ed.), *Economic and Philosophic Manuscripts of 1844*, Foreign Languages Publishing House, Moscow.

Georgescu-Roegen, N. (1971). *The Entropy Law and The Economic Process,* Harvard University Press, Cambridge, Mass.

Maler, K.G. (1990). *Sustainable Development,* Stockholm School of Economics, Stockholm, Mimeo.

Malthus, T.R. (1798). *An Essay on the Principle of Population as it Effects the Future Improvement of Mankind,* 1976, Norton, New York, J. Johnson, London. Reprint 1976, Norton, New York.

Meadows, Dennis, H., Ronders, J. and Behrens, W. (1972). *The Limits of Growth: A Report for the Club of Rome's Project on the Predicament of Mankind,* Universe Books, New York.

Pearce, D.W. (1998). *Economics and Environment: Essays on Ecological Economics and Sustainable Development,* Edward Elgar, Cheltenham, UK and Northampton, MA, USA.

Pearce, D.W., Barbier, E.B. and Markandya, A. (1990). *Sustainable Development: Economics and Environment in the Third World,* Edward Elgar, London, and Earthscan, London.

Rawls, J.R. (1971). *A Theory of Justice,* Harvard University Press, Cambridge, MA, USA.

Ricardo, D. (1817). *The Principles of Political Economic and Taxation*, Reprint, 1955, Dent, London.

Solow, R.M. (1956). A Contribution to the Theory of Economic Growth. *Quarterly Journal of Economics*, 70, 65-74.

Tietenberg, T. (2003). *Environmental and Natural Resource Economics*, 6th Edn., Addison Wesley, Boston.

Tisdell, C.A. (1999a). Conditions for Sustainable Development: Weak and Strong, pp. 23-36 in A.K. Dragun and C.A. Tisdell (Eds.), *Sustainable Agriculture and Environment*, Edward Elgar, Cheltenham, UK and Northampton, USA.

Tisdell, C.A. (1999b). *Biodiversity Conservation and Sustainable Development*, Edward Elgar, Cheltenham, UK and Northampton, MA, USA.

Tisdell, C.A. (2005). *Economics of Environmental Conservation*, 2nd Edn., Edward Elgar, Cheltenham, UK and Northampton, MA, USA.

Tisdell, C.A. (2009a). *Resource and Environmental Economics: Modern Issues and Applications*, World Scientific, Singapore, New Jersey, London.

Tisdell, C.A. (2009b). Economic Reform and Openness in China: China's Development Policies in the Last 30 Years, *Economic Analysis and Policy*, 39(2), 271-94.

Tisdell, C.A. (2011). Core Issues in the Economics of Biodiversity Conservation, *Annals of the New York Acadamy of Sciences*, In Press.

Chapter 2

Environmental Ethics

Some Obiter Dicta

B.N. Ghosh

"Earth provides enough to satisfy every man's need, but not every man's greed".

—Mahatma Gandhi

In recent years, analysts and policy-makers have become concerned about the appropriate use of our environmental resources. The environment is under constant attack and is being assaulted by factors and forces that accompany the rapid rate of industrialisation and the escalating rate of human poverty. Both poverty and growth do have convoluted interactions with environmental resources, and in many respects, they are the primary causes of environmental damage and degradation. Whilst a moderate rate of growth is desirable and often harmless, too much of it at too fast a rate impinges on the eco-system and environmental tranquility. Innumerable *ethical issues* are involved in the management of human environment. The present chapter aims at analyzing some of these issues and many other related problems including the policy of sustainable development and proper management of environmental resources.

POVERTY, RESOURCES AND ENVIRONMENT

It should be appreciated that poverty arises because of lack of basic resources. Many of the resources that can alleviate poverty are natural resources that one can gather from forests, water, mountains, land, mines and the nearby environmental endowment. There seems to be a vicious circle of relationship working here. In order to reduce poverty, more of natural resources will be depleted which, in future, will aggravate poverty. Thus one starts from a situation of poverty and again lands into poverty, and the ***poverty cycle*** continues. One of the reasons for the growth and sustenance of poverty is the degrading of natural resources or environment. Too much of use of natural resources due to excessive growth of population, or due to lack of proper policy to regenerate or renew the natural resources leads to environmental fall out. The natural resources have a natural power to regenerate themselves, but it takes time and if it is constantly assaulted without giving them the necessary time to regenerate, the depletion is inevitable. Over-use and wrong use of resources are contributory factors for the damage and depletion of our natural resources.

Can a payment for the environment lead to sustainable resource use and poverty alleviation? This is a complex question which does not have a straight-forward answer. If the user is a rich person or a multinational agency, some payment will be made for the use of resources. In the process of exploitation of environmental resources, the user may derive some values (benefits) which may remain unknown to the owner, and the actual value may be several times higher than the amount paid for the use of those resources. This is particularly the case for many plants having medicinal value. In such a case, there will be tendency on the part of the user to over-exploit the natural resources. More often than not, the exploiter is receiving enormous benefits and he will go the whole hog to maximize his gains. In such a case, even if payment is made, it will bring more harm than good to the environment.

If the owner is an individual person, his poverty may be alleviated but at the cost of the country's natural resources. There may be private benefit but the public loss may be enormous. Moreover, some of the benefits of our natural resources cannot be evaluated by the market price. The indirect benefits of natural resources are in particular not amenable to market valuation.

The fact that over-use of natural resources can lead to more of poverty is substantiated by the case of many developing countries. In Lao PDR, more than 50% of the GDP is derived from agriculture and 80% of the people rely on natural resource base to eke out living. This practice has been going on since long, and as a consequence, agriculture and other natural resources have become overburdened and less productive. The situation has been aggravated by deforestation, illegal logging, and the conversion of cultivable land to industrial areas, decline in non-timber forest products, loss of biodiversity and water resources and depleting fish stock and so forth. Clearly, the poverty of the people is again raising its ugly head. So the government of Lao PDR has to take a number of practical steps to not only reduce poverty but at the same time to ensure sustainable development. Sustainable development is that development which does not harm the environment or distorts the ecological balance of the society. It needs to be noted that both poverty reduction and sustainable development are simultaneously possible with good policies and their rational implementation. In most of the developing countries, there is an urgent need to minimize the impact of environmental changes.

The case of Philippine is getting more publicity now as the country where profitable agricultural production is destroying the eco-balance. In Philippines, for increasing palm oil production, the plant and animal diversity has been seriously damaged, and carbon emissions, however, have not been reduced. All this is for the benefit of a handful of business people and commercial farms. The period of globalization has witnessed a structural change in the agrarian development of many countries. Agricultural land is becoming more and more an ordinary commodity. The commodification of land in many countries including Africa, India and Pakistan has been contributing substantially to the degradation of environment. It is high time for less developed countries to find out some viable alternatives to land degradation and soil erosion problems. This may make land more fertile once again and alleviate the poverty of the cultivators. It is not impossible to enhance the resource use efficiency and transform traditional agriculture, and at the same time, keep intact the ecological balance.

Population growth and poverty are responsible for land and forest degradation. The poor people have to depend on the nearby forest to collect fuel-wood, fodder and get some extra income for the expanded family. All these are important cause of deforestation. However, legal and illegal logging and exploitation of forest resources with the connivance of forest officers are also equally responsible for this state of affairs in many developing countries. The programs for afforestation and regeneration have never been a howling success in these countries due to a variety of reasons.

Be that as it may, the poor people suffer disproportionately from the resources imbalance created by various factors. Pollution of all types affect the poor most in terms of loss of income, sickness and loss of amenities that are possible from the nearby forest. Population growth and poverty go hand in hand.

As suggested earlier, poverty and environmental deterioration are mutually reinforcing, and so is the population growth. Poverty is correctly blamed for both excessive population growth and environmental pollution It is rightly said that "poverty is the worst pollutant". Poverty induces parents to have more children so that the total family income can be maximized as each child in a poor family becomes an earning member at a very low age, and the marginal cost of rearing children in a poor family is also very low. Thus, since the benefit of a marginal child is higher than the cost of having an extra child, the rate of growth of population particularly in poor families and rural areas is very high. The poor people are rational in the calculation of economic benefits and costs of extra children.

But the relevant issue is: if poverty is reduced, wherefrom the resources are coming? Are those natural resources being replenished? In many cases, one of the contributory causes of environmental degradation is the pressure of population, which is again the cause of poverty. In the mainstream thinking particularly stemming from the western countries, however, the basic advice is that for the sake of development, the concern for environment must be relegated to the background. This type of wrong advice is not called for. The world institutions, in general, are concerned more with quantitative economic growth and do not care for the quality of human life. Since life is cheaper in developing countries, the developed capitalist countries often transfer their dirty industries to the former type of countries. Many of the world institutions do recommend the entry of such bad industries to less developed countries in the name of globalization and liberalization. In fact, one of the basic purposes of liberalization in LDCs is to open their gates to all the polluting industries of the developed capitalist countries. While permitting these industries, the LDCs do not strictly adhere to the international environmental norms.

Is Economic Growth Conducive to Environment?

There is no inherent contradiction between growth and environment if the rate of growth is moderate and not competitive. A growth rate based on the domestic saving (investment) and the existing capital-output ratio is supposed to be moderate in nature as it is based on the carrying capacity of the economy. Technological improvement can also raise and sustain this moderate rate of growth. Such a rate of growth is not necessarily harmful for the environment.. However, the craze for a high rate of quantitative growth by hook or by crook can endanger the environmental equilibrium. The problem is that this craze very often dominates the policy makers and planners. A high rate of growth is looked upon as an evidence of a high level of achievement by the political authorities. So, eventually, growth becomes a competitive game. The greatness of a country is measured in terms of its quantitative growth rate.

When important national problems are discussed in a forum, economic growth comes at the top, and environment does not capture the mind of the people or planners. It is believed by politicians that public vote is based on per capita income growth and not on the clear sky and the clean air. When questions were asked in a recent survey in America, people showed more concern for economic growth and its sustainability rather than environment. Moderate growth is a necessary but not a sufficient condition for clean environment. Economic development is definitely more conducive to environmental balance. This is substantiated by the fact that a sustainable type of economic development is quite compatible and consistent with environmental development. In fact, economic development includes environmental development.

Economic growth entails many types of activities that may endanger environment. In a bid to achieve a high rate of growth, different types of polluting industries may be set-up in areas that were earlier devoted to agricultural works and forestry. In such a case, growth disturbs the environment and the ecological balance. A high rate of economic growth is often found to be positively correlated with environmental pollution. It is believed that much of the current urban environmental degradation has been underpinned by a long-lasting trend of uncontrolled quantitative growth which has resulted in ill-health and pollution. Economic growth of course brings many new opportunities but it creates many problems at the same time or in the process of growth, which are detrimental for the environment. Perhaps it is not so much the growth itself, but the composition or content of growth that is more decisive for the environment. A growth in a clean way does not have to necessarily jeopardize the environment. A growth exercise that is based only on the establishment of dirty industries and the exploitation of our ***natural capital*** is detrimental to environment. Promiscuous and rampant growth can put pressure on the ecological balance.

It is touted by western capitalist economists that economic growth can correct many environment-retarding factors such as population and poverty by increasing the growth of per capita income. There is no doubt that these two problems are the worst enemies of environmental and sustainable development. Indeed higher income level can reduce the level of poverty and in that case, the exploitation of forest resources and pollution of environment by the poor classes of population will be substantially reduced. The poor will then not make the environment dirty and polluted. For instance, they will not defecate near the road side, or will not cut the forest for fodder and fire wood. They will have more decent ways of performing their daily chores. Moreover, at a higher level of per capita income, the parents will not have to depend so much on the additional income from additional children. In other words, when the parental income level is higher, the dependence on children will automatically reduce.

In the early stages of development, or in the traditional sector, the cost of rearing a child is very low, while the expected benefit out of him is too high. However, at the advanced stage of industrialization, the situation is quite opposite. As a result of increased income per capita, birth rate tends to fall. At a higher level of income, standard of living becomes a more dominant consideration. At this stage, many parents will like to have a car rather than a baby. Urbanization will also help reducing the birth rate. In an urban area, a big family is a liability. Lack of accommodation, higher rents and higher cost of rearing of children help to reduce the tendency towards higher birth rate. Moreover, birth is reduced by late marriage, which is a trend in many urbanized societies. Birth rate can also be reduced if the death rate is low. In fact, with the rise in the level of income, it is possible to have better medical care and public health services, and all these will reduce the death rate. Since the death rate is lower now, the birth rate also becomes lower than before. In the same way, since poverty is a negative function of income, a rise in the level of income, will help to reduce poverty. But the level of income that is needed to reduce the rate of growth of population, will have to be critical minimum. If the rise in income is very low and slow, population and poverty will rather increase than decrease. Thus, people say that a good rate of growth is the best contraceptive.

However, there is a note of caveat. It is often found from experience in less developed countries that it is during the period of fairly good growth rate that an economy experiences environmental hazards and pollution. The point is made by the South against the limitless growth philosophy of the North. The North maintains that in LDCs, the pollution is not created by economic growth as such but by the teeming millions of people. The North goes so far as to say that the LDCs is suffering not from high growth rate but from low growth rate syndrome. Some researchers cite results of cross-national regressions showing that there is an inverted U-shaped relationship between per capita income and the correlates of environmental degradation in a country like India (see Rao, in Chary and Vayasulu, 200). This implies that at a lower per capita income growth, the environmental pollution is showing an upward trend, and at a still higher level of income, the level of pollution is going to be lower. If this is true, one should specify the exact level of income that is expected to show such a declining trend of national pollution.

But in spite of the fact that a very high level of income growth is a method to bring down the pollution level, a number of cogent points can be made here. First, a higher level of per capita income growth is often associated with a higher level of inequality in income distribution. This is true for many LDCs including both India and China (Ghosh, 2009, p. 28). It is evident from the following table. As the table shows, in India and China, the levels of income inequality were lower in 1986 when these countries had lower rate of economic growth. However, with higher levels of economic growth, income inequalities were exacerbated in both these countries

It should be pointed out that growth and environment are not necessarily competitive issues. The one is interacting with the other. So the choice cannot be an absolute one for all the time to come. There are in fact *many points of trade-off* : there can be higher growth rate with a tolerable rate of environmental damage, or a lower level of growth rate with maximum possible level of environmental resources and natural capital. The pertinent question is: does the society want at the present stage of development? It is a choice between the present and the future. If the society puts more premium on the present, then higher level of growth is the answer, but if the society discounts the present and puts a premium on the future, then lower rate of economic growth will have to be associated with a higher level of environmental growth and protection.

GLOBAL ENVIRONMENT : SOME BASIC ISSUES

The issues involved in the study of global environment (GE) will be analyzed in three parts. First, a list of major global environmental problems will be provided. The second part of the discussion will be devoted to the global inequalities in handling GE problems and the last part will analyze a few serious problems of less developed countries including India. The impact of different types of pollution on human health is discussed in the next section of this chapter.

Major Global Environment Issues

The major global environmental problems include:

Global climate change
Ozone layer depletion
Acid rains
Noise pollution

Pollution of international waters
Desertification of land
Deforestation
Soil erosion, radioactive pollution
Loss of valuable species
Loss of biodiversity

In the year 1991, the World Bank, UNDP and UNEP identified the flowing four major global environmental issues. These issues are: biodiversity, climate change, depletion of the ozone layer and the problems of international waters. Various environmental treaties have been trying:

(1) To conserve biological diversity and to fairly and equitably share the benefits of genetic resources.
(2) To control and reduce the harmful effects of desertification and deforestation
(3) To protect and enhance wetlands
(4) To conserve and rationally use the marine living resources
(5) To protect the endangered species of flora and fauna from over-exploitation
(6) To conserve and effectively manage migratory species.
(7) To protect the ozone layer, and phase out the ozone depleting substances
(8) To stabilize the emission of greenhouse gases
(9) To prevent the dumping of hazardous waste
(10) To regulate the quantity/quality aspects of international waters.

Global Inequalities in the Management of Environmental Problems

Some developed countries (DCs) are unnecessarily internationalizing the global environmental problems to gain some advantages and to put the LDCs into trouble. The following discussion will make the point clear. The DCs are interested only in those problems the solution of which will give them some advantages and will lead to an effective control over the LDCs. In Africa, for a long time, desertification was a global problem but no attention was paid by the DCs until recently. While LDCs are trying to phase out the harmful CFCs, the DCs under Montreal protocol have been given the power to increase the production of CFCs. The LDCs are not, however, getting any special financial assistance for such an attempt, although the production of CFCs may reduce the tempo of industrialization of these countries. CFCs are often necessary for industrialization and growth. The DCs are mostly the polluters in developing countries which are accommodating more and more multinational corporations. The principle that "*polluters must pay*" was adopted in the Rio Declaration; but the MNCs do not bother to pay any heed to such a treaty.

The DCs dump hazardous wastes to LDCs to reduce cost and evade laws. Sometimes the Northern companies make deals with Southern companies, or villagers in the coastal areas for dumping hazardous waste. In 1996, more than 289 ships came to India for the disposal of such harmful waste from America [Gupta, J., 2000 in Cary and Vayasulu (eds.), book, p. 259].

Western researchers very often use genetic resources from developing countries without paying any compensation. The patent system especially in the field of bio-technology exploits the LDCs. Since the products of biotechnology industries earn a huge amount of profit, the North is interested in patent protection, although the basic inputs are obtained from developing countries.

The international agencies are interested in making international treaty on forestry. This is objected by LDCs on the basis of the fact that forests are national resources and no international treaty is needed for the protection of forests. The above points will show that the DCs are trying to internationalize every environmental issue in their favour for imposing control and gaining international supremacy.

Problems before Developing Countries : International Inequalities

In the area of environment management, LDCs confront with many serious problems. Some of the major problems are:

First, in the long-run, it is believed that reduction in emission will reduce the tempo of economic growth in such countries.

Second, the western polluters in such countries cannot be compelled to pay for the pollution created by them. The polluters must pay paradigm has been changed to Leadership paradigm. The western polluters are accorded the role of leaders who will try to control pollution in all countries. The principle of "grandfathering" is getting the upper-hand.

Third, while the DCs have got the *de facto* property rights to pollute more, the LDCs are punished for that. The international agencies are following a double standard, and the interests of LDCs are consistently overlooked.

Fourth, the technology transfer agreements are increasingly using the provision of certain environmental criteria. In that case, it will be difficult for LDCs to absorb western technologies.

Fifth, global power politics is being used in the case of global environment management. The western investors are asking for a complete control over their own investment and its consequences in the developing host countries. This will intensify the problems of environment in LDCs.

Sixth, current international trade from the South involves a form of invisible subsidy to the North as the raw materials exported by the South are not priced on the basis of ecological value. The eco-labelling and packaging rules imposed by importing DCs may reduce the trade from the South.

Seventh, international debt problems make the LDCs economically and environmentally vulnerable as many of these countries have to exploit their forest resources to a level which is unsustainable.

The aforesaid problems are common to many of the developing countries, and unless the international economic and power structure is based on equity principle and the Southern countries develop a strong countervailing power, such a deplorable situation is likely to continue forever.

THE IMPACT OF ENVIRONMENTAL POLLUTION ON HUMAN HEALTH

Neo-liberal globalisation that spreads the message of increased tempo of industrialisation through capital and technology intensive methods of production has exacerbated the environmental pollution (EP) several times in recent years. The EP is the common name given to a syndrome of different types of pollution including air pollution, water pollution, dust pollution, soil pollution, radioactive pollution, noise pollution and thermal pollution. The modern world has been witnessing various types of environmental pollution. Population growth and environmental pollution go hand in hand. Population growth directly or indirectly affects the environmental tranquillity and ecological balance. Run-away world population figures continue to exacerbate environmental problems. In the year 1999, the world had 6 billion human beings, by the year 2110, it will have 10 billion people. These additional people need food, fuel and other necessities that the earth is already straining to provide. The world already has 500 million undernourished people, and this number will increase 10 percent in the next decade. Along with population growth will come different types of environmental degradations. Environmental pollution has been affecting the people of both developed and underdeveloped countries. However, the nature of environmental pollution is different in these two types of countries. In developing countries, urban slums and pollution are the outcome of excessive population growth and uncontrolled migration of people. The threat to human life has become so imminent that concerns have been expressed by experts and even by world bodies to control the environmental problems at the earliest opportunity.

Linked with globalisation is the ecological system of the world. Globalisation may not be the casual factor but it is associated with the declining state of our eco-system. The ecological development of the world has a definite impact on human health and well-being. The UN Report on the State of the World's Cities (2008/9) puts the current ecological footprint of humanity as 2.2 hectares per person, while the earth's biological capacity remains at 1.8 hectares. The two Asian superpowers, China and India have ecological footprints that are twice their bio-capacity. That implies that what the population consumes in a year, their area of earth will take two years to produce. This seems to be consistent with the Malthusian legacy: the growth of population is going to outstrip the growth of food supply. If this is true, it is sure to impinge on human health and happiness. There are of course other challenges on which the Report has elaborated. These challenges relate to mobility, waste management and environment. Some of the natural catastrophes like the climate change, triggered partly by irresponsible human actions during the period of globalisation, will not be amenable to proper management and control. The magnitude of the climate change in the years to come will be simply incredible. There could be long spells of heavy rains, prolonged dry weather, powerful cyclones or tsunamis. The glaciers of the Himalayas may start melting and the sea levels may rise. But the rub is that it will be difficult to reverse the trend, even if the whole world starts taking measures to protect the environment now (*The Hindu*, December 3, 2008). The impact of the climate change will not be confined to one or two countries, but will be felt across the world. The disaster will have direct deleterious impact on human and animal health.

Climate Change and Other Agents of Pollution

The indirect impact of climate change on human health will be through its damaging effect on the production of food and fibre. In a report by the Food and Agricultural Organisation (FAO, 2008), it is said that ocean warming, frequent tropical cyclones, flash floods and droughts are very likely to bring a devastating impact on food production system in the Pacific Island countries. Climate change-related disasters have already seriously constrained the development of these islands. The climate change is also going to reduce food security (especially for households). Moreover, increasing coastal inundation, salinization and erosion as a consequence of the rise of the sea level, and human actions may indeed contaminate and reduce the area of agricultural lands for the purpose of food production. This is sure to impinge on the problem of household and national food security. The FAO has advised the world countries to build resilience of the food systems to avoid enormous future economic losses in food and agriculture, fisheries and forestry. In order to fight the menace of climate change-related disasters, countries will need to assess the degree of vulnerability of their food security systems, and accordingly can make viable plans for agriculture, forestry and fisheries. It becomes clear from the report of the FAO that the indirect effect of climate change on human health will be serious and unpredictable.

Because of excessive concentration of population and industrial activities, enormous amount of domestic and industrial waste is discharged into rivers, which creates water pollution. The deposit of chemicals may also pollute the air. Water may also be polluted by cadmium, which causes impairment of the central nervous system. Nitrate pollution of water supplies, which causes methaemoglobinaemia, particularly among infants, is reported in Northern American countries. Water pollution may also cause hepatic disorders, typhoid and intestinal diseases. Water pollution may be biological, radio-active or chemical. The constant use of synthetic detergents leads to considerable river pollution. The presence of detergents renders life impossible for some micro-organisms which affect the natural biological purification of water. Detergents also affect the re-oxygenation capacity of river water. Excessive fluoride in the water may cause dental fluorosis or hearing deficiencies in children. Chemical contamination of water may be enough to produce toxic effects. The biological contamination of water is responsible for many types of endemic infectious diseases like typhoid, cholera and dysentery. Epidemiological studies carried out in many countries have established that soft water may produce cardiovascular and cerebrovascular diseases.

Only 10 percent of the world's fresh water is polluted in terms of diminished oxygen content. However, while most rivers are fairly clear of organic pollution, some contain high concentration of pesticides and/ or polychlorinated biphenyls. These are dangerous carcinogens. A more serious problem regarding fresh water is that its distribution is uneven. Less than 50 percent of the world's rural population and nearly 75 percent of its urban population have access to clean water. Adequate sanitation is available only to 15 percent of the rural population and 59 percent of the urban population. The UN hoped that by the year 2000, everyone could have clean water and adequate sanitation. But the hope has not been a reality. A high rate of population growth prevents adequate investment for the availability of pure water and adequate sanitation in developing countries. Salt water pollution remains a concern especially after the outbreak of algae blooms along the southern Scandinavian and the US east coasts. The North Sea seal and dolphin deaths and the appearance of medical rubbish have polluted the salt water. One hopeful sign was the fact that in 1987, eight North Sea countries agreed to reduce waste incineration at sea by at least 65 percent by the end of 1990, and phase it out completely by 1994. However, it has not been possible to do so.

Noise is also capable of producing damage to human beings. Noises produce congestion, visual distraction and so on. Noise at high pitch (over 80 decibels) may cause a temporary decrease in the size of blood vessels and produce high pulse rate, hearing damage, cardiovascular problems and the constriction of muscles. The blaring of loud speakers and radios in urban areas may create noise over 70 decibels which in many cases can cause nervous breakdown and disturbance of sleep.

It is necessary to dispose of hazardous wastes. Most developed countries send their waste to the less developed countries of Africa and elsewhere. OECD countries generate between 300 and 800 million tonnes of hazardous waste a year. The US alone contributes 88 per cent of this amount. The developing countries of India, Brazil, South Korea and China also contribute substantial hazardous wastes. A large proportion of this waste is produced by chemical and mineral industries. The cost of disposal of hazardous waste is pretty high and several countries flush their wastes into oceans, even though there are international and regional conventions to control such dumping. Land disposal is the most popular method of waste disposal. The Kommenekimi method of Denmark destroys more than 90 percent of the country's hazardous waste and at the same time supplies 35 percent of the heating needs. However, because of many difficulties, the developed nations continue to transport their hazardous waste across borders for someone else to deal with. Some of the third world countries accept the waste in exchange for hard cash. But this has triggered widespread concern recently in African countries. In 1988, the Organisation of African Unity adopted a resolution condemning the use of Africa as a garbage heap.

Another important issue in the environmental balance is the fact that the earth is gradually burning due to human activities that artificially increase the amount of carbon dioxide and other greenhouse gases in the atmosphere. The burning of fossil fuels releases an extra 5000 million tonnes of carbon every year and the destruction of forest and other vegetation creates another 1600 million tonnes of carbon annually. The other gases that are harmful for the environment are nitrous oxide, methane, chlorofluorocarbons and ozone, which, when trapped below about 12 Km, are damaging to the human body. According to one estimate, within a century, there will be a warming of the earth by between 1.5°C and 4.5°C. This will have far-reaching environmental and socio-economic implications. The wind pattern and rainfall will change, making some regions wetter and some regions drier. In some regions, there may be tropical storms. In the event of the greenhouse effect, food production will fall marginally, because some areas may not be able to adjust to the climate changes.

Air Pollution

In the present section of the book, I will concentrate on the health damaging effects of only air pollution. Of all the agents of environmental pollution, air pollution (AP) is the most serious one that affects health. There are, of course, ozone depletion, global warming and climate change. Air pollution refers to the presence of many types of obnoxious substances and chemicals in solid, liquid or gaseous forms in the air. Air becomes polluted when the concentration of sulphur and nitrogen-based compounds and other harmful particles like dust, smoke, ash and gas, and suspended particulate matter are accumulated in the air beyond the acceptable standard for good health. The World Health Organisation estimates that more than a billion people in Asia are exposed to air pollution levels that exceed its guidelines. This is considered to be a reason for the premature death of half a million people every year. Researchers at Brigham University and Harvard School of Public Health have recently found that average life expectancy had increased by three years between 1980 and 2000 in those cities in America where pollution had been kept at bay. Reducing pollution produces measurable gains in terms of human health and life expectancy. Cleaner air has lengthened life expectancy by five months in fifty-one US cities, as a news report revealed in January 2009.

Air pollution may be caused by the combustion of fuels, incineration and the fumes from different types of transportation. Nearly 70 per cent of air pollution is caused by the exhausts from automobiles which have become a part of modern life. The clean air is contaminated by carbon dioxide, sulphur compounds, carbon monoxide, nitrogen compounds, hydrocarbons and particulate matters. Different types of compounds become highly toxic when they are mixed in the air. Air pollution largely affects the industrial sector of both the developed and the developing countries. A high concentration of carbon monoxide in the blood obstructs the supply of oxygen necessary to maintain the vitality of the body's cells. Tetraethyl mixed with automobile fumes and lead poisoning become very harmful to the human body. All these may cause anaemia, nervous disorders and muscular problems. Air pollution is responsible for various types of respiratory diseases. Air pollution in some countries has, of course, decreased since the 1970s, particularly in some western countries. The Protocol of the 1979 Convention on Long-Range Trans Boundary Air Pollution which came into force in 1987 is further expected to reduce it. However, according to the data supplied by the NEP (1984), out of 54 major cities, 27 were found to have unacceptable or just marginal air according to the standard of the World Health Organisation. Extrapolated world-wide, this means that some 990 million people (almost 50 percent of world's urban people) breathe marginal or unacceptable air. The cities surveyed which had marginal or unacceptable air included Delhi, Dublin, Hong Kong, Shanghai, New York, London, Milan, Tehran, Seoul, Rio De Jenero, Sao Paulo, Paris, Beijing, Madrid and Manila.

The World Development Report (2007) has studied the seriousness of the AP problem in different countries of the world and finds that the most polluted city in the world in terms of particulate matter (PM) is Cairo (Egypt) followed by Delhi (India), and the least polluted is the Shenyang city in China. (Vide the Table 1).

TABLE 1

Most Polluted Cities by Particulate Matter (PM)

Particulate Matter $\mu g/m^3$ *(2004)*	*City*
169	Cairo (Egypt)
150	Delhi (India)
128	Kolkata (India)
125	Tianjin (China)
123	Chongqing (China)
109	Kanpur (India)
109	Lucknow (India)
104	Jakarta (Indonesia)
101	Shenyang (China)

Source : *World Development Report,* 2007.

A large number of epidemiological studies have provided quite authentic evidence that AP is associated with increased morbidity and mortality (WHO, 2002). AP mostly affects the cardiovascular and respiratory systems. A study by Gregory A. Wellenius *et al.* (2007) finds a small but statistically significant inverse association between ambient particles and blood oxygen saturation, but no association between short-term fluctuations in ambient pollution levels and functional status as assessed by circulating levels of BNP. However, in this case, the sample size was too small to give the correct result.

The Reports of the WHO disclose that each year, more than 2.5 million people die of air-borne diseases. EP can be an indoor or outdoor phenomenon, or both. Tobacco pollution can lead to indoor or outdoor pollution. The problem of air pollution is more or less affecting all the countries in the world, and it has a marked deleterious effect on human health. The National Resources Defence Council of America reported some years ago that every year due to AP, nearly 64, 000 people die of cardiopulmonary diseases in the United States, and nearly 30,000 people die prematurely every year due to coal-fired power plants. It is estimated that for every one per cent loss of ozone, the risk of skin cancer in human beings will increase by about six per cent (GEO Year Book, 1997). The increasing demand for energy in the modern world leads to the burning of fossil fuels, which is a contributory factor to the aggravation of AP.

A study by Beatrix Groneberg-Kloft *et al.* (2006) has established that AP remains a leading cause of many respiratory diseases, including chronic cough. The authors observe that a long-term exposure of children to nitrogen oxide is associated with increased incidence of chronic cough and decreased lung function parameters, and there is some evidence that chronic inhalation of diesel can also lead to the development of the cough. Sulphur dioxide is also known as a major respiratory irritant. Another powerful oxidant is ozone that affects the functionality of the respiratory tract. An exposure to a slightly elevated ozone concentration produces many unwanted respiratory syndromes. A correlation is found between indoor AP and chronic cough; arsenic contaminated well water in Bangladesh has been responsible for chronic respiratory diseases too. Depleted uranium, which is a radioactive heavy metal, is also associated with chronic cough.

AP consists of particulate matter, sulphur dioxide, nitrogen oxide, carbon monoxide, hydrocarbons, volatile organic compounds and ozone. Except ozone, all other chemicals, having the value of the air quality index exceeding 150, can affect human health. Lead and arsenic, very common chemicals found in water, soil, air, paints and children's toys, may cause serious damage to health. Arsenic is associated with diseases of the skin and of the nervous system. It also triggers the risk of certain types of cancer including kidney, lungs and bladder cancers. Lead poisoning, which is a common type of poisoning in less developed countries, is associated with brain damage, mental retardation, stunted growth and reduced I.Q. Very often, indoor AP is many times more serious than outdoor AP in terms of health risk Indoor AP in India causes nearly half a million of death of women and children every year (Mukherjee and Hazra, 2008, p. 14). Apart from all these, there are hazardous chemicals all around us that have deleterious effects on human and animal health. The International Environmental Organisation has identified more than 180 air pollutants and some of them are so poisonous that they can have a devastating effect on human health including skin, brain, nervous system, reproductive system, and respiratory and circulating channels.

Continued exposure of workers in asbestos factories and to silica dust in various parts of the world, particularly in LDCs, leads to different types of serious diseases. Asbestosis causes pleural and parenchymal changes and may also cause broncogenic carcinoma and malignant mesothelioma among the workers. Silicosis is caused by exposure to silica dust and it involves diseases mainly of the lungs. It is accompanied by cyanosis and inflammation of lungs, shortness of breath and fever. In the Western countries, the incidence of silicosis has substantially reduced over the years due to the use of many protective gadgets and respirators. However, in LDCs, the multinational factories and local producers do not often provide the workers with protective devices. Hence, the death rate due to silicosis could not be reduced. This Grinder's disease is taking a large toll of human lives. Occupational hazards become more pronounced in LDCs because of poor working environments and absence of necessary safety nets. Globalisation has indeed increased the risk of lives of poor factory workers. The bad news is that the green house gas emission has increased in transitional economies by 7.4 per cent, while it was reduced by 5 per cent in the industrialised countries during the period 1990-2006 (UN Framework: Convention on Climate Change).

Motivated by the philosophy of maximisation of profits, neo-liberal globalisation has witnessed an expansion of polluting industries in different parts of the world, particularly in the Third World countries. Table 2 reveals the global trend of carbon concentration, ozone depletion and global temperature over a period of four decades. The Table clearly shows that while carbon concentration in the air has been increasing with the rising temperature, the concentration of ozone is gradually declining. This is pretty bad news for human health and wellness. This may, in the longer run, jeopardise human well-being, declining stock of ecological capital and inferior quality of human life. A study by Phillip O'Hara (2006) finds that there is a contradiction between environmental protection and the expansion of business/profit in our times. The temptation for profit in the days of globalisation is so great that the world's most polluting country, the United States, refuses to sign the Kyoto protocol to minimize the level of pollution. What has become clear by now is that there is a positive trade-off between business profits and environmental pollution in the world today; but the problem is that the capitalist countries do not appear to understand this empirical truth.

TABLE 2

Global Environment (1960-2000)

	1960	*1970*	*1980*	*1990*	*2000*
Air Concentration of CO_2	317	326	339	354	369
Global Temperature Change (Compared with 1961-90 average)	0	-0.05	+0.12	+0.26	+0.38
Ozone Concentration (Dobson Units)	293	276	227	172	135

Source : Compiled from various sources, as given in O'Hara (2006).

The level of pollution varies from country to country, and even in the same country, there may be different cities with different degrees of AP. Generally speaking, urban areas are more affected by AP than rural areas. Urban areas have higher concentration of motorised vehicles and various types of industrial activities. The Central Pollution Board of India measured the air quality in four Indian cities in 2007 and found that in terms of nitrogen dioxide and respirable suspended particulate matter (RSPM), the city of Mumbai was the most polluted one, followed by Delhi. The level of AP in these two cities was beyond the nationally acceptable average (vide the following Table 3). However, due to heavy

TABLE 3

Air Quality in Major Indian Cities (13-14 February, 2007)

City	*Sulphur Dioxide*	*Nitrogen Dioxide*	*RSPM*
National Standard	80	80	100
Chenai (Adyar)	7	12	94
Delhi (BSZ Marg)	7	70	133
Mumbai (Sion)	35	103	293
Vadodara	NA	NA	NA

Source : Govt. of India, Central Pollution Control Board (www.cpcb.nic.in).

rains, no comparable data could be collected for another city (Vadodara).

In all the cities in the world, due to a large number of vehicles and burning of fossil fuel, the AP is on the rise every day. Whereas in developed countries old vehicles are discarded or destroyed after a specified number of years, in LDCs, they are repaired and used on the road, and this increases the concentration of pollutants in the air. Of course, there are official provisions to check pollutants from vehicles and punish/suggest to the owners to remedy the situation. However, due to corruption and regulatory capture, the system of checks and balances does not work well in LDCs.

The damage done by AP in many countries is well-known. Who is not aware of the Chernobyl nuclear power plant disaster of 1985? This has left a horrible impression in the minds of the people of Ukraine. In Kolkata, the winter season experiences deaths of many old people and children every year due to chronic obstructive pulmonary diseases and respiratory distress syndrome. The city of Kolkata has over 12,000 small and big registered factories, and there are over 60,000 registered vehicles operating daily (Mukherjee and Hazra, 2008, p. 45). The damage to health in Kolkata is caused by high concentration of respirable suspended particulate matter and nitrogen dioxide in both industrial and residential areas. However, the Central Pollution Control Board of New Delhi (India) finds that the level of concentration of sulphur dioxide is not that alarming in Kolkata. AP is very high in the thermal power stations in India. Sulphur dioxide emitted from these thermal plants is the major source of AP in many parts of India. A report by the WHO discloses that about10 to 15 per cent of the population in India is suffering from asthma, bronchitis, hay fever and common cold, and these are all air-borne diseases.

The BBC report in the year 2002 disclosed that in the United Kingdom, the Great Smog of 1952 which spread over London for six days was the main cause of death of 12,000 people. In the history of India, the worst type of air pollution was caused by what is known as the Bhopal Gas Tragedy. The leakage of a hazardous gas called methyl isocyanate from the Union Carbide factory, a multinational pesticide company, was responsible for the death of about 4,000 people and caused permanent or temporary injury to about 5,00,000 which led to the death of another 6,000 people. The gas leakage not only disabled the factory workers but also affected the members of their families and the residents who were staying nearby including children and women. In the United States of America, in one of the worst incidents of AP, 20 people died and 7000 people were injured in Donara, Pennsylvania in

1948. The leakage of a poisonous gas, anthrax spores, from a biological warfare laboratory in the former USSR led to the death of hundreds of civilians including men, women and children in 1979. There are numerous examples to prove that no place in the world is safe from AP. The Environmental Protection Agency reported in 1985 that indoor air pollution was three times more dangerous than outdoor air pollution in American homes, and in such a situation, the members of a family are susceptible to serious health problems including cancers. This shows that ladies who most of the time stay at home are prone to some type of air-borne diseases. The chances for such diseases are far higher for housewives of LDCs than those in DCs, as in the latter countries, most of the female members also work outside homes. In fact, housewives, who are constantly exposed to cooking fuel in developing countries, have higher possibilities of health damage from indoor air pollution.

In China, the major cities like Beijing and Shanghai are affected by the over-accumulation of carbon monoxide, nitrogen oxide and photochemical smog. However, China has been trying to reduce the AP by following the standard set by the European Union. In an empirical study conducted in Ukraine, it is estimated by Elena Strukova *et al.* (reported in Mukherjee and Hazra, 2008, Ch. 8) that in Ukraine, AP-related mortality represents about six per cent of total mortality in the country; in Russia, it is about four per cent. But morbidity represents about thirty per cent of total air pollution health load.

The monetary cost of mortality and morbidity due to AP is around 2.6 billion USD per year. This cost of AP in Ukraine is too high and in the near future, as the authors maintain, it will offset the economic growth rate.

It needs to be borne in mind that the actual effects of AP on human health are not properly appreciated in LDCs because most of the minor ailments produced during the short period or due to short-term exposure are neither medically consulted nor registered on medical records. However, the acute effects of AP are produced rather dramatically and can be easily identified. Short-term elevations in atmospheric particulate matter have been often associated with the triggering of acute cardiovascular syndromes which include myocardial infarction, ventricular arrhythmias and ischemic strokes. Studies have found a positive association between short-term increases in ambient particles and the risk of heart ailments. The long-term effect of AP is also ignored as it cannot be separated when it is mixed up with the basic syndrome of another acute or chronic disease. Moreover, the effect of AP on plants and animals is neglected in such studies. Thus, the actual impact of AP remains underestimated. The authorities very often forget about human health but are excited about the progress of urbanisation, rapid rate of industrialisation and a high rate of quantitative economic growth. Quantity of material gains often overshadows the quality of human life. This is indeed the travesty of modern age and civilisation.

Although there are different issues involving exhaustible and renewable resources, there is no denying the fact that polluters should be made to pay for the amount of pollution or misuse of natural resources, particularly when negative externality is created by polluters.

SOCIAL COST OF POLLUTION

Some authorities favour the idea of calculating costs and benefits for the purpose of proper management of environment. However, it is indeed very difficult to calculate the cost and benefits of environmental control. In the next section, we explain the meaning and impact of externalities (external factors or influences).

What is an Externality ?

Externality is an external influence that does not come from the institution of market but affects the welfare of a person either adversely or favourably. Externalities adversely affect economic efficiency.

Externalities can be positive or negative. Air pollution is an example of negative externality. A negative externality is a cost which is not reflected in the market place. A positive externality is a benefit accruing to a person without his own effort and it is not generated from market transaction. Education is a good example of a positive externality. An educated person is likely to produce many benefits to the society as a better citizen, as a source of information advice and so on.

Positive and negative externalities can be found both in the case of consumption and production. An example of positive externality in consumption is vaccination. By vaccination not only is the individual vaccinated protected from certain diseases, but the whole society is protected. An example of negative externality in consumption is the noisy motorbike which creates disturbances to the others. An example of positive externality in production is beehives which not only produce honey to the owner but would also help the pollination of plants. Another example of negative externality in production is the riverside chemical factory which dumps its chemical wastes in the river and pollutes its water.

As already been pointed out a positive externality is a benefit and it is not, therefore, a social problem because it may simultaneously improve the economic welfare of everybody (e.g. a good climate). Since negative externality create real problems, the subsequent discussion would be devoted to negative externality.

Negative Externality and Market Failure

In the case of negative externality, the social cost is higher than the private cost. To be more precise, the marginal social cost is higher than the marginal private cost, and the actual output produced by the negative externality creating industry (or film) does not take into account the social cost that it creates. Thus competitive equilibrium will result in : (i) more than efficient output and, (ii) less than efficient price. The market price falls short of the actual marginal social cost of the product. Under such a situation, an excessive amount of resources is allocated to produce the output which involves externality (vide Figure 1). The market fails to achieve allocative efficiency.

Negative Externality

The Figure assumes that the demand curve for the product (D) which reflects marginal benefits (MB). When the social cost (equivalent to MD) is neglected and the production consideration is based on the marginal cost (MPC) alone, the output would be OM and the price \$ 100 per unit of output. However, if social cost is added to the marginal private cost, then one can get the actual marginal social cost (MSC). Thus, the correct equilibrium point is not A but B, and the correct output is not OM but ON. The actual output OM is not the correct output and the actual price (OP_0) is not the correct price because at this level of output and price, the marginal serial cost (MSC) is higher than MSB. The efficient output corresponds to point B where MSC= MSB. Thus, in the case of negative externality, actual output is higher than efficient output and actual price is lower than the efficient price (P_1 = \$ 110).

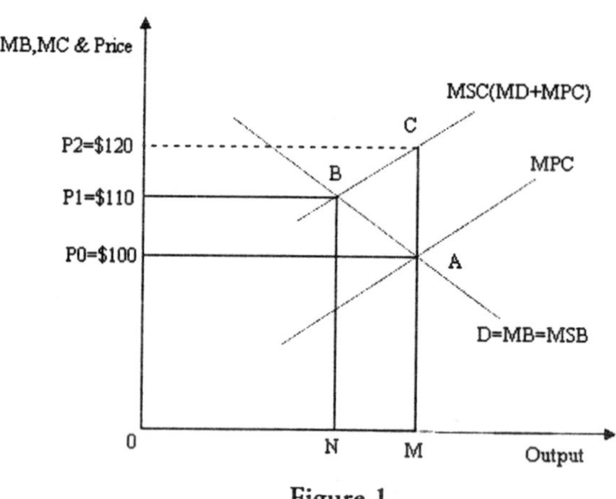

Figure 1

Why is Negative Externality a Problem?

The problems of negative externality in production, discussed earlier, can now be summarised:

(i) Price mechanism fails to allocate resources.
(ii) Actual competitive output is more than efficient output.
(iii) The actual competitive price is lower than the efficient price.
(iv) The society is at a loss because the competitive productions relations and conditions to losers.
(v) Since externality is not transmitted through the market and there is no market for negative externalities, market solution does not seem to be possible.

ETHICAL ISSUES IN ENVIRONMENTAL POLLUTION/ DAMAGE

Environment pollution (EP) is associated with many types of ethical problems, some of which are briefly discussed below:

1. As pointed out in the foregoing analysis, negative environmental externalities (external influence) impose extra cost to poor people who are affected by the pollution. This leads to lower benefits and loss of utility. Very often, the social cost is not paid by the polluter but has to be borne by the sufferer. This is ethically unjust and wrong. Negative externalities thus lead to *market failure* because market mechanism cannot account for them, and as a result, the market price does not show the real worth of a commodity. In such a situation, there are misallocations of resources.
2. As most of the poor people depend on forest for environmental resources for their livelihood, it is morally unjust to destroy or damage these resources. As discussed earlier, poverty is also responsible for environmental degradation. However, once the level of poverty is reduced through economic development, there is a possibility of less of environmental damage.
3. It is the basic human rights to enjoy a clean and healthy environment. EP deprives human beings of such natural rights and hence EP is unethical.
4. EP endangers the health of poor people who reside in vulnerable areas. EP imposes higher burden on poor people because sickness leads to loss of income and productivity and mandays.
5. Pollution increases the cost of production (if social cost is

included) and reduction in output. Both are socially undesirable, particularly for the poor people.

6. Justice needs that polluters must be punished and they should pay for the damage .But very often they go scot free because of legal loopholes and *regulatory capture.*
7. More often than not, EP and damage is due to over-use of natural resource, which is made possible by increasing domination over nature. Too much domination over nature is unjust and unethical in the sense that it is exploited for personal or selfish ends.
8. The damage to environment and are unethical and unjust for the future generations who have equal rights to enjoy the benefits and gifts of nature. Human rapacity is responsible for the degradation of environmental resources. The depletion of these resources is unjust and unethical for the posterity.
9. EP leads to loss of life or extinction of certain species of animals, fish and plants and the loss of *ecological balance.* Non-human creatures have also the right to live in a natural environment along with human beings. The harm caused to them is unethical and unjust.
10. Environmental pollution leads to *hedonic injustice* by distorting the prices of land, houses and other habitats. In the case of EP, the prices of houses and lands in the polluted areas go down abnormally and the owners suffer unjustly without any fault of their own.

MANAGEMENT OF ENVIRONMENTAL POLLUTION

1. Corrective Pigovian Tax

A tax equivalent to the marginal damage created by the polluter can be imposed by the government on the polluter. Such a market-internalizing tax is able to secure the position of Pareto optimality.

2. Government Regulation

Government regulation may be resorted in two cases: (i) to impose the correct dose of taxation on the polluter, and (ii) the merger of firms (the polluting firm and the victimized firm). However, such a merger is not possible for many reasons. But nevertheless, the imposition of tax is likely to reduce the output of the polluting firm, which would be a gain to the victim because less output means less damage.

3. Market for Pollution Rights

Some economists feel that it would be better to allow the market to solve the negative externality problem by permitting the polluters to pay fees for the pollution. Such fees can be correlated to the pollution damage. The creation of a market for buying and selling polluting rights will have several advantages (e.g., the right can be given to the highest bidder), identification of polluters would be possible and emission standard can be fixed and controlled.

4. Internalization of Externalities

The internalization of externalities requires that the producers (polluters) are forced to take account of the costs they impose on others. In the case of negative externality, internalization requires that the social cost of negative externality be incorporated into the market price of the product.The increase in the market price of the good (as a result of marginal social cost incorporating marginal private cost plus the cost of damage) would mean that the equilibrium quantity demanded and supplied of the good would come down, and therefore, negative externality would be internalized.

5. Creating Property Rights

According to Ronald Coase, well-defined property rights are sufficient to internalize any negative externality when there are small numbers of affected parties and when the transaction cost is very low. Government intervention is essential because it can not only establish rights to use resources but also encourage the trading of those rights, and minimize the transaction costs involved.

SUSTAINABLE DEVELOPMENT

Sustainable development (SD) is now regarded as one of the best policies for the management of environment *vis-à-vis* human consumption needs.

Ever since the publication of the Brundtland Commission Report in 1987, the concept of sustainable development (SD) has been the subject of a new understandings and many misunderstandings. The United Nations World Summit in 2005 reiterated the views of the Brundtland Commission for the popularization and practice of SD among all countries in the world for the sake of meaningful human existence with ecological balance. Sustainable development is a type of development that aims to use natural capital and environment in a way that can meet the present and the future human needs. Thus, the use of these resources by the present generation will not deplete the stock of these resources for the posterity. Thus, SD incorporates an element of intergenerational equity. However, SD needs to be distinguished from *Green Development* which basically aims at preserving and promoting the environmental resources. The concept of green development is an overly protective idea, and has nothing to do with economic and social development.

For understanding the core idea of SD, one needs to know the level of human consumption and the replenishment capacity of the stock of natural capital. In this context, we can come across the following three situations.

CASE ONE : Human consumption exceeds the replenishment capacity. So there is depletion of natural capital. This, then, does not constitute sustainable development.

CASE TWO: Human consumption of natural capital is just equal to the replenishment capacity. This is, therefore, the steady state growth.

CASE THIRD: Human consumption of the natural capital is much lower than the replenishment capacity. This is precisely the case of sustainable development.

SD recognizes the constraints and limits to development, and believes in the inter-linkages that exist among the economy, society and environment. It is based on the recognition of the nexus that exists between socio-economic development and environmental development. The theory of SD works through inter-linkages, inter-generational equity and operational efficiency. It is a common experience that economic growth in the days of globalization impinges on natural environment and disturbs the ecological balance. The details about the type of nexus between growth and environment and the impact of growth on environment have been discussed in the last section of the present chapter.

In recent years, the craze for quantitative growth in developing countries under the influence of globalization does not pay much heed to environmental protection, which slowly destroys the stock of natural capital and is detrimental to both the present and the future generations. However, a contrary idea held by many economists is that, natural capital, human capital and knowledge capital do not diminish over time. However, although these capitals are not completely destroyed in the long-run, their replenishment capacity many not be the same as their levels of wear and tear and destruction due to over-use. But it must be conceded that by following the policy of sustainable development strategy, the planners are able to make sure that there will not be any natural capability failure in the area of natural resource use. SD is essentially a policy-oriented strategy of development.

***Sustainable environmental development** (ED)* has to take into account the carrying load on the environmental resources. As hinted earlier, if the load is such that it cannot replenish the damage or loss over a period of time, then the ED does not qualify for sustainable development. However, if the damage or loss done to the environment can be replenished over time and if the consumption level of the natural stock of capital is much less than the replenishment power of the environment, then it is a case of sustainable ED. The problem with this type of analysis is that it considers ED in isolation and does not take into account many types of inter-linkages and feedbacks. It also assumes away the problematics of measuring the extent of real environmental damage or losses.

References

Chary, S.N. and Vyasulu, Vinod (Eds.), (2000), *Environmental Management*, Macmillan, Delhi.

Ghosh, B.N. (2010), *Rich Doctors and Poor Patients : Market Failure and Healthcare Systems in Developing Countries*, Wisdom House, U.K., pp. 1-17.

Ghosh, B.N. (2001), *From Market Failure to Government Failure*, Wisdom House, Leeds, UK, Ch. 1.

Ghosh, B.N. (2009), *Understanding Engineering Humanities*, Trinity Press, UK, Chs. 10 & 11.

Mukherjee, Debashree and Hazra, Somnath (eds.) (2008), *Air Pollution and Health*, The ICFAI University Books, Hyderabad (India).

O'Hara, Phillip (2006), "The Contradictory Dynamics of Globalization" in B.N. Ghosh and H. Guven (Eds.), *Globalization and the Third World: A Study of Negative Consequences*, Palgrave-Macmillan, London and New York, *World Development Report* (2007).

Part II

Natural Resources

Chapter 3

Ecological Economics of Shifting Cultivation in an Indian Ecosystem

NIRMAL CHANDRA SAHU AND CHANDRA DHWAJ PANDA

ABSTRACT

Shifting cultivation is a traditional practice involving field (not crop) rotation, which, in policy circles, is held as eco-damaging through loss of forests and erosion of soil. However, it is a major food production system of the tribal ecosystems in India. In this context, the research problem here is: why the tribal communities, even those in transition, continue this practice in spite of policy incentives against it? We respond to this question through a case study of 60 Lanjia Saura families of Gadiabanga village in Rayagada District of Orissa. The primary data collected through a census survey over one year have been meticulously processed.

It is found that the practice of shifting cultivation is a part of culture rather than commerce. Forest land and human labour are the only two inputs into its production function. Through a series of Probit and Logit models, the factors that influence the probability of a tribal household to cultivate Podu (shifting cultivation) area beyond the per capita median size are determined. It is observed that the likelihood for a family to extend the area is increased by incomes from Podu and the forest, poverty distance and the outstanding debt burden. Of all the variables, debt is the most important determinant. In the face of low carrying capacity of Podu land and low productivity of labour, the incentives to continue the practice come from the socio-cultural milieu, intensity of poverty and debt, natural insurance against uncertainty and the notion of complementarity of economic activities. In terms of variety and certainty, shifting cultivation has an edge over the other socio-economic sectors. The entire process of tribal development over the last six decades is yet to introduce so diverse and so stable a source of livelihood. The policy prescription from the study is that any alternative to the practice should be capable of meeting the resilience condition of the system.

1. INTRODUCTION

Shifting cultivation is one of the most important food production systems of the tribal communities. It is a system of field (not crop) rotation. Preliminary clearings of the forests are done by 'slash and burn' procedure. Short periods of cropping are alternated with long fallow periods. The practice involves spatial and temporal impermanence with regard to the structure of land occupation. There occurs great diversity in the practice among different regions and tribal communities. The variety of social institutions, cultural milieu, heritage, traditions, religious beliefs, rituals, soils, hydrological systems, climatic conditions and vegetations has yielded and preserved this diversity (NCA, 1976, CSE, 1985, Fernandes *et al.* 1988, Pal, 1989, Sachidananda, 1989, Ramakrishnan, 1992). Even though there is a wide diversity, the process moves through the main route of forest land occupation, site selection, land preparation, sowing, weeding, watching, harvesting, consumption and storage with a long series of follow-through rituals (Ramani, 1988).

Forests and tribals interact through a symbiosis in an ecological economic system. Sustainability of tribal well-being hinges on the health of forests, which is a function of the scale of shifting cultivation. If Podu (local name of shifting cultivation in Orissa) practice remains within the stability and resilience of the forest-linked tribal ecosystem, sustainability can be promoted. The system of cultivation is observed sustainable under low population density and long fallow periods, with systematic investment in natural capital through micro-level sustainable management and planning.

The central theme of this paper is that why the tribals who have experienced some amount of transition and transformation continue with shifting cultivation. Are there inherent ecological and economic advantages for the tribal families and the village environment? Through a case study we respond to this question. The case under study hovers around the Podu practice of 60 tribal families belonging to the Lanjia Saura ethnic group, who live in a village called Gadiabanga. The village is located near Gunupur town in Rayagada District of south Orissa, India.

In the literature and materials we have surveyed, accurate data are not available on the state and extent of shifting cultivation in Orissa. There is no unanimity in the estimates (Sahu, 1986, Sachidananda, 1989,

Patnaik, 1993, Mohapatra (C. R.), 1997, Samal, 1994, CPSW, 1996). But all studies indicate that Podu has a predominant presence in Orissa. It is concentrated on the northern and southern districts of the state. The Podu property rights of the tribals are peculiar. In the past, the rulers had permitted the Adivasis for the practice. The statutes recognised the privileges of the tribals after independence in the Protected Forests. In the Reserves, the tribals are prevented to resort to the practice. But it continues as a traditional right. In Orissa, the Podu cycle is now reduced from 15-20 years to about 2-7 years due mainly to population pressure. For this, it is alleged that the practice is responsible for a number of evils, mainly associated with deforestation and soil erosion. However, this view is contested with arguments from anthropology, sociology, economics and ecology. One well-recognised positive feature of Podu is associated with simultaneous sowing of a variety of seeds for sequential harvesting. As against monoculture, this practice has been a traditional coping strategy of the tribals against uncertainty and vulnerability.

One of the most frequently articulated views on shifting cultivation relates to its dynamic linkage with the forest. The process is considered as a contributor to deforestation. Several works including Sahu (1986) and Jyotishi (2000 and 2001) have summarized the frequently attributed evil effects, such as destruction of valuable forests for the sake of less valuable grains, acute soil erosion, drying up of springs below the hills, heavy floods and siltation of dams, Mohapatra (C.R., 1997) has considered Podu as an ecological hazard. Contrary to this Fernandes *et al.* (1988) and Mohapatra (L.K., 1997) have attributed the real devil to the population pressure, circumscription of forest area for reservation, location of mining, multipurpose river valley and other development projects and the claims of commerce and industry for short-term profit.

As elsewhere in the world and India, the policy of government of Orissa has been to persuade the Adivasis to give up shifting and take to settled cultivation. The forest and tribal development strategy has been to provide some better economic alternatives so that the tribals cease the practice. The policy efforts in Orissa have been resettlement in colonies, perennial cropping system, alternation of agriculture with silviculture (*Taungya* cultivation), rational land use on watershed basis, alternative-employment in forestry and subsidiary occupations, household level poverty alleviation schemes and building up of income generating assets. Notwithstanding all these, shifting cultivation continues in tribal pockets of Orissa. The people have responded to the environmental stress through out-migration (Chopra and Gulati, 2001). The regularity and promotional efforts of the Forest and Soil Conservation

Departments could not bring in discipline and change in the practice. Why? The problem under study hovers around this question.

2. METHODOLOGY

The secondary information collected from a wide number of sources are used to appreciate the nature and extent of shifting cultivation, the strategy of tribal development and build a helpful profile of Gadiabanga village. But the core of the research problem is addressed with a large amount of primary information gathered from the village using a specially designed questionnaire-*cum*-schedule. However, as the research problem warranted systematic field observation, a method of maintaining a diary for each family developed. A census approach was followed to collect the data from each of the sixty households. The village was visited at least once a week through a period of more than one calendar year during 1999-2000. Even after recording the data for one full year, the village has been visited again and again as per the requirements of the study. The collected information were collated through several stages of tabulation to build household-wise datasets on the general particulars, property, shifting cultivation, agriculture, home garden, miscellanies crops, livestock production, forest-related activities, labour inputs to each production sector, daily wage income, loan and debt, and the links with the market. During the process of data collection, information on production, consumption, exports and imports of each of the item were maintained in local measures. Conversion factors were developed by physical weighing of most of the items to make quantitative analysis with standardized data. The prices of the different commodities and items were ascertained from Gadiabanga, Gunupur and Brahmani weekly market (*Hata*). Later, after thorough discussion with the tribals the ex-village price set of the year was finalised. For the Purpose of understanding the process of shifting cultivation and other production activities and finalisation of prices, a series of PRA (Participatory Rural Appraisal) type surveys were conducted in the village (Panda, 2003). The total value of property is estimated as the sum of the money value of land, house and cattle owned by a household. The value of annual production, income, consumption, exports and imports were determined from the weekly and monthly diary records maintained in physical terms by using the final price set. For each family these variables were estimated as, the sum of the value of the output from Podu, agriculture, home garden, miscellanies crops, forests, livestock and daily wage sectors. Human energy inputs were recorded in terms of the hours of work devoted to a particular process or activity. Then for estimating employment, a human day is calculated taking 8 and 12 hours of work for the adults and children respectively.

In order to ascertain the variables that influence the probability that a household would cultivate Podu land above the per capita median size, discrete dependent variable models are tried. Out of several types of binary choice models, Probit and Logit models are estimated in this study. The Probit model emerges from the assumption of the normal cumulative distribution for the probabilities, which has the following functional form (Hill *et al.*, 2001, pp. 370-74; Green, 2000, pp. 811-20):

$$\text{Prob}\,(y=1) = P_i = F(I_i) = F(\alpha + \sum_{j=1}^{K} \beta_j\, x_{ij})$$

where,

$$P_i \begin{cases} = 1 \text{ for the households cultivating per capita Podu area above the median, and} \\ = 0 \text{ otherwise.} \end{cases}$$

I_i = The index function of the probability

F = The following Probit function for a standard random variable z:

$$F(z) = \int_{-\infty}^{z} \phi(t)\, dt$$

The function ϕ (.) is the commonly used notation for the standard normal distribution. The Logit model uses logistic cumulative distribution function. In this model, the probability (P) that the observed value y takes the value 1 is

$$P = F(\alpha + \sum_{j=1}^{K} \beta_j\, x_{ij}) = \frac{1}{1+e^{-(\alpha + \sum_{j=1}^{K} \beta_j x_{ij})}}$$

Both the binary choice models are tested with ML estimators.

3. THE VILLAGE ECONOMY : A BRIEF PROFILE

Gadiabanga is located 5 km. away from Gunupur town in the Rayagada district of south Orissa, India. The village occurs on a small compact patch of land on the foothills of mountains, which contain the Pedakonda and Rangamati Reserved forests. The area has a tropical monsoon type of climate with pronounced summer. On an average, the area receives 1228 mm of rainfall in about 100 days of rain. All the 60 households of the village belong to the Lanjia Saura tribe. During the last 15 years they have taken to Christianity. In the process, some of the traits of the Lanjia Saura community are no more prominent. Dress and food habits are on the change. The state of education is very poor, even though the village has a Sevashram. Only 11% of the population is literate, which is much lower than the District and State literacy rates. Male literacy is 9 percent, which is far higher than the female literacy rate of only 2 percent. The village is Malaria prone. Digestive disorders are a common ailment of the people. Marriage and festival habits linger as a part of the tradition and heritage. The traditional beliefs and Church going habits co-exist. Celebration of the Christmas Day is an additional festival of the village, but it is gradually becoming the most important one. So far as the state of the overall development is concerned, the village is now served by a road, a recently dug tank, a Catholic Church house, a Sevashram and a Kanyashram. Three to five years old forest plantations are visible. In these, cashew occupies a prominent place. Notwithstanding the induced change and development, all the households are below the poverty line (Panda, 2003).

Forest is the predominant claimant of the Gadiabanga land. The cultivable area is about 23% of the total territory, but the low land paddy is only a small fraction. The water bearing capacity of the soil is moderate. The forest vegetation is of the southern dry mixed deciduous type. The Pedakonda has been encroached for cultivation since long. Under a Long-Term Action Plan of the JFM programme, cashew plantations have been raised. The economic activities of the households include shifting and settled cultivation, raising of home garden and miscellaneous crops, collection of forest produce, livestock production and daily wage engagement. All the households depend on Podu cultivation for a number of subsistence crops. The village is not free from middlemen and moneylenders.

In Pedakonda, forest encroachments are of two types. These are for permanent and shifting cultivation. Under the first type, the encroacher completely occupies a part of the forest. He erects a permanent structure on the land, uses it for cultivation, and tries to settle there by building a house. In case of the latter, the encroacher clears a patch of forest land, cultivates it for some years and after that leaves aside that patch as fallow for some years. This type of shifting cultivation practice is quite prevalent in the entire Rayagada Division (GOO, 1989). To protect and preserve the Pedakonda Reserve Forest from such evil practice, a Vana Surakshya Samiti (VSS) was formed in 1990. But shifting cultivation could never be stopped. Later on, however, growing conflicts of interests among the villagers led to the collapse of the Samiti.

All the 60 households practice Podu on 86.11 Acres of forest land on Pedakonda. The mean Podu area per family and per capita are 1.44 (± 0.07) and 0.30 (± 0.02) Acres (± SEM) respectively (Table 1). The

TABLE 1

Distribution of Families According to the Range of Podu Land

Sl. No.	*Podu area range (Acres)*	*Households*		*Total Podu land*		*No. of Members*		
		No.	*%*	*(Acres)*	*%*	*Male*	*Female*	*Total*
(1)	*(2)*	*(3)*	*(4)*	*(5)*	*(6)*	*(7)*	*(8)*	*(9)*
1.	Up to 0.50	7	11.66	3.50	4.06	11	10	21
2.	0.51 to 1.00	9	15.00	9.00	10.45	20	22	42
3.	1.01 to 1.50	20	33.34	28.20	32.75	52	43	95
4.	1.51 to 2.00	21	35.00	37.96	44.08	58	63	121
5.	2.01 to 2.50	1	1.66	2.10	2.44	2	3	5
6.	2.51 to 3.00	2	3.34	5.35	6.22	3	2	5
7.	Total	60	100.00	86.11	100.00	146	143	289

Source : Panda, 2003.

tribals raise seven crops on Podu lands. They are Great millet, Fox tail millet, Red gram, Horse gram, Cowpea, Castor and Niger. Land and labour productivities are highest for Great millet, which is the staple food of the people in the village. Per family and per capita mean productivities are also high for the millets (Table 2). The SEMs are found very low, which indicates that the crops are stable in terms of their physical contribution. In terms of diversity of items, the forest economy is neatly predominant. The two other main economic sectors, such as shifting and settled cultivation, together provide the same number of products, which the forests contribute. So far as food variety is concerned, the contribution of Podu crops is high. Because of simultaneous sowing and sequential harvesting of several Podu crops, the Gadiabanga economy has relatively lower degrees of the problems of risk, uncertainly and vulnerability in comparison to regular agricultural ecosystems.

In respect of physical production from all activities, the millets occupy a very good position next only to Paddy. Red gram and Horse gram do not indicate substantial productivity advantage under settled in comparison to shifting cultivation. There occurs a reiterated observation in several studies that the Podu productivity is very low. But an agriculture that receives modern chemical, biological and mechanical technological inputs stands no comparison with shifting cultivation. Against the low capital-intensive tribal agriculture of a mountain or a forest-based ecosystem, Podu productivities are not worse. It is, however, not possible to provide conclusive evidence from this study, because the survey period was the first year of the current Podu cycle. Moreover, when all products under settled cultivation are considered, the paired t-test procedure does not indicate advantage under shifting cultivation.

The contribution of the forest sector is as high as about 47%, followed by agriculture. Shifting cultivation occupies the third position accounting for about 16% of the total income. It has been observed that, as Podu land increases, income from Podu increases. But there is an inverse relationship between income from forest and Podu land under cultivation. To some extent, therefore, there is an element of competitiveness in respect of the claims of the sectors on the human effort. So for as demand on human labour is concerned, claim of the Podu sector is very high (about 45%). The forest sector takes the third position after agriculture. In terms of human energy, it has been estimated that the overall Podu activities such as the forest clearing, forest burning, land preparation and crop watching claim about 33% of the total human energy inputs. Crop specific activities claim only 6% of

TABLE 2

Productivity Statistics of Podu Crops in Gadiabanga

Sl. No.	*Crop*	*Families (No.)*	*Area (acre)*	*Total production (kg.)*	*Land productivity (kg. per acre)*	*Labour productivity (kg. per humanday)*	*Price per kg (Rs. of 1999-2000)*
(1)	*(2)*	*(3)*	*(4)*	*(5)*	*(6)*	*(7)*	*(8)*
1.	Great millet	60	86.11	5272.50	61.23	1.32	6.00
2.	Fox tail millet	60	86.11	3035.00	35.25	0.84	2.00
3.	Castor	57	78.61	2080.00	26.46	0.60	7.00
4.	Redgram	60	86.11	2067.50	24.01	0.55	12.00
5.	Horsegram	40	55.26	1523.00	27.56	0.58	4.00
6.	Cowpea	60	82.26	295.50	3.59	0.68	8.00
7.	Niger	38	58.76	106.60	1.81	0.34	12.00

Source : Panda, 2003.

energy. In economic activities, the share of women is less in terms of labour contribution. When total household well-being is considered including the domestic work burden, the true contribution of women becomes prominent.

Forest land and human labour are the only two inputs into the Podu production function in Gadiabanga. A Cobb-Douglas production function type of analysis shows that the returns to scale parameter is 0.97. The output elasticity of land and labour are respectively 0.42 and 0.55. Thus the Podu production function exhibits constant returns to scale. However, it is marginally less than unity, which indicates that the swiddeners have entered the stage of marginal diminishing returns. In other words, the scale of Podu production should decrease, rather than increase in Gadiabanga. The position of Podu practice in the village yields a different picture under law of variable proportions. Keeping the human labour constant, if Podu area increases these is a scope to increase output. The most efficient production point is at 2.37 acres of Podu land which is beyond the current median range of the people. There is, however, no prescription here that the Gadiabanga tribals should encroach into a larger per capita forest land area from Pedakonda. Because, the benefits from the present state of the forest is far higher than what it would be when it is converted into Podu land.

4. BINARY CHOICE MODELS: RESULTS AND DISCUSSION

The fundamental issue addressed here relates to the question as to why the Savaras of Gadiabanga continue the Podu practice. The village economy is at the threshold of a transformation. The process of transition is at its last phase. It is pretty close to an urban center. It is not located in the interior of a forest or on a hill top. A good road connects the village with the sub-division headquarters Gunupur on one side and the Panchayat headquarters Chinasari on the other. Development activities including forest rehabilitation are under implementation since some time, even though the impact is marginal. Because of contact with the people and market in urban center, the Savaras have better exposure to information and modern world, the state of literacy is poor though. The links with Christian Missionaries have induced changes in thinking and attitudes. Notwithstanding all these, the socio-economic and cultural milieu have not changed. The practice of shifting cultivation continues. At the end of the present Podu cycle at Pedakonda, there is an urge to explore Rangamati Reserved Forest. Why? Is it simply a tendency to free ride on the public natural forests?

In order to find clues to this question, binary choice models are considered as the best econometrics approach. Since, all the families of

Gadiabanga resort to shifting cultivation, it was difficult to get an appropriate limited dependent variable. But it was observed that the per capita Podu area is different for the different families. Even though the areas earmarked in Pedakonda, for each family have been decided on the basis of ability since long, the present households have taken larger plots at a distance. It was decided to follow this habit in the form of a dummy variable. The median per capita Podu area in Gadiabanga is 0.296 areas. This being the divider, a dummy variable, *Podu*, is constructed. All those who have per capita Podu area above the median are taken as 1 and 0 otherwise. In a series of simple and general Probit Model and a final Logit model, the dependent dummy variable is regressed with the number of explanatory variables. The objective is to find what explanatory variable increases the probability that a household will take per capita Podu area larger than the median.

The definitions of the variables used in the models are given in Table 3. In Table 4, the coefficients of the index function of the

TABLE 3

Definition of Variables

Sl. No.	*Name*	*Explanation*
1.	Podu	Dummy for shifting cultivation (1 = Per capita Podu area above the median of 0.296 acres, and 0 otherwise)
2.	Income	Total annual income of the household (Rs.)
3.	Pcincom	Per capita annual income (Rs.)
4.	Poduy	Annual income from Podu (Rs.)
5.	Forsty	Annual income from forests (Rs.)
6.	Wagey	Annual income from daily wage (Rs.)
7.	Poarea	Podu area of the household (Acres)
8.	Property	Total value of property (Rs. x 10^3)
9.	Fsize	Family size (No. of members)
10.	Litrat	Literate members in the family (No.)
11.	Polab	Labour input to total Podu production (Humandays)
12.	Loan	Amount of loan availed (Rs. x 10^3)
13.	Debt	Amount of loan outstanding (Rs. x 10^3)
14.	Povdy	Distance of per capita annual income (Rs.) from poverty line (=Rs. 4160.40, as defined under SGSRY)
15.	Povte	Distance of per capita total annual expenditure (Rs.) from poverty line (Rs. 4160.40)
16.	Povfc	Distance of per capita annual food consumption (Rs.) from poverty line (Rs. 4160.40)
17.	Calorie	Distance of per capita per day food consumption in terms of energy from poverty line (2400 Calorie)

Source : Panda, 2003.

TABLE 4

Coefficients of the Index Function of Probability of *Podu* in the simple Probit Models

Sl. No.	*Constant*	*Explanatory variable*		*Log L*	χ^2	*Mc-Fadden*	*Rsqd_ML*	*Correct prediction (%)*
		Name	*Coeff.*					
(1)	*(2)*	*(3)*	*(4)*	*(5)*	*(6)*	*(7)*	*(8)*	*(9)*
1.	-1.646 (-3.098) *	Pcincom	0.0008(3.240) *	-35.70	11.77 *	0.1415	0.1781	71.67
2.	-0.864(-1.767) J	Poduy	0.0006(1.886) J	-39.78	3.61 J	0.0434	0.0584	63.33
3.	-0.552(-1.699) J	Wagey	0.0005(1.988) #	-39.53	4.12 #	0.0496	0.0663	56.67
4.	1.199(2.348) #	Fsize	-0.2532(-2.438) #	-38.23	6.72 *	0.0808	0.1059	53.33
5.	0.178(0.966)	Litrat	-0.5919(-1.919) J	-39.39	4.39 #	0.0528	0.0706	58.33
6.	-0.302(-1.457)	Debt	0.1535(2.262) #	-38.69	5.80 #	0.0698	0.0922	56.67
7.	1.619(3.034) *	Povdy	0.0008(3.240) *	-35.70	11.77 *	0.1415	0.1781	71.67
8.	1.952(2.951) *	Povte	0.0009(3.065) *	-36.50	10.17 *	0.1223	0.1559	70.00
9.	3.091(3.060) *	Povfc	0.0011(3.112) *	-36.33	10.51 *	0.1263	0.1607	78.33
10.	0.507(2.136) #	Calorie	0.0007(3.105) *	-36.41	10.36 *	0.1246	0.1586	70.00

Notes : 1. Estimated from survey data using Limdep econometric software (Version 7.0.3: EA/Limdep 1.0.2 of August 17, 1999).
2. Figures in the parentheses are 't'-ratios.
3. Rest. Log L is -41.59 in all models.
* Significant at 99% or higher level of confidence (a = .01 or lower).
\# Significant between 95% and 99% level of confidence (.05 > a > .01).
J Significant between 90% and 95% level of confidence (.10 > a > .05).

Source : Panda, 2003.

probability in the simple Probit Models are presented. It shows that variables like per capita income, wage income, family size, literacy, outstanding debt and the poverty distance variables have regression coefficients that are significant at the acceptable levels of confidence. In all the models, the estimated chi-squared statistics is also found significant. Even though the R-squared ML is not very high, the percentage of correct prediction is satisfactory remaining in the ranges of 53% to 78%. The Mc-Fadden statistics, measured by

$$LRI = 1 - \frac{\ln L}{\ln L_0}$$

which is an analog to the R^2 in the conventional regression, is low in most of the models. It is found that per capita income, Podu income, daily wage income and the degree of indebtedness positively influence the likelihood that a Savara household will take more land for shifting cultivation. Similarly, literacy tends to decrease that probability. The signs of the regression coefficients in all these Probit Models are as per our expectations. It is however, surprising that family size tends to decrease that probability. Similarly, it was also not our expectation that some poverty distance variables will have a negative influence on that probability. It is to be noted that the poverty distance variable will have decreasing influence on probability, when the regression coefficient is positive. In term of quantitative significance, *Fsize*, *literacy* and *debt* are relatively meaningful. However, because of the low model fit measures, implying inadequate explanation in the index function probability, a set of general Probit Models are formulated.

Table 5 furnishes the estimated coefficients of the index of the probability of the Podu dummy dependent variable of the general Probit and Logit Models tested with ML estimators. Everywhere, Poduy and family size have significant influence among all the variables. The qualitative and quantitative significance of wage income and debt are consistent. But literacy, even though has appropriate sign, has t-values below the acceptable range. The variables that surprise us are poverty distance considered in term of per-capita income and minimum calorie requirement. The influence of poverty distance in terms of income *(Povdy)* is consistently negative, which indicates that as per capita income decreases the probability increases. But the calorie deficiency as a determinant is found to have positive regression coefficients implying a tendency to reduce the likelihood of increasing Podu area. With one of these two poverty distance variables, the fit measures are observed to be quite ugly. When the two are considered in one model, then net interactive effects seems indeterminate.

TABLE 5
Coefficients of the Index Function of Probability of *Podu* in the General Probit and Logit models under ML Estimation with t-ratios in Parentheses

Sl. No.	*Variable*	*GBPM1*	*GBPM2*	*GBPM3*	*GBPM4*	*Logit*
(1)	*(2)*	*(3)*	*(4)*	*(5)*	*(6)*	*(7)*
1.	Constant	-2.2306 (-1.643)J	-2.1043 (-1.559)	-5.0699 (-2.523)#	-5.1497 (-2.548)#	-8.8305 (-2.546)#
2.	Poduy	0.0017 (2.990)*	0.0017 (2.961)*	0.0018 (2.981)*	0.0018 (3.002)*	0.0032 (2.867)*
3.	Wagey	0.0008 (1.651)J	0.0009 (1.807)J	0.0012 (2.100)#	0.0013 (2.255)#	0.0022 (2.212)#
4.	Forsty	—	—	0.0008 (1.933)J	0.0008 (2.045)#	0.0014 (1.976)#
5.	Fsize	-0.8175 (-2.586)*	-0.8884 (-2.833)*	-1.5153 (-2.907)*	-1.602 (-3.183)*	-2.723 (-3.065)*
6.	Litrat	-0.3222 (-0.836)	—	-0.1929 (-0.485)	—	—
7.	Debt	0.2606 (2.088)#	0.2515 (2.027)#	0.2321 (1.724)J	0.2267 (1.681)J	0.4078 (1.665)J
8.	Povdy	-0.0016 (-1.859)J	-0.0016 (-1.840)J	-0.0029 (-2.578)*	-0.003 (-2.630)*	-0.0051 (-2.587)*
9.	Calorie	0.0015 (2.304)#	0.0015 (2.220)#	0.0022 (2.746)*	0.0022 (2.723)*	0.0037 (2.620)*

10.	Log L	-24	-24.38	-21.88	-22	-22.05
11.	χ^2	35.17*	34.42*	39.42	39.17*	39.07*
12.	McFadden	0.4228	0.4138	0.4739	0.471	0.4697
13.	Rsqd_ML	0.4435	0.4366	0.4816	0.4795	0.4786
14.	Correct prediction (%)	80	81.67	83.33	83.33	83.33

Note : As in Table 4.

Source : Panda, 2003.

Nevertheless the quantitative significance of both these variables is relatively low, with the absolute coefficient of the distance in terms of income being relatively high. One reason for this could be traced to the consumption habit of the tribals. Since the calorie variable is constructed from food consumption and as the tribals sale away a good quantity of Castor and Red gram, which are high energy intensive products, the calorie deficit might be having a negative influence on the desire to extend Podu cultivation area. But per capita income is a more wholesome measure. Its distance from the poverty line is quite likely to induce the Savaras to continue with the practice.

Out of several General Binary Probit Models (GBPMs) considered in Table 5, GBPM4 is considered as the best. The estimated equation is

$$
\begin{aligned}
\hat{I} = &-5.15 + 0.0018\, Poduy + 0.0013\, Wagey + 0.0008\, Forsty \\
&(-2.55) \quad (3.00) \qquad\qquad (2.66) \qquad\qquad (2.05) \\
&-1.60\, Fsize + 0.23\, Debt - 0.003\, Povdy + 0.002\, Calorie \\
&(-3.18) \qquad\quad (1.68) \qquad\qquad (-2.63) \qquad (2.72)
\end{aligned}
$$

The estimated index function has a significant Chi-squared and Mc-Fadden statistics. The model also explains more then 83% of the probability. The numerical weight of the most of the regression coefficients and their partial derivatives (marginal effects) are low for most of the explanatory variables. However, family size has a significantly determining influence for reducing the probability that a family will have the desire to take Podu land more than the median.

The influence of debt burden is quite opposite. It increases the likelihood of taking more Podu land. If the debt burden of the tribal households increases by Rs. 1000 over and above the present burden of Rs. 2100, the estimated probability is found to be 0.557. This is calculated using the cumulative distribution function of F $(\hat{I})$, where F represents the standard normal distribution (N ~ 0, 1) and I is the index function, in which all variables are measured at their means, but debt is raised by Rs. 1000. This influence is also shown in Fig. 1 drawn as per Greene (2000, pp. 817-8). Since the estimated probability that an individual family will choose to take Podu area beyond the median is greater than 0.5, it is predicted that, when debt burden increases the family will choose to claim more forest land .

This model is reworked with a Logit formulation. The characteristics in the numerator of the odd ratio (Podu = 1) is given in the 7th row of Table 5. It is found that the qualitative and quantitative significance of the characteristics are similar to that of the Index function of probability of the Probit model. The coefficients are 1.6 to 1.8 times more in the Logit function than in the Probit function, as it is expected from standard literature (Greene, 2000, p. 817). Given the similarity in the coefficients of the Probit and Logit function the usual interpretation remains valid. Thus, all the models taken together stand to justify that the complementarity in the components of income such as those from Podu, daily wage and forest, and debt burden and poverty distance tend to increase the likelihood that a household of Gadiabanga will claim an area larger than the median for shifting cultivation.

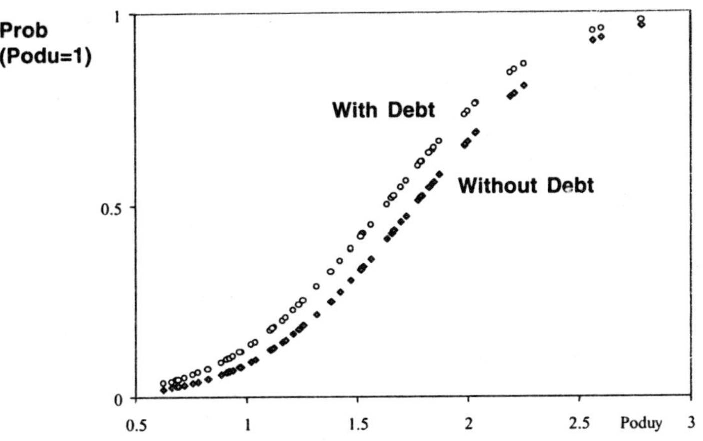

Figure 1 : Effect of *Debt* on predicted probabilities of the GBPM4 Model

5. CONCLUSION

Shifting cultivation is a part of culture rather than of commerce. The Savaras of Gadiabanga do not compete for Podu land. Forest land distribution is socially determined by convention. The current generation just honours the tradition. There is no fear or care for the regulation of the Forest Department. The Podu land remains earmarked for each family in three patches of Gadiabanga. If somebody has the capacity to cultivate a larger forest area, he gets it at a distance. The follow period is 4 to 5 years, contrary to what is reported by the Forest Department.

Through a series of Probit and Logit Models, the influence of the factors on the tendency of the tribal households to cultivate Podu area beyond the median size is estimated. It is found that the likelihood that the households would extend the Podu area is increased by the income from Podu, daily wage and the forests, and the poverty distance and the outstanding debt burden. Of all the variables which tend to increase that probability, debt is found to be the most important variable. In the face of low carrying capacity of the Podu land and low productivity of labour, it is observed that the socio-cultural tradition and milieu, the intensity of poverty and debt, a natural insurance against uncertainty and the notion of complementarily of economic activities are found responsible for the continuity of the Podu practice.

It is observed that in terms of variety and certainty, shifting cultivation has an edge over the other socio-economic sectors. The entire process of tribal development is yet to introduce so diverse and so stable a source of livelihood. The policy prescriptions from the analyses of the study is that by way of an alternative to Podu, there should be a source of income, which is stable and complementary to other activities and capable of meeting the resilience condition of the system. The non-institutional indebtedness should not be a burden on the tribal life. There is also a need to make the tribals free from the poverty trap. As long as these prescriptions are not met, the Podu practice is bound to continue. So far as its possible damage to the ecosystem is concerned, there is a need to integrate the practice with the demands of the ecosystem by making good amount of investment in natural capital. Natural capital here refers to the vegetation cover and the soil quality. In Gadiabanga it has been seen that Cashew plantation are increasingly accepted as an economic and ecological asset of the village. It is, therefore, necessary to approach the Podu practice with micro plans for ecological and economic promotion, rather than control it by law and regulation. As it is observed that Gadiabanga is a *typical* tribal village in transition, the policy implications of this case study are expected to be valid in a far wider context.

References

Chopra, K. and S.C. Gulati (2001), *Migration Common Property Resources and Environmental Degradation: Interlinkages in India's Arid and Semi-arid Regions*, Sage Publications New Delhi, 163 p.

C.P.S.W. (1994), "Orissa's Environment—Citizens Report", Bhubaneswar.

C.S.E. (1985), *State of India's Environment. The Second Citizens Report, 1984-85*, CSE, New Delhi, pp. 167-72.

Fernandes, W., G. Menon, L.P. Viegas (1988), *Forests, Environment and Tribal Economics: Deforestation, Impoverishment and Marginalisation in Orissa*, Indian Social Institute, 363 p.

Government of Orissa (GOO) (1989), Working Plan for the Reserved and Rayagada Forest Division, 1989-90 to 1998-99 for ten years, District Forest Office, Rayagada.

Greene, W.H. (2000), Econometric Analysis, 4th Edition, Prentice Hall Inc., New Jersey, 1004 p.

Hill, R.C., W.E. Griffiths and G.G. Judge (2001), *Undergraduate Econometrics*, John Wiley & Sons Inc, New York, 402 p.

Jyotishi, A. (2000), "Swidden Cultivation: A review of concepts and issues", *Working Paper*, Institute for Social and Economic Change (ISEC), Bangalore.

Jyotishi, A. (2001), "Institutional pluraism: Case of swiddeners in Orissa, *Working Paper*, The Institute for Social and Economic Change (ISEC), Bangalore.

Mahapatra, C.R. (1997), Podu: An Ecological hazard, in P.M. Mohapatra and P.C. Mohapatro, (ed.) *Forest Management in Tribal Areas: Forest Policy and People Participation*, Concept Publishing Company, New Delhi, pp. 70-85.

Mahapatra, L.K. (1997), Parameters of Forest Policy and Tribal Development in P.M. Mohapatra and P.C. Mohapatro (ed.), *Forest Management in Tribal Areas : Forest Policy and People Participation*, pp. 21-26.

N.C.A. (1976), *Report of the National Commission on Agriculture, Part IX: Forestry*, Ministry of Agriculture and Irrigation, Government of India, New Delhi.

Pal, B.N. (1989), "Some economic aspects of tribal agriculture" in Vidyarthi and Sahay (ed) in *Applied Anthropology and Development in India*, Delhi, p. 117.

Panda, C.D. (2003), *Ecological Economics of Shifting Cultivation*, Unpublished Ph.D. Thesis, Berhampur University, Orissa.

Patnaik, N. (1993), *Swidden Cultivation amongst two tribes of Orissa*, CENDERET, Bhubaneswar, SIDA, New Delhi, ISO/SWEDFOREST, Bhubaneswar.

Ramakrishnan, P.S. (1992), "Jhum: Is There a way out?" in A. Agrawal (ed.), *The Price of Forests*, Centre for Science and Environment (C.S.E.), New Delhi, pp. 304-11.

Ramani, V.S. (1988), *Tribal Economy: Problem and Prospects*, Chugh Publications, Allahabad , pp. 99-103.

Sachidanada (1989), *Shifting Cultivation in India*, Concept Publ., New Delhi, p. 15.

Sahu, N.C. (1986), *Economics of Forest Resources: Problems and Policies in a Regional Economy*, B.R. Publishing Corporation, New Delhi.

Samal, J. (1994) , "Role of forest in a poor Resource-base Tribal Economy", *Journal of Labour and Social Research*, New Delhi, Vol. 2, No. 2, April-June, pp. 23-47.

Chapter 4

Economic Analysis on Potentiality of Motra (Clinogyne Dichotoma) Cultivation in District Coochbehar

B. CHATTERJEE AND A.K. MAITI

ABSTRACT

Motra-mat, i.e. "Sital Pati" produced from perennial plant 'Motra' (*Clinogyne dichotoma* Salisb) is densely associated with the cultural heritage of West Bengal. Though initial investment is quite higher as compared to conventional crops, but due to the ratooning nature of motra the recurring cost is quite less as compared to other crops and yield can be obtained during 30 years or more. It also provides a good employment opportunity to the men as well as women family labour. The Motra cultivation and Sital pati preparation gives a serious clue in the policy making perspective looking at rural development in Coochbehar district of North Bengal which may improve economic and social life of a specific group of people -the rural poor. In this study an attempt is taken to study the economic potentiality of cultivation of motra and marketing of Sital pati in Coochbehar-1 block of Coochbehar district which is the predominant motra producing area. It is found that the benefit-cost ratio is quite marked which implies that motra production as well as motra-mat weaving is profitable. It is also observed that it holds an important rank when single crop situation as well as the existing crop rotations are analyzed. This clearly indicates the potentiality of motra cultivation in the study area.

INTRODUCTION

Motra-mat, i.e. "Sital Pati" is deeply associated with the cultural heritage of West Bengal. It is used to get relief from severely hot summer season as it easily soaks our thermal perspiration and gives a cooler feeling . This pati is produced from the plant 'Motra' (*Clinogyne dichotoma* Salisb) belonging to the family Marantaceac. It is an hydrophilous plant especially perennial is nature. It can thrive well under in undated as well as acidic soil condition that is a rare crop as per as choice of crop in this critical condition of North Bengal area is concerned. Motra involves though higher investment but provide good return during a fairly long. But, due to the ratooning nature of motra the recurring cost is quite less as compared to another crops and yield can be obtained during 30 years or more. Period of time (average 30 years) due to its ratooning nature. It also provides a good employment opportunity to the men as well as women in an immaculate fashion. The men collect the culms and thereafter complete stripping and drying. Women perform mainly the weaving function as they are very skillful and specialized in this aspect.The Motra cultivation and Sital pati preparation gives are serious clue in the policy making perspective looking at rural development in this area which likely improve economic and social life of a specific group of people—the rural poor.

In this study an attempt is taken to study the economic potentiality of cultivation of motra in Coochbehar-1 block of Coochbehar district because of dominance of motra cultivators, pati weavers, local wholesale market and accessibility. In spite of the fact that Motra is a perennial but expensive crop it is also remunerative crop as opined by sample farmers. Hence, it is deemed essential to analyze the feasibility of cultivation of this crop *vis-à-vis* competing crops practiced by the sample farmers.

Maiti, Mukherjee and Banerjee (1986) have observed that mat-stick once sown gives production for five consecutive years. Naturally cost of production is expected to be more in the first year and minimum in the fifth year. On an average for five years, surpluses have been found to be positive and maximum in second year. Cost of production of mat has remained more or less constant over five years duration.

Debnath (1989) has observed that, about 84 per cent operational area, in the selected farms, devoted to Motra cultivation by the smallest size groups. It is an indicative of the fact that the Motra growers having relatively small area used to take the cultivation of Motra and its weaving of mat by family members as the main source of income through self-employment proposition. It has also been estimated that total quantity of labour required for cultivation of Motra and weaving mat from Motra of one acre in a year is 4.5 times higher than the employment generated from best alternative crop combination fitted in an annual rotation.

Chatterjee and Sengupta (1991) have stated that, after the complete harvest of the motra crop, ratoon crops are generated from the stubble's and even more production (quantitatively and qualitatively) could be obtained from the ratoon . A huge quantity of bio mass was likely to be available from the crop-residue. Also the unused portion of motra sticks after motra mat-weaving is used as fuel purpose.

METHOD AND MATERIALS

Motra, as a specialized enterprise, is not cultivated throughout the State uniformly. It's cultivation is restricted to a very few districts of West Bengal and also in restricted blocks and villages of a district. Due to this reason, it needs to identify the tract where motra is cultivated on a bulk-scale and covers the bulk of the motra-mat (Sital Pati) growers. After identifying the tracts, multi-stage sampling technique has been adopted to select the ultimate sample unit.

The present study is confined to Coochbehar district of the State of West Bengal. Motra is cultivated in Coochbehar and a few pockets of Jalpaiguri and North Dinajpur districts. The district Coochbehar has been selected purposively for the study because of it's leading position in respect of area coverage and household associated with cultivation and weaving.

Out of 12 blocks in the district Block Coochbehar-1 is selected purposively. According to District Agriculture Office Report, this block possesses the lion's share in respect of area coverage under this crop. Majority of households who migrated from East Bengal do reside in this block and are skilled in pati weaving. Hence, this block has been selected.

Motra is not uniformly cultivated all along the sample block. It's cultivation is limited within a few villages of the block. Following the method of Probability Proportional to Area under motra, three villages are selected from the sample block—the villages selected are Dhaluabari, Gangaalerkuti and Ghaghirghat.

A list of farmers has been prepared and motra growers were identified from each selected village. On the basis of operational holding the motra cultivators are classified and stratified into 3 groups. During survey period it is observed that motra cultivation is dominated in less privileged group and their operational holding range up to 1.60 ha. Only a few cultivators, whose inclusion will affect statistical test because of want of minimum degree of freedom, possessing more than 2.00 ha of land. Hence, the sample farmers are stratified into three groups, i.e. up to 0.50 ha. 0.51 to 1.00 ha. and 1.01 ha. and above. Then from each group 15 per cent motra cultivators are randomly selected. In such a way, 200 farmers from Coochbehar-1 are selected for the study.

In the fulfilment the objectives of the study, the data have also been collected with the help the of a pre-tested survey schedule prepared for this purpose. Data have been collected by visiting and interviewing each farmer personally. Further a monthly visit to each market intermediary was made for the collection of market information.

To fulfil various objectives set out tabular method of analysis is followed.

RESULT AND DISCUSSIONS

Cost of cultivation and returns are important for economic analysis of a crop or a farm. It is one of the criterion to judge the profitability of an enterprise. This helps to draw important conclusions and policies to be implemented. Hence, this section is dedicated to look at the cost and returns of Motra cultivation in the study area. In this study cost and return are worked out on the basis of plant age as there is distinct year-to-year variation noticed.

TABLE 1

Age Group-wise and Item-wise Cost and Return Per ha. of Motra Cultivation (Saf-cut) in Block -Coochbehar-1 at Current Year Price of 2004-05

(Rs. per year Rs.)

Inputs	*Age group*				*Cumulative of 6 Years*
	1	*2-3*	*4-5*	*6*	
1. Levelling and land preparation	2625.00	0.00	0.00	0.00	2625.00
2. Seed	22500.00	1500.00	0.00	0.00	25500.00
3. Transplanting	2062.50	375.00	0.00	0.00	2776.50
4. Manures and fertilizers	5250.00	3187.50	1800.00	0.00	15225.00
5. Intercultural operation	7125.00	2625.00	2250.00	3000.00	19875.00
6. Harvesting	0.00	0.00	13942.50	10334.13	38219.13
7. Intrest on working capital	2707.03	479.53	1360.42	935.05	7321.98
8. Misc. cost	450.00	3157.03	3417.48	3699.42	17298.44
Total cost	42719.53	11324.06	22770.40	17968.60	128877.05
Total return	0.00	0.00	107250.00	99435.00	313935.00
B.C. Ratio	—	—	4.71	5.53	2.43

Table 1 represents the age groupwise and itemwise distribution of cost and return fetched per ha. of motra cultivation (Saf-cut) in Block-Coochbehar-1 at current year price of 2004-05. In an overall analysis it is found that harvesting expenses (29.65%) is the highest of all the cost items though expenditure on seed expense (19.78%), intercultural operations (15.42%) and expenditure on manures and fertilizers (11.81%) have considerable importance It is also clear from the table that the highest proportion (33.14%) of total cost has been incurred in the initial year of the plant followed by the age group of 4-5th year (17.66%) and 6th year (13.94%). It is observed to be minimum (8.78%) in 2-3rd year Except the miscellaneous Cost, all other costs are noted to be maximum in the initial year resulting highest cost of cultivation. This reveals that motra cultivation is expensive which restricts the allocation of area under this crop. From table it is observed that during the 6years of life span the maximum cost is incurred for the harvesting which is only performed during the last 3 years. The next important cost item is noted to be the cost of seedlings, which is occurred in the initial year and a few rhyzomes, are replaced in the second year Intercultural operation is done in all the years. The beginning and ending of harvesting are found to be at the age of 4th year and 6th year with a good number of human labour is required as it is performed by the human beings only But in case of return, it is found to be maximum in the age of group of 4-5th yearwhich is reduced in 6th year. It is also observed that motra is quite remunerative as investment of Rs. 1.00 gives a return of Rs. 2.43 to the farmer when the total life span of the crop is considered. The benefit-cost ratio is worked to be maximum during the 6th year of plant age because of much lower cost of cultivation. In spite of higher return per rupee investment, the most critical condition is that the farmer has to meet a good amount without any return for the first three years. This is supported by the study of ......

TABLE 2

Benefit Cost Ratio of Motra Cultivation (Saf-Cut) in Block Coochbehar-1 in the Year of 2004-05

Age Group	*Cumulative Cost for the age group (Rs ./ ha.)*	*Cumulative Return for the age group (Rs. / ha.)*	*Benifit-Cost Ratio for the age group (Rs. / ha.)*
13 years	65368	—	—
4th 5th years	10909	214500	1.93
6th years	128877	313935	2.44

Table 2 exhibits the benefit cost ratio of motra cultivation in block Coochbehar-I in the year of 2004-05 for the farmers who are specialized in total harvesting (saf-cut) practice. It is noted from the table that the cumulative cost, cumulative return and benefit cost ratio are found to be maximum in the age group of 6th year followed by 4-5th year of age group. In the first three years cumulative cost is noted to be to the magnitude of Rs. 65367.65 whereas no return has been fetched by the farmer during this period of time. The cumulative cost for the total life span of six years is to the tune of Rs. 1,28,877.05, i.e., Rs. 21479.51 per ha. per year. This indicates that motra cultivation is expensive. In spite of its expensive nature, the rate of return per rupee is also lucrative. As by investing of Rs. 1.00 farmer gets Rs. 2.44.

After a detailed analysis of cost and returns from motra-stick cultivation it is felt necessary, to analyse the cost and returns from motra-mat weaving. It is the final product of motra stick and farm family members (specially female members) are very much engaged in this practice. This provides a regular flow of income to the family.

Table 3 presents the size-group-wise and item-wise cost and return of medium-sized motra-mat weaving in district Coochbehar. An inverse relationship between total cost and size-group is noticed which is mainly due to the cost of raw motra-stick. This may be due to scale economy. It is further observed that labourers engaged in motra-mat weaving are more efficient in case of large farmers as the operations performed require less time and total cost of motra-mat is also least. Total return

TABLE 3

Size-wise and Item-wise Cost and Return of Medium-sized Motra-mat

(Rs. Per motra-mat)

Items	*Size Group*		
	Small	*Medium*	*Large*
1. Raw motra-stick	41.05	40.24	37.96
2. Motra-stick splitting	10.27	10.03	9.44
3. Fibre Extraction	25.72	25.58	25.05
4. Motra-mat Weaving	25.78	25.18	25.08
Total Cost	102.82	101.03	97.53
Motra-mat Value	125.00	126.40	127.00
By-product Value	29.72	29.67	28.71
Total Return	154.72	156.07	155.71
Benefit Cost Ratio	1.50	1.54	1.59

from motra-mat weaving is found to be highest in case of medium size of farm group. However, the inter sizegroup variation in total receipt is not marked. In case of B-C ratio a direct relationship between farm size-group and the ratio is noted which ranges from 1.50 to 1.59. The ratio clearly shows the remunerativeness of motra-mat weaving in the sample area.

Table 4 reveals the size-group-wise and item-wise cost and return of large sized motra-mat in district Coochbehar. An inverse relationship between total cost and size-group is noticed which is mainly due to the cost of raw motra-stick which may be due to scale economy. It is again observed that labourers engaged in motra-mat weaving are more efficient in case of large farmers. Total return from motra-mat weaving is found to be highest in case of large size group. However, the inter size-group variation in total receipt is not marked but a positive relationship is also noted. In case of B-C ratio a positive relationship between farm size-group and the ratio is noted which ranges from 1.30 to 1.38. The ratio noticeably shows the remunerativeness of motra-mat weaving in the sample area. An inter-motra-matsize comparison shows the remuneratives of medium sized motra-mat in comparison to large-sized motra-mat as indicated by B-C ratio. It is also observed that the cost of weaving as well as return from motra-mat praparation are more in case of large-sized motra-mat than the medium sized motra-mat. May be because of remunerativeness, farmers are more engaged in medium-sized motra-mat weaving.

TABLE 4

Size-wise and Item-wise Cost and Return of Large sized Motra-mat

(*Rs. Per motra-mat*)

Items	*Size Group*		
	Small	*Medium*	*Large*
1. Raw motra-stick	49.10	48.17	47.02
2. Motra-stick splitting	10.34	9.99	9.90
3. Fibre Extraction	39.51	38.80	38.40
4. Motra-mat Weaving	39.45	38.70	40.20
Total Cost	138.40	135.66	135.52
Motra-mat Value	164.00	168.60	171.20
By-product Value	16.60	15.90	16.60
Total Return	180.60	184.50	187.80
Benefit Cost Ratio	1.30	1.36	1.38

The feature of crop rotations in the sample area are as follows:

Abbreviations/Crop Rotations:

R_1 : Jute, Pulse (Lathyrus), Mustard
R_2 : Jute, Winter-paddy, wheat
R_3 : Jute, Winter-paddy, Pulse (Lathyrus Late varieties)
R_4 : Aman-paddy, Winter-paddy, Pulse (Lathyrus Late varieties),
R_5 : Aman-paddy, Winter-paddy
R_6 : Jute, Winter-paddy
R_7 : Motra

Hence, while examining the feasibility of Motra cultivations, these crop rotations are studied and compared with Motra cultivation.

The overall comparison clearly exhibits the superiority of Motra crop over the existing crop combinations or rotations followed by the sample farmers of this study area. While considering the crop rotations followed, it is observed that Motra is not too expensive in nature. But it requires higher capital for initiating it's cultivation. Another point that goes against this crop is that it yields from the 4th year of plant age and the farmer has to incur cost for more than three years after planting while for the other combinations farmers get return after a few months interval only.

Table 5 succinctly expresses the feature of cost , return and benefit-cost ratio of motra and other competing crops in single crop situation per year. Cost A_1 and cost D are observed to be highest in tobacco followed by motra and jute and lowest in mustard. In case of total return as well as net return perspective Tobacco holds it's predominant position followed by motra and winter paddy. It is also observed that the benefit-cost ratio is highest in motra followed by Tobacco and wheat.

Table 6 expresses that the cost D under perennial motra is lower than all other existing crop rotations in this block. In this respect it is observed that R_2 is found to be in the top most position followed by R_3 and R_6 respectively. It is also found that the gross return is highest in case of R_2 followed by R_3 and R_4. In case of R_5 the gross return is lowest. The net return is also found to be highest in R_2 followed by R_3 and R_4 respectively the lowest net return is found to exist in case of R_1. It is also noted that the benefit cost ratio under perennial motra crop holds it's predominant position over the other six existing crop combinations which are practiced as rotational crop cultivation in block Coochbehar-1 which clearly indicates the potentiality of motra cultivation in the study area.

TABLE 5

Per Hectare Cost and Return of Existing Crops Practiced by Sample Farmers in Block-Coochbehar-1 (2004-05)

Cost and Return (Rs.)	*Crops (Single)*							
	Winter Paddy	*Aman Paddy*	*Wheat*	*Pulse*	*Mustard*	*Tobacco*	*Jute*	*Motra*
(1)	*(2)*	*(3)*	*(4)*	*(5)*	*(6)*	*(7)*	*(8)*	*(9)*
Cost A_1	11091.65	7015.68	8536.55	5283.26	5323.62	19557.00	12900.72	18895.54
Cost D	13506.53	8641.39	10637.13	6660.93	6633.59	24656.72	15890.15	21479.50
Gross Return	25470.21	13885.09	21635.48	11653.18	10016.72	52765.38	27807.76	52322.50
Surplus Over Cost A_1	20378.56	10126.41	17098.93	10370.22	4693.10	33208.38	14907.04	33426.96
Surplus Over Cost D	17963.68	8500.70	14998.35	8992.55	3383.13	28915.38	11917.61	30843.00
B.C. Ratio at Cost A_1	2.29	1.98	2.53	2.20	1.88	2.69	2.15	2.77
B.C. Ratio at Cost D	1.88	1.60	2.03	1.75	1.51	2.14	1.75	2.43
Labour Utilization	147.00	95.00	76.00	73.00	75.00	175.00	184.00	240.00

TABLE 6

Hectare-wise Cost and Return of Perennial Motra and of Prevailing Crop Rotations of Block Coochbehar-1 in the year 2004-05

Cost & Return	*Crop Rotations*						
	R_1	R_2	R_3	R_4	R_5	R_6	R_7
(1)	*(2)*	*(3)*	*(4)*	*(5)*	*(6)*	*(7)*	*(8)*
Cost D (Rs.)	29184.70	40033.80	36057.60	28808.90	22147.90	29396.70	21479.50
Gross Return (Rs.)	53477.70	84913.50	74931.20	64265.50	48612.30	59278.00	52322.50
Surplus over Cost D (Rs.)	24293.00	44879.60	38873.50	35456.60	26464.40	29881.30	30843.00
B-C Ratio at Cost D	2.33	1.98	2.41	2.35	1.51	2.14	2.43

CONCLUSION

It is observed that motra is quite remunerative as investment of Re. 1.00 gives a return of Rs. 2.43 to the farmer when the total life span of the crop is considered. The benefit-cost ratio is highest during the 6th year of plant age. However, the farmer has to bear a considerable amount during the first three years. The feature of cost, return and benefit-cost ratio of motra and other competing crops in single crop situation, Cost A_1 and cost D are observed to be highest in Tobacco followed by Motra and jute and lowest in mustard. In case of total return as well as net return perspective Tobacco holds it's predominant position followed by motra and winter paddy.

Cost D under perennial motra is lower than all other existing crop rotations in this block. It is also noted that the benefit cost ratio under perennial Motra crop holds it's predominant position over the other six existing crop combinations which are practiced as rotational crop cultivation in block Coochbehar-1 which clearly indicates the potentiality of Motra cultivation in the study area. The study highlights the dominance of Tobacco cultivation in the district but also suggests the necessity of encouraging Motra cultivation.

REFERENCES

Chatterjee, B. and Sengupta, K. (1991), "Sital Pati Gach," Byabaharik Udvidbidya, West Bengal State Book Board, 1(b) pp. 382-83.

Debnath, P.K. (1989), "Some Economic Aspects of Motra (Clinogyne Dichotoma Salisb) Cultivation in Coochbehar District", M.Sc. (Ag.) Thesis, B.C.K.V.

Maiti, A.K.; Mukherjee, A.K. and Banerjee, B.N. (1986), "Economics of Production and Marketing of Mat in West Bengal—A Case Study", *Jorn. of Agril. Marketing,* July-Sept. 1986; pp. 28-32.

Chapter 5

Environmental Perils Due to Intensive Agricultural Practices

K.U.K. Nampoothiri and T.K. Hrideek

1. INTRODUCTION

The adage that "nature has enough for every ones need but not for their greed" is becoming more relevant in the present scenario of resource exploitation to meet the growing demand of mankind. In the long past men were satisfied with the forest products including meat to satisfy their quench, their hunger and thirst. As demand increased, crops were domesticated and grown initially in the river banks from the sixth millennium BC (Mehra, 1997). The details are available in Vedas particularly Rigveda and in Atherva Vedas, indicating that traditional varieties, country ploughs, cow dung, cow urine, sheep excreta and seed extracts formed the essential components of cultivation, none of which hampered the nature to any appreciable extent (Aiyar, 1952). This was followed by the awareness on the importance of good seeds through selection, preservation and storage which are reported to have been practised in the deltas of Godavary and Krishna rivers in the 1st century AD (Randhawa, 1980). Thereafter human population started to get worried about the out-break of diseases and pests and vagaries of nature which now we call as abiotic and biotic disorders. It was only natural since causal agents such as fungi, bacteria, virus, nematodes and insects were present for millennium. What is significant is that the ancient scientists recognized and used only organic agents to control the pests and diseases with very little adverse impact on the environment (Randhawa, 1980).

2. THE FOOD CRISIS AND INDIAN SOCIETY

Early in the 20th century India faced many famines and deaths due to starvation which was a national danger since we could not have depended upon imported food grain at that level especially when it came from 12,000 miles away. At one point we reached a stage when the stocks existed for only two weeks and there was nothing in the pipe line.

3. GREEN REVOLUTION

The term green revolution refers to the renovation of agricultural practices beginning in Mexico in the 1940s, significantly increasing the amount of calories produced per hectare The beginning of the Green Revolution are often attributed to Norman Borlaug, an American scientist interested in agriculture, who in the 1940s developed new disease resistant high yielding varieties of wheat in Mexico. Countries all over the world in turn benefited from the Green Revolution work conducted by Borlaug and his research institution. India for example was on the brink of mass famine in the early 1960s because of its rapidly growing population. After introducing new seeds of wheat from Mexico and rice seeds from the International Rice Research Institute (IRRI) in the Philippines, India produced 17 million tons of wheat in 1967-68 and 71 million tons in 2003. The major benefits of the green revolution were experienced mainly in Punjab and Haryana between 1965 and the early 1980s. By 1980, almost 75 percent of the total cropped area under wheat was sown with high-yielding varieties. For rice the comparable figure was 45 percent. The plan was implemented only in areas with assured supplies of water with means for large inputs of fertilizers and adequate farm credit.

The crops developed during the green revolution were high yielding varieties—meaning they were domesticated plants bred specifically to respond to fertilizers and produce an increased amount of grain per hectare planted. There were three basic elements in the method of the green revolution: 1. continued expansion of farming areas; 2. double-cropping in existing farmland; and 3. Using improved seeds.

4. IMPACT OF GREEN REVOLUTION (GR)

4.1. Fertilizer

While green revolution was a need of the hour to save India from starvation, it was also accompanied by many adverse impacts. Large quantities of fertilizers were used to increase yield. This caused the deficiency in secondary and micronutrients, i.e. Sulphur, Zinc, Boron, Iron, Manganese and Molybdenum including universal deficiency of

NPK. Carbon/nitrogen balance was upset causing metabolic problems. Intensive cultivation of land without conservation of soil fertility and soil structure lead ultimately to the increase in soil salinity and soil fatigue due to intensive cultivation. The high yielding crops gobble up nutrients like nitrogen, phosphorous, iron and manganese, making the soil anemic.

Many people suppose that the excess run-off of nitrates comes from farmers heaping on too much nitrate fertiliser which is then washed-off by the rain. Researchers on the experimental site at Rothamsted have been studying the nitrate leached from plots of land, some of which have been left bare and have not been fertilized since 1843, yet continue to leach nitrates in the water.

4.2. Pesticides

The most serious environmental impact was due to the use of very strong pesticides at higher doses. Indiscriminate use of pesticides, fungicides and herbicides caused adverse changes in biological balance. Due to intensive use of pesticides, pests became tolerant to pesticides. Presence of pesticide residues in food materials and even in the breast milk has been reported. Aerial spray of endosulphan to control pests of cashew in Kerala resulting in serious human maladies is a glaring example. Pesticide exposure can have chronic and acute impacts on human health. Long-term low dose exposure to pesticide causes immune suppression, hormonal disruption, diminished intelligence, reproductive abnormalities and cancer (Gupta, 2004).

Another problem which occurs due to pesticide is the conversion of pesticides into the obsolete form, which may even show more harmful effects than the former. When the pesticides are not used within the prescribed time of their efficacy, they become obsolete. "Obsolete pesticides" are defined as those pesticides that can no longer be used for their intended purpose or wanted to be used and therefore must be disposed-off. They are decomposed into other chemical components, which are sometimes even more toxic than the original pesticides. Most pesticides expire two years after production, meaning they cannot be used unless they are tested and proved stable. A recent report by the National Consumer Council mentioned that, despite severe restrictions, many older chemicals such as the insecticides lindane and dieldrin are still found in relatively high concentrations in groundwater.

4.3. Irrigation

Irrigation also played a large role in the green revolution and this forever changed the areas where various crops can be grown. For instance, before the GR, agriculture was severely limited to areas with a significant amount of rainfall, but by using irrigation, water could be stored and sent to drier areas, putting more land into agricultural production—thus increasing nationwide crop yields. Irrigation especially flood irrigation without proper control resulted in vast areas of soils getting alkaline or saline making them uncultivable. Unscientific tapping of underground water lead to the rapid exhaustion and declining water table in many areas to the extent of one metre per year. Nitrate contamination in ground water and accumulation of heavy metals like Arsenic, Lead and Cadmium is also reported.

4.4. Varieties

The rapid replacement of numerous locally adapted varieties with one or two high yielding strains in large contiguous areas resulted in the spread of serious diseases capable of wiping out entire crops, because by having increased crop homogeneity there were not enough varieties to fight pests and disease. Diversity is a central principle of traditional agriculture in India. Such diversity contributed to ecological stability, and hence to ecosystem productivity. The lower the diversity in an ecosystem, the higher its vulnerability to pests and disease. The development of high yielding varieties meant that only a few varieties are grown. In India for example there were about 30,000 rice varieties prior to the green revolution, whereas today there are only around 1000 thus depleting available genetic base. In order to protect these few varieties then, pesticide were used increasingly as well.

4.5. Social

The rapid modernization of agriculture and the introduction of new technologies such as those that characterized the green revolution have had a differential impact on rural populations by both class and gender. Indian farmers started growing crops the American way—with chemicals, high-yielding seeds and irrigation thus loosing the uniqueness of Indian cultivation practices.

The major technological thrust of the GR was the development by agricultural research centres of high yielding varieties of rice and wheat which, under favourable conditions increase grain yield considerably over indigenous varieties. But increase in grain yield is not the only desired criteria of preference for women farmers who also value biomass and other components of the crop or plant. To a small producer, rice is not a just grain: it provides straw for thatching and mat-making, fodder for livestock, bran for fish ponds and husks for fuel. These products not only have a role in the domestic economy but are often a valuable input to other income generating enterprises that provide a livelihood for many of the rural poor, especially women.

4.6. Ecology

The GR was based on the assumption that technology is a better alternative for nature's limits. However, the assumption of nature as a source of shortage, and technology as a source of plenty, leads to the creation of technologies which create new scarcities in nature through ecological destruction such as reduction in availability of fertile land and loss of diversity of crops.

The green revolution has reduced genetic diversity at two levels. First, it replaced mixtures and rotations of crops like wheat, maize, millets, pulses and oil seeds with monocultures of wheat and rice. Second, the introduced wheat and rice varieties came from a very narrow genetic base. On this narrow and alien genetic base the food supplies of millions are precariously perched. Another range of problems relates to wildlife. Intensive farming is often blamed for its destruction, again mainly through excessive use of pesticides, but also through removal of habitat.

5. EVERGREEN REVOLUTION/SECOND GREEN REVALUATION

Realizing the problems enumerated above, Professor Swaminathan suggested for an evergreen revolution. The evergreen revolution means increasing agricultural productivity in perpetuity without associated ecological harm. Technological advances must be combined with ecological thinking in order to create true sustainable benefits. It integrates human and ecological objectives to create sustainable solutions to the most pressing global challenges—climate change, food security, and ecosystem degradation. The path way of evergreen revolution as outlined by Prof. Swaminathan (See Figure 1).

5.1. Organic Agriculture

Organic agriculture is a holistic production management system that promotes the health of the agro-ecosystem related to biodiversity, nutrient biological cycles, soil microbial and biochemical activity. By definition, organic farming does not involve the use of expensive agrichemicals—they are not permitted. Fertilizers are either created *in situ* by green manuring and leguminous crop rotation or on-farm via composting and worm farming. Organic manure covers manure made from cattle dung, excreta of other animals, rural and urban composts, other animal wastes, crop residues and green manures.

Most of the farmers choose to use farmland manure because of its easy availability, the ability to improve the soil, tilth and aeration, increase the water holding capacity of the soil and stimulate the activity of micro-organisms. Composting is the process of reducing vegetable

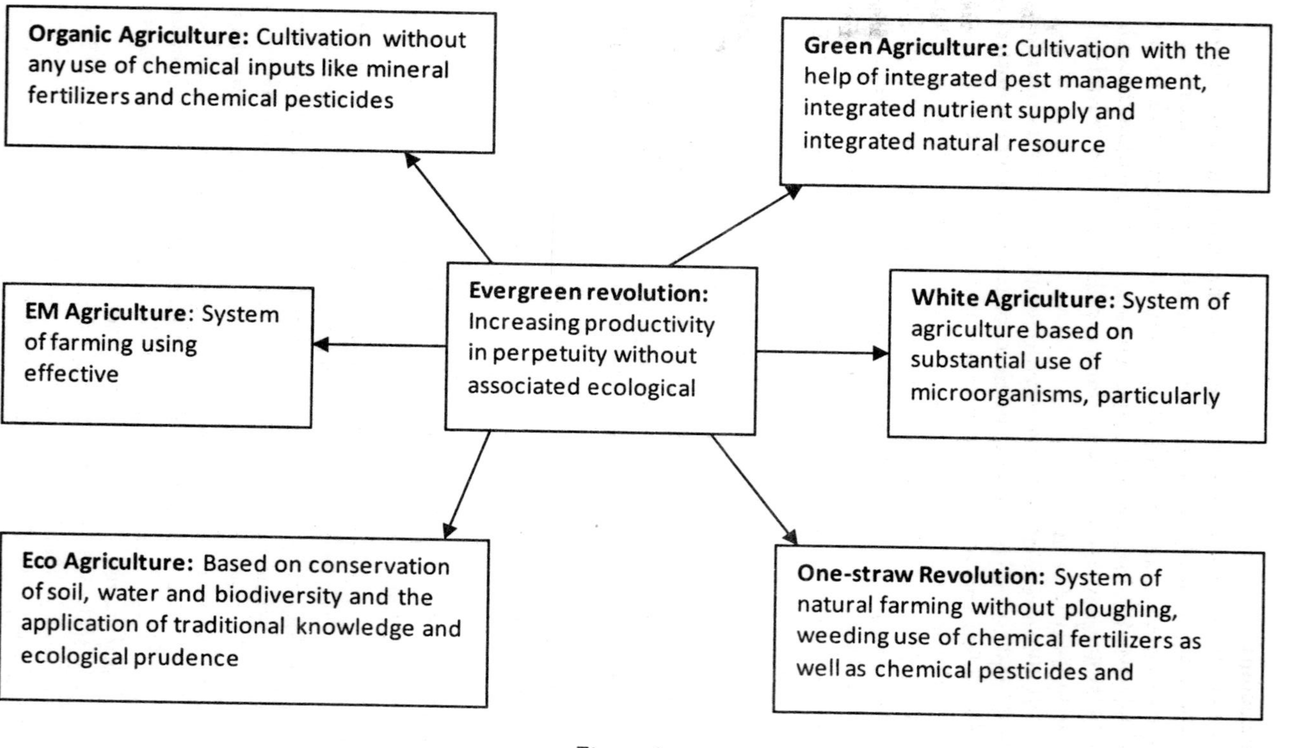
Organic Agriculture: Cultivation without any use of chemical inputs like mineral fertilizers and chemical pesticides
Green Agriculture: Cultivation with the help of integrated pest management, integrated nutrient supply and integrated natural resource
EM Agriculture: System of farming using effective
Evergreen revolution: Increasing productivity in perpetuity without associated ecological
White Agriculture: System of agriculture based on substantial use of microorganisms, particularly
Eco Agriculture: Based on conservation of soil, water and biodiversity and the application of traditional knowledge and ecological prudence
One-straw Revolution: System of natural farming without ploughing, weeding use of chemical fertilizers as well as chemical pesticides and

Figure 1

and animal waste to a quickly utilizable condition. This is done through the action of microorganisms on the wastes. These wastes may include leaves, roots, stubbles, crop residues, straw, hedge clippings, weeds, water hyacinth, saw dust, kitchen wastes and human habitation wastes. The waste materials undergo intensive decomposition under medium-high temperatures in heaps or pits with adequate moisture for around 3-6 months. The finished compost is an amorphous, brown to dark brown mix of humified materials. Green manuring involving the cultivation of leguminous plants is of advantage because of their symbiotic nitrogen or N fixing capacity.

The greater resistance of crops to pests and the diseases save farmers significantly in expensive fungicides and pesticides.. Biological pest management is the process of using the natural predators of pests like birds and parasites instead of chemical pesticides. Mechanical methods involve the mechanical removal of the pests by hand picking the larva and grubs, removing eggs from the tips of the leaf by pitching of the terminal portion, warding-off birds that damage grains using effigies or by producing noise using drums, controlling pests by dusting ash on plants, etc. Agronomical methods include various methods like intercropping, trap cropping, border cropping, crop rotation, fumigation, use of lights traps, use of bird perches etc. Biocontrol is possible by the use of parasites, predators, botanical pesticides, pheromones, etc.

While establishing new organic production system, attention has to be paid to certain aspects so that the basic requirements of certified organic farming could be achieved. Preferably, the entire farm unit should be converted to organic in a phased manner and the grower should present a conversion plan to the certification body when applying for certification. The varieties selected for organic crop production must be well adapted to local conditions and tolerant/ resistant to pests/diseases. Seeds for raising nursery should be collected from organic estates/blocks only. Advantages of the system are that organic food is normally priced 20-30% higher than conventional food. This premium is very important for a small farmer whose income is just sufficient to feed his/her family with one meal.

Organic farming also normally does not involve capital investment compared chemical farming. Further, since organic fertilizers and pesticides can be produced locally, the yearly costs incurred by the farmers are also low. The creation of living, fertile soil conditions through early corrective soil re-mineralization and strategic Keyline chisel ploughing are significant establishment costs that, however, reap ongoing benefits to production at minimal maintenance.

In cases of natural calamity, pest or disease attack, and irregular rainfall, when there is a crop failure, small farmers practicing organic farming have to suffer less as their investments are low. This system involves synergy with various plant and animal life forms. Small farmers are able to understand this synergy easily and hence find it easy to implement the same

Small farmers have abundance of traditional knowledge with them and within their community. Under organic farming, the farmers can make use of the traditional knowledge. Further, in case of organic farming, small farmers are not dependent on those who provide chemical know-how. The organic gardens do not release harmful non-natural pesticides or herbicides into the environment; this has proved immensely helpful in preserving the diverse ecosystems of the earth.

Compared to most popular practices of agriculture, organic farming is the safest and most economical form of farming. All you need is awareness of the utility of naturally available products and natural ways to solve problems common to gardening while increasing production. Through organic farming we are able to maintain biodiversity, conserve soil and water, producing nutritious food which has more longevity. This way, you will be able to enjoy tasty and healthy organic vegetables, fruits, simultaneously upholding the cause of ecological conservation.

A major benefit to consumers of organic food is that it is free of contamination with health harming chemicals such as pesticides, fungicides and herbicides. As you would expect of populations fed on chemically grown foods, there has been a profound upward trend in the incidence of diseases associated with exposure to toxic chemicals in industrialized societies.

5.2. Climate Friendly

The synthetic inputs upon which conventional agriculture is so dependent are energy expensive to mine and manufacture. Today the embodied energy of industrial agriculture uses up 9 calories for every one calorie of food that it produces. Organic agriculture with its low input needs of naturally derived substances produces less greenhouse gas emissions and is considerably more climate friendly.

5.3. Ecology Friendly

Farmers pour tons of phosphate and nitrogenous fertilizer on their cropping lands every year. Because it is soluble, much of this fertilizer is either washed-off the soil surface and into waterways (especially phosphates) or leaches through the soil profile beyond the reach of plants and finds its way less directly into waterways (especially nitrates).

Nitrate contamination of groundwater (indicated by > 10 mg/L nitrate) in Australia is widespread in every state and territory, occurring over regional and local scales (LWRRDC, 1999). In many areas, the concentration is greater than the recently revised Australian Drinking Water Guidelines level of 50 mg/L nitrate, resulting in groundwater that is unfit for drinking. In some of the more contaminated areas, the concentration is in excess of 100 mg/L (LWRRDC 1999). With fresh water reserves under increasing pressure from climate change this is a grave situation for humanity.

5.4. Biodynamic Farming

Biodynamic farming is also a part of organic farming where chemical fertilizers are replaced by microbial nutrient givers like algae, fungi, bacteria, microhiza and actinomycetes. Biodynamic farmers use a low cost microbial solution sprayed onto their crops farming.

6. ORGANIC FARMING—DISADVANTAGES

6.1. Productivity

Proponents of industrialized agriculture point out its superior productivity. In the short term, this yield is possible by expending massive inputs of chemicals and machinery, working over bland fields of a single crop (monoculture). However, over the longer time frame, productivity advantages dwindle. Organic agriculture is often criticized for its comparatively lower yield. There are adequate studies to indicate that the yield reduction, if at all any, is only for the initial 2-3 years after which the yield stabilises.

6.2. Cultivation

While their conventional counterparts may sow by direct drilling of seed into herbicide treated soils, organic farmers are usually at least partly dependent on cultivation to remove weeds prior to sowing. In contrast to cultivation, direct drilling does not mechanically disrupt soil structure and removes the risk of exposed soil being lost to wind or water erosion. This is a valid argument where farmers are working with marginal quality soils. However, the structure of agrichemically-deadened soils is weakened by the corresponding loss of soil life and thus unable to maintain its integrity under occasional cultivation. So it's a circular argument. Despite many criticisms, the GR has forever changed the way agriculture is conducted worldwide, benefitting the people of many nations in need of increased food production.

6.3. GM Crops

Organic growers do not use genetically modified or engineered food crops, some of which are engineered to tolerate herbicides (e.g. "Roundup Ready Canola") or resist pests (e.g. Bollworm resistant cotton). Conventional growers, on the other hand, are free to "take advantage" of GM crops. According to a report from the Directorate-General for Agriculture of the European Commission, productivity gains attributed to GM crops are usually negligible when growing conditions, farmer experience and soil types are factored in and are often in fact negative. The main advantage farmers using such crops gain is convenience only. There are worrying indications that GM crops may be associated with harm to both human health and the environment. The main concern is that once they are released it is nigh impossible to "un-release" them. Therefore, long-term effects are to be studied before GM crops are used widely

6. CONCLUSION

Population rich but land hungry countries like China and India have no option except to produce more foodgrains and other agricultural commodities per units of land. With more than 400 million Indians going to bed hungry every day, the Government of India was committed to declare agriculture as the first priority in 1998. The seriousness of the problem can be realized by the fact that price of pulse—the poor man's protein—has gone up by 34 percent and that of milk by 25 percent over the last year.

The initiation of exploitative agriculture without a proper understanding of the various consequences of every one of the changes introduced into traditional agriculture and without first building up a proper scientific and training base to sustain it may only lead us into an "era of agricultural disaster in the long-run, rather than to an era of agricultural prosperity." Such a challenge can be met only by harnessing the best in frontier technologies and blending them with our rich heritage of ecological prudence. Eco-technologies for an ever-green revolution should be the bottom line of a strategy to shape our agricultural future.

References

Aiyar, V.V.S., 1952. The Kural or the Maxims of Tiruvalluvar, Third Edition, Dr. V .V. S. Krishnamurthy, Tiruchirapalli, India. 287 p.

Gupta, P.K., 2004. Pesticide Exposure-Indian Scene, Toxicology, 198: 83-90.

Mehra, K.L., 1997. Biodiversity and subsistence changes in India: The Neolithic and Chalcolithic Age. Asian Agri-History 1:105-26.

Randhawa, M.S., 1980. A History of Agriculture in India, Vol. 1, Indian Council of Agricultural Research, New Delhi, India, 541 p.

Sridevi and Subhashini Sridhar, 2000, Traditional Agriculture practices for crop protection—Testing and Validation, *Tred. Knowledge systems of India and Sri Lanka.*

Swaminathan, M.S. (2000). An Evergreen Revolution, *Biologist*, 47(2): 85-89.

Chapter 6

Water Pollution in India

A Status Analysis

K. SADASIVAM

Water is essential for human, animal and plant life. Without water life is not possible on earth. It is a part of all organisms some of which contain even more than 90 per cent water. Water plays a key role in photosynthesis. It is a medium for transport of photosynthesis. Crop production depends on the availability of water. Water is necessary for industrial production, energy generation and navigation. Water is a renewable natural resource produced by the hydrologic cycle. All the items used by man are known as resource. Water resource is quite essential for the human society along with the development of other resources as well. Human beings and all other living creatures, plants and trees are made up of 60 per cent water content. Hence, it has been truly said that "nothing will survive in this world without water" and "wars of the 21st century will be fought for water only not for oil".

Water as a resource is under made relentless pressure due to population growth, rapid urbanization and large-scale industrialization. Decreasing per capita water availability and increasing water pollution are serious issues. The annual availability of water resources in India is estimated as 1869 Billion Cubic Metres (bcm). However, due to hydrological, topographical and geological constraints, only 1122 bcm of water can be utilized through development of 690 bcm of surface water and 432 bcm of replenishable ground water.

The availability of safe drinking water becomes a problem throughout the world. Water resource is a saviour of human life. Today due to overwhelming growth of population, about 2 billion populations living in 80 countries of the world face the problem of water crises. Lack of clean drinking water brings in its wake health problems. Diseases like typhoid, cholera, diarrhea, jaundice spread mainly due to contaminated water. Eleven million children under the age of 5 die each year of easily preventable diseases like diarrhea.

Lack of clean drinking water most affects the health of children. But among adults, women are more exposed to hazards of polluted water than men. Firstly, they are the primary carriers of water, and secondly, they wash clothes and utensils. Moreover, since childcare is primarily the women's responsibility, when children get infected women are more likely to catch the infection than men. Poorer households will be clearly affected the most, as they have less access to clean water sources and are less likely to have help in domestic chores.

Water constitutes 70 per cent of the total earth surface and is 1384.12 MKm^3. Out of this 97.16 per cent occurs on oceans as salt water and rest 2.6 per cent is fresh liquid and gaseous.

With 2.4 per cent of world's land area and 4 per cent of its fresh water, India has to support 17 per cent of the world's human population and 18 per cent of its cattle population. India is situated in the monsoon region. But it has to face a paradoxical situation of scarcity amidst plenty. The per capita annual availability of water is 1869 cum considering all the river systems. The bulk of the flows are contributed by thinly populated basins like Brahmaputra and Barak. Many basins like Pennar and Sabarmathi are severely water stressed already.

The water use in 1997-98 was estimated as 629 bcm of which 524 bcm was for irrigation alone. By 2050, the demand for water has been projected as 1180 bcm. This indicates that all the utilisable water resources will have to be put to use by 2050 to meet the demand. This calls for exploitation of all potential storage sites and replenishable ground water resources, which is indeed a tall order. Although about 173 bcm of storage has been constructed, the per capita storage in India is very low compared to many other countries like USA and China

During the last 50 years, the share of urban population in the country has increased from 14 per cent to 32 per cent. While the population of India has grown two and half times, the Urban India has grown by nearly 5 times. It has been assessed that the urban population may reach 50 per cent of the total population by the middle of this century. In cities like Bangalore, Delhi, Mumbai, Hyderabad and Chennai, there is already an acute shortage of drinking water supply and water is being transmitted from long distance to cater to needs. Between 2000 and 2050 fresh water withdrawals by urban areas will rise from an estimated minimum amount of about 15 bcm to a projected maximum of about 60 bcm. The options like watershed management, rain water harvesting, ground water exploitations, which create spatially distributed resources, are unable to meet these concentrated demands. Supply of safe drinking water to such a large urban population is proving a Herculean task for creating concentrated sources of water.

The growth of urban megalopolises, increased industrial activity and dependence of the agricultural sector on chemicals and fertilizers has led to the over-loading of the carrying capacity of our water bodies to assimilate and decompose wastes. The increasing discharge of domestic and industrial wastes has also led to the contamination of ground water, making it unfit for human consumption at many places and over one-third of deaths are caused by consumption of contaminated water and on an average as much as one tenth of each person's productive time is sacrificed to water-related diseases.

EFFECT OF FLUORIDE ON HUMAN HEALTH

Fluoride is often called a "two-edge sword". It helps in mineralization of bones. Presence of small quantities of fluorides, less than one part per million in drinking water prevents dental decay and helps enamel formation. However, fluoride, when consumed more than one mg/l can cause several kinds of health problem, namely, skeletal fluorosis, dental flyorosis, osteofluorosis and get-fluorosis. The effect of fluoride on human health is given in Table 1.

TABLE 1

Effect of Fluoride on Human Health

Amount of fluoride in groundwater (ppm)	*Adverse effect on Human Health*
0	Adverse effect on pregnancy power and can also damage a foetus
0.5 – 1.5	Helps in teeth-enamel formation
1.5 – 4.0	Dental Fluorosis
4.0 – 10.0	Dental Fluorosis, Osteo-Fluorosis, Severe pains in joints, back bones and other disorders.

Source : Rajiv Gandhi National Drinking Water Mission, New Delhi.

SITUATION OF WATER POLLUTION IN INDIA

The indiscriminate disposal of water after use in the form of wastewater causes water pollution. The tragic incident of Minnamata in Japan is well known. A paper factory using mercury compounds carelessly dumped its waste effluents into the sea, which formed

$(CH_3)_2Hg$ and $(CH_3CH_2)_2Hg$, which in turn were consumed by the sea fish. The Japanese people who consumed such fish indicated symptoms of mercury poisoning like gingivitis, vomiting, fever, diarrhoea, paralysis of extremities, etc. There were several instances of marine flora and fish dying in the sea, on account of deoxygenating of water, perhaps due to thermal pollution. Most of the rivers in India are polluted due to industrial activity. Thus in Bombay, Ulhas River is polluted due to disposal of effluents from rayon and dyestuff industries. The rivers in India with polluting industries shown in parenthesis, are listed as follows: Ganga (Jute, sugar), Sone (paper pulp), Gomati (paper), Yamuna (insecticides), Chaliyar (rayon waste), Kaveri (rayon, sugar), Godavari (paper, small-scale industries), Mahi (dyestuff), Mullamutha (antibiotics), Brahmaputra (black liquor, paper industry), Juhari (fertilizer waste), Patalganga (organic chemicals) and Valdhun (disposal of dye intermediates). It is paradoxical to note that holier the river, the more it is polluted. It is bound to be polluted due to disposal of industrial effluents in the streams that feed the river. The Government of India has fortunately now constituted the Ganga River Authority to keep the Ganges clean. We need such authority for all principal rivers in India. There exist similar organizations in most developed cities of the world, e.g. Thames River Authority in England.

Industrial Effluents or Waste Water

The water after it is used once for industrial purpose cannot be reused for the same purpose without treatment. Such water which emerges out after use from industries is called as the industrial effluent. Such effluents have no definite composition, as anything which is not required is carelessly dumped into its stream. Such unwanted disposable material is contributed by chemical firms, food and beverage industry, textile and apparel industries, electronics and electrical material industries, or thermal power plants. The quality of such water is characterized by the study of its various physical, chemical and biological properties. The disposal of arsenic compounds by paper industry, or toxic chemicals by fertilizer plant had caused serious problem in western India. In many places in India, town authorities have not hesitated to let up sewage into streams or brooks or rivers, thinking that dilution would solve the pollution problem. Unfortunately, biological matter and bacteria such as coliforma, sterpococcii, fecal coliforma, crenthronix, are not eliminated by mere dilution. Such water needs chemical treatment, like disinfection. This would also kill pathogenic bacteria. The projected population and respective wastewater generation in India from 1977-78 to 2051 is presented in Table 1 (Appendix). During 1977-78 the 60 crore population generated 116 lpcd/day and gross wastewater generated was 7007 which increased respectively to 119 lpcd and 12,145 mld during 1989-90 with 102 crore population. It is projected that per day wastewater generation in 2051 will be 1093 lpcd and gross wastewater generation will be 1,32,253 mld. The state-wise domestic wastewater generation and treatment (2001) is given in Table 2 (Appendix). It shows that the waste water generated is highest in Maharashtra (4692 mld) of which 4193 mld of water untreated and discharged into the environment. Tamil Nadu occupies 9th place in domestic waste water generation and treatment.

Pollution Due to Sewage and Sludge

Very few cities in India have sewage treatment plants. Even major cities like Bombay, Calcutta do not have plants to treat all sewage and sludge. Unfortunately, most of our villages are fitted with antiquated treatment plants or septic tanks. The growths of slums in bigger metropolitan cities like Bombay, Calcutta, Delhi are posing alarming problems of sewage disposal. In fact, the greatest problem Ganga Water Authorities faced during their initial working was the tackling of untreated and indiscriminately disposed-off sewage.

Pesticide Pollution

Pesticide residue is the source of greatest pollution of land and soil. The most important ones like DDT, malathion, para-malathion, aldrine, dialdrins, etc., cause a serious problem of land pollution. The characterization and determination is not simple. These pesticides have deleterious effect on health. Several of them are carcinogenic and cause long-term harmful effects upon health and hence need analysis.

Solid Waste Problem

The liquid waste was thoroughly investigated by environmentalists, but this is not the case with solid wastes. As a matter of fact with greatest industrialization, the production of colourful cartons, boxes and packing for food, beverages, edible oil, we will have an alarming problem of solid waste disposal. The problem is further aggravated due to non-degradable nature of plastic and polymeric materials. A survey was made in the city of Bombay on a particular month for solid waste composed of garbage, paper, glass, plastic, etc. It was found that animal waste (1,500 tons) constitutes the largest share followed by mineral waste (1,100 tons), agricultural waste (550 tons), with household waste of 250 tons per year. In comparison, paper (30 tons) and plastic waste (4 tons) are very less. The methods for disposal of such solid waste are equally tedious. They are physical composition, incineration and land filling. All these methods available for their disposal are expensive. Table 2 gives percentage of solid waste in different towns in India as well as in typical cities of America, England, Japan and Switzerland. Such waste has all components of nature. From

TABLE 2

Characterisation of Water from Powai Lake, Wells and Municipal Supply

Source	*pH*	*Hardness $CaCO_3$ ppm*	*Dissolved Oxygen*	*NO_2*	*F*	*MPN per 100 ml*
Powai	8.1	74.4	7.0	0.6	2.0	2400
	7.9	71.6	7.2	0.8	1.0	430
	8.3	80.0	8.1	0.6	2.0	430
	8.2	74.4	5.6	1.3	1.0	2400
	8.0	74.8	8.0	0.6	2.3	2400
	8.2	78.2	8.2	0.7	1.6	1100
	8.2	76.8	8.4	0.6	2.4	2400
Potable water						
Well water	7.0	12.5	12.5	0.1	0.2	100
	7.2	301	1.4	1.2	—	1000
	7.9	95.2	1.4	0.7	—	1000
	7.1	124	1.4	1.1	—	1000
	7.4	251	1.8	0.9	—	1000

Source : Environmental Pollution Analysis, S.M. Khopkar, p. 83.

Table 3 (Appendix) it is noted that the highest garbage was encountered in Pune; paper and glass was disposed heavily in American cities; rags found extensive usage in Japanese solid waste, while Bombay headed the list having maximum of plastic waste.

Water Pollution Analysis

First we would consider analysis of potable or drinking water. Nowadays fresh water cannot be used directly for drinking purposes due to pollution. Some kind of treatment is essential before it is used for drinking. The quality of water depends upon its source or history. The world history of water signifies the terrain through which water is flowing its origin and most important the extent to which it is contaminated on its way by industrial waste water. The solubilisation and fragmentation of the terrain through which it flows determines its quality. The extent of dissolved solid materials present in water decides its quality. Such quality is also decided by the solubility of the geological deposits, contact of water with sediments, time of interaction and special factors related to environment. To ascertain suitability of water for consumption, it is necessary to undertake examination of quality of water. Such quality is ascertained in three ways, viz.

(a) Physical examination of water
(b) Chemical characterization of water
(c) Biological investigation of water

Table 3 indicates the characterization of water from Powai Lake, wells and municipal taps in the city of Bombay, while Table 3 describes important characteristics of well water in typical coal fields in Eastern India, where most of the water comes from wells. Both the tables give prevailing status about the quality of water. For instance, there is too much of chromium, hardness and dissolved solids in Powai Lake water. Also the MPN count is alarmingly high, viz. 430-2,400. This shows Powai water is unpotable or unfit for drinking. The second (Table 2) throws much light on the quality of well water in east India.

TABLE 3
Characteristics of Well Water of Coalfield Area in Eastern India

Parameters	*Average values*
pH	7.0
Dissolved solids	672.0 mg/litre
Total hardness	617.0 mg/$CaCO_3$/litre
Nitrate	78.0 mg/litre
Sulphate	47.0 mg/litre
Chloride	62.0 mg/litre
Oil and Grease	0.038 mg/litre
D.O.	6.0 mg/litre
B.O.D.	0.82 mg/litre
C.O.D.	4.5 mg/litre
Iron	0.10 μg/litre
Manganese	0.003μg/litre
Fluoride	0.23 μg/litre

Source : Environmental Pollution Analysis, S.M. Khopkar, p. 84.

Physical Examination of Water

Much information on the quality of water can be ascertained by physical examination (Table 4). We consider then such examination, as study of colour, conductivity, temperature, odour, turbidity and hardness.

TABLE 4
Physical Characteristics of Water
(Average Value in Brackets)

Sl. No.	*Physical characteristics*
1.	Acidity – no definite data
2.	Alkalinity – 10-500 mg/litre as $CaCO_3$ (100) (total)
3.	Colour – 50 platinate units (10)
4.	Hardness – 5-1000 mg/litre $CaCO_3$ (150)
5.	pH 6.0-9.0(7.5)
6.	Specific conductance 30-1000 micro ohms (400)
7.	Temperature 1-30°C (20°C)
8.	Turbidity 0-1000 JU (100 JU)
9.	Odour – 3 TON

Source : Environmental Pollution Analysis, S.M. Khopkar, p. 84.

CHEMICAL CHARACTERIZATION OF WATER

Most important part of water quality assessment is the chemical characterization of water. Table 5 lists important chemical constituents of water, which are generally analysed to assess its quality. In addition to chemical parameters, this table also lists important physical properties of water.

TABLE 5
Physical, Chemical and Biological Examination of Water

Chemical Characteristics	*Physical properties*	*Biological investigations*
Ca^{2+}	Colour	Total coliforma
Mg^{2+}	Conductivity	Faecal coliforma
Na^{+}	Temperature	Faecal streptoccii
K^{+}	Odour	Crenothrix
$Cl^{?}$	Turbidity	Plankton
SO_4^{2-}	Hardness	Algae
CO_3^{2-}		Nitrosomonas
HCO_3^{2-}		
Total dissolved solid		Diatomaceous algae

Al^{3+}	
Ba^{2+}	Protozoa
Boron	Rutifers
F?	Escherichia Coli (i.e. E.)
Coli)	
$NO?_2$	
$NO?_3$	
PO_4^{3-}	
Fe^{3+}	
Mn^{2+}	
SiO_2^{2-}	

Source : Environmental Pollution Analysis, S.M. Khopkar, p. 89.

TABLE 6

Major Elements Present in Drinking Water (mg/litre)

Element	*PHS–limit*	*WHO–limit*
As	0.05	0.20
Ba	1.0	None
Cd	0.01	Nil
Cr	0.05	0.05
Cu	1.0	1.5
Fe	0.03	1.0
Pb	0.05	0.10
Mn	0.05	0.50
Se	0.05	0.05
Ag	0.05	None
Zn	5.0	15.0

Source : Environmental Pollution Analysis, S.M. Khopkar, p. 90.

Table 6 has prescribed WHO limit of metals, which is in agreement with that prescribed by Public Health Services (PHS) in India. This gives guidelines for the environmental engineer to control purity of drinking water, by devising proper water treatment process. It is worthwhile at this stage to consider two important aspects of chemical constituents. First is their significance in determining quality of water and second is their analysis in the laboratory.

BIOLOGICAL INVESTIGATION OF WATER

Apart from direct bacteriological examination, indirect proof for the presence of microorganism and biodegradable organic matter can be found out from the knowledge of dissolved oxygen (DO) and biochemical oxygen demand (BOD).

DISSOLVED OXYGEN OF WATER

All living organisms are dependent on one another, in order to maintain the metabolic processes that produce energy for growth and reproduction. N_2 and O_2 dissolve poorly in water, but their solubility increases with temperature as per Henry's law. At saturation, dissolved gases in water have 38 per cent oxygen. The solubility is 14.6 mg/litre at O°C and it rises to 7 mg/litre at 35°C at one atmosphere pressure. Further, the rate of dissolution of O_2 in polluted in water is much less than in pure water. Thus, DO determines whether biological changes are brought out by aerobic or anaerobic organisms. Former use free oxygen for oxidation to produce innocuous end products. DO is one singlé test to indicate how aquatic life is supported; as all aerobic process depend upon DO.

BIOCHEMICAL OXYGEN DEMAND (BOD)

BOD is defined as the amount of oxygen required by bacteria while stabilizing decomposable organic matter under aerobic conditions. In other words, BOD represents quantity of dissolved oxygen in mg/litre required during oxidation of decomposable or biodegradable organic matter by aerobic biochemical action. The decomposable organic matter at a time serves as the food for bacteria and energy is obtained due to such oxidation. Thus, BOD gives an idea about the extent of pollution. It serves as a test to evaluate the purifying capacity of receiving bodies of water. BOD test is a bioassay test devised to measure oxygen consumed by living organisms, while using organic matter present as waste in water. While evaluating this, samples are protected from sunlight, excessive agitation or shaking, and kept at a fixed temperature in an incubator. This also favours uniform bacterial growth. To complete such growth 20 hours are required. Typical BOD values of different samples of water are listed in Table 7.

TABLE 7

BOD of Various Samples of Water

Type of sample	*BOD, mg/litre*	*Remark*
H_2O	1.0-3.0	Reasonable
River	5.0-20.0	Tolerable
Sewage	50.0-100	Very bad
Industrial Waste	100-10,000	Extremely poor

Source : Environmental Pollution Analysis, S.M. Khopkar, p. 99.

The state-wise biochemical oxygen demand (BOD) in 2001 is presented in Table 4 (Appendix). The BOD load before treatment is found to be larger in Maharashtra (1937 tonne/day) followed by Uttar Pradesh (1699 tonne/day and Tamil Nadu (1215 tonne/day. The selected state-wise average biological water quality of wetlands in wildlife habitats of India in January 2005 is given in Table 5 (Appendix). The states like Meghalaya and Tamil Nadu have been affected with heavy pollution.

BACTERIOLOGICAL EXAMINATION OF WATER

The disease producing organism, i.e. pathogens should be removed from water. Those producing infectious disease like typhoid or cholera are viruses. The common indicator organisms are from coliforma group present due to sewage contamination. By special test, their number is determined. Fecal coliforms and fecal streptococci are also isolated. They are generally determined by various methods. Their population is usually expressed in terms of MPN, i.e. most probable number. For example, for untreated water, coliform density should not exceed 5000/ml. The MPN or plate count throws light on the quality of water.

WATER POLLUTANTS

The main sources of water pollution are domestic sewage, industrial effluent and agricultural runoff. The lack of investment in appropriate infrastructure to manage these wastes has been a major factor in deterioration of water quality across the country.

About 75 per cent of the waste water produced is from domestic sector, but the sewage treatment facilities are inadequate in most cities and almost absent in rural India. Only 25 per cent of class I cities (population more than 100,000), have wastewater collection, treatment and disposal facilities. And less than 10 per cent of the 241 smaller towns have wastewater collection systems. Some 20 per cent of all the wastewater generated in class I cities and only 2 per cent of all wastewater generated in class II towns is treated. In 1988, Mumbai and Delhi individually generated more wastewater than all the class I towns put together.

Estimates of wastewater generated in rural India are not available, but only 3.15 per cent of the rural population had access to sanitation services in 1993. According to Central Statistical Organization, in 1993, about 75.7 million people living in cities and 563.6 million people living in rural areas do not have access to toilets of any type. As a result, huge quantity of organic waste finds its way into water bodies exposing the population to disease. Some of the water-related diseases are diarrhea, hepatitis, roundworm, hookworm infection, quinea worm, schistosomiasis, leishmaniasis and lymphatic fillariasis. It has been estimated that India loses about Rs. 36,600 crore per year due to water related diseases.

Indian industry has grown substantially in the last four decades. According to CPCB, as of 1995, out of 8,432 large and medium industries in the country, only 4,989 (59%) have installed appropriate measures (adequate and operating effluent treatment plants (ETP)) to treat wastewater before discharge. ETP in 1,233 units are not adequate. There are over two million small-scale industries in the country, which also produce considerable effluents. Policy-makers have so far ignored pollution contribution from small scale industries.

With the advent of the Green Revolution, agriculture in India has become very resource intensive. High Yielding varieties require high chemical and water inputs, and these are often applied indiscriminately. Excess chemicals find their way into water sources, thus polluting them. Pollution from agricultural chemicals is difficult to monitor and control. In comparison to the developed countries, over all amount of chemicals does not seem to be large, the manner in which it is applied, where it is applied (upstream) are causes for concern. Excess fertilizer enriches the receiving water bodies resulting in proliferation of biota, which eventually use up all the exygen in the water. Pesticides in the water accumulate with increasing concentration in the food chain, thus affecting the various animals and plants in the food chain, including humans.

PRIMARY WATER QUALITY CRITERIA

The primary criteria of water quality in drinking water source without conventional treatment and drinking water source with conventional treatment are given in Table 8.

TABLE 8

Primary Water Quality Criteria

Criterion	*Drinking water source without conventional treatment*	*Drinking water source with conventional treatment*
Dissolved Oxygen(mg/l) minimum	6	4
Biochemical Oxygen Demand (BOD) maximum (mg/l)	2	3
Total coliform count (MPN/100 ml) maximum	50	5,000
pH	6.5 – 8.5	6 - 9

Source : Clean Water Environmental Governance-1, Indira Gandhi Institute of Development Research, p. 7.

Even after enactment of Water (Prevention and Control of Pollution) Act as early as in 1974, water quality continues to deteriorate in the country. Rivers, streams and ground water are severely polluted. The water quality in most of the country's rivers is below CPCB prescribed criteria. The amount of pollutants that find their way into water sources are way above receiving water body's natural assimilative capacity. Surface water pollution is increasingly becoming a source of conflict among upstream and downstream water users because the later suffer the effects of upstream pollution. Water quantity available for specific uses will decline with pollution. When quality deteriorates, water loses its economic value. For example, with progressive quality deterioration, water uses may successively shift from drinking water to bathing water, water for livestock, agriculture and industrial uses and so on. Pollution also creates water scarcity in regions which otherwise have abundant water resources.

References

Clean Water Environmental Governance-1, Indira Gandhi Institute of Development Research, pp. 6 -9.

Khopkar, S.M., Environmental Pollution Analysis, New Age International (P) Ltd. Publishers, New Delhi, June 1995, pp. 7-10

Tamil Nadu—An Economic Appraisal, 2001-02, pp. 227-28.

The Hindu, September 27, 2007, p. 1.

www.indiastat.com

APPENDIX

TABLE 1

Projected Population and Respectively Wastewater Generation in India (1977-78, 1989-90, 1994-95, 2001, 2011, 2021, 2031, 2041 and 2051)

Year	*Urban Population*	*Litres/Capita/ Day (lpcd)*	*Gross Wastewater Generation (mld)*
1977-78	60	116	7007
1989-90	102	119	12145
1994-95	128	130	16662
2001	285	—	—
2011	373	—	—
2021	488	121 (Assumed)	59048 (Projected)
2031	638	121 (Assumed)	77198 (Projected)
2041	835	121 (Assumed)	101035 (Projected)
2051	1093	121 (Assumed)	132253 (Projected)

Source : Ministry of Environment & Forests, Govt. India.

TABLE 2

State-wise Domestic Wastewater Generation and Treatment (2001)

States/UTs	*Wastewater Generation MLD*	*Wastewater Treatment MLD*	*Untreated Wastewater Discharge MLD*
Maharashtra	4692	499	4193
West Bengal	2113	372	1741
Uttar Pradesh & Uttaranchal	2292	772	1520
Bihar & Jharkhand	1363	135	1228
Andhra Pradesh	1271	208	1063
Rajasthan	1055	27	1028
Gujarat	1709	701	1008
Madhya Pradesh & Chhatisgarh	1159	227	932
Tamil Nadu	1094	290	804
Delhi	2700	1927	773

Karnataka	1036	387	649
Punjab	616	0	616
Kerala	428	0	428
Orissa	374	0	374
Assam	222	0	222
Chandigarh	272	91	181
Pondicherry	36	0	36
Meghalaya	30	0	30
Haryana	330	303	27
Manipur	24	0	24
Tripura	22	0	22
Goa	20	0	20
Nagaland	20	0	20
Himachal Pradesh	13	3	10
Andaman & Nicobar Islands	8	0	8
Mizoram	4	0	4
India	22903	5942	16961

Source : Central Pollution Control Board, Ministry of Environment & Forests, 2001.

TABLE 3

Disposal of Solid Waste (%)

	Garbage	*Paper*	*Glass*	*Rags*	*Plastic*
USA	5.0	54.4	9.1	2.6	1.7
UK	13.0	50.0	6.0	3.0	—
Switzerland	14.5	33.5	8.5	3.0	2.0
Japan	36.9	24.8	3.3	7.1	2.2
India	31.7	0.25	1.0	3.6	0.2
Nagpur	31.4	0.2	0.07	0.3	0.2
Poona	67.6	8.7	0.6	1.6	0.7
Kanpur	35.8	4.5	1.0	5.7	0.8
Calcutta	45.1	3.2	0.4	3.6	0.6
Bombay	52.3	12.0	—	4.3	2.7

Source : Environmental Pollution Analysis, S.M. Khopkar, p. 10.

TABLE 4
State-wise Biochemical Oxygen Demand (BOD) Load Generated (2001)

(*Load in Tonne/day*)

States/UTs	*BOD Load*	
	Before Treatment	*After Treatment*
Maharashtra	1937	275.0
Uttar Pradesh	1699	262.0
Tamil Nadu	1215	211.0
Gujarat	639	190.0
Andhra Pradesh	578	163.0
Madhya Pradesh	1116	145.0
Karnataka	610	106.0
Punjab	383	94.0
West Bengal	205	78.0
Haryana	227	61.0
Rajasthan	118	50.0
Bihar	243	45.0
Kerala	122	28.0
Orissa	138	26.0
Himachal Pradesh	37	10.0
Jammu & Kashmir	59	7.0
Assam	59	5.0
Delhi	8	5.0
Pondicherry	25	5.0
Goa	14	4.0
Daman & Diu	22	3.0
Sikkim	14	1.0
Nagaland	8	0.4
Dadra & Nagar Haveli	1	0.3
Manipur	1	0.3
India	9478	1775

Source : Central Pollution Control Board, Ministry of Environment & Forests, 2001.

TABLE 5
Selected State-wise Average Biological Water Quality of Wetlands in Wildlife Habitats of India (January, 2005)

States	*Saprobic Score*	*Diversity Score*	*Biological Water Quality Class*	*Biological Water Quality*
Andhra Pradesh	5.22	0.536	C	Moderate Pollution
Arunachal Pradesh	6.41	0.450	C	Moderate Pollution
Assam	5.01	0.498	C	Moderate Pollution
Bihar	4.94	0.516	C	Moderate Pollution
Delhi	4.16	0.563	C	Moderate Pollution
Haryana	5.08	0.603	C	Moderate Pollution
Punjab	5.77	0.598	C	Moderate Pollution
Rajasthan	5.00	0.72	C	Moderate Pollution
Uttar Pradesh	4.90	0.628	C	Moderate Pollution
Meghalaya	4.41	0.344	D	Heavy Pollution
Jammu & Kashmir	5.21	0.485	C	Moderate Pollution
Tamil Nadu	4.34	0.400	D	Heavy Pollution

Source : Ministry of Environment and Forests, Govt. of India.

Chapter 7

Use and Misuse of Freedom of Man

Case Study of Hydropower

BHARAT JHUNJHUNWALA

1. THE FRAMEWORK

The purpose of this universe appears to be 'evolution of consciousness' as seen in the meaning of the word 'Brahman'—to grow. It is obvious that man, being part of the Brahman, should act in a way that furthers the purpose of Brahman—to grow. Since Brahman is consciousness, it follows that growth must be defined as growth of consciousness. Brahman may here be defined synonymously with 'Nature' or 'Universal Consciousness'.

Man has higher degree of freedom than other living beings. It is man's solemn responsibility to exercise his freedom in a way that leads to evolution of his consciousness. Man is part of Brahman hence evolution of his consciousness spontaneously becomes evolution of Brahman.

There is a caveat, however. Man's consciousness can evolve in two ways:

- *Evolution*: While helping others in raising of their consciousness.
- *Devolution*: While hitting at others and reducing their level of consciousness.

An example will clarify the matter. Planting of trees on barren land leads to all-round increase in consciousness. The barren land has low level of consciousness. This is increased by growing of trees, animals, insects, etc. On the other hand, planting of the same trees in a beautiful garden would lead to lowering of level of consciousness. The scented flowers and flowing waterfalls will be replaced with a jungle of trees.

Man must exercise his freedom to modify nature in a way that leads to an overall evolution of consciousness and not to its devolution.

This seems to be the difference between Asuras and Devtas. Both are *bhogi.* But Asuras undertake bhoga while hitting at others while Devtas undertake bhoga by helping others evolve along with. Cutting of a forest for making charcoal may be done both by Asuras and Devtas. The Asuras will cut the forest and leave the land barren. Devtas will cultivate the land and grow wheat and paddy. Thus, Rig Veda mentions cutting of forests in a positive sense:

> "Goddess of the wild and forest... seemest to vanish from sight. And, yonder what seems a dwelling-place appears... another there hath felled a tree" (10.146).

The Asuric use of freedom leads to reaction from other elements of Universe. Let us say the civilization of Indus Valley collapsed because of over-exploitation of the forests. As a result, the silt got deposited in the river beds and led to floods which destroyed the civilization. We may conclude from this that the loss of consciousness in the forests that were cut was more than the gain of consciousness by use of the forest resources. In the result nature—silting and floods—took revenge. The loss of consciousness due to the cutting of the trees was not compensated by an increase in consciousness brought forth by the use of that timber, etc.

The main problem is that it is difficult to understand and comprehend whether consciousness is increasing or decreasing by a particular action. It requires one too connect with various players and appreciate the increase or decrease in their consciousness. People of the Indus Valley failed to connect with the consciousness of the forest and rivers and were, therefore, punished. They probably thought they were doing a great service to the Universe by cutting the forests for providing fuel wood to the artisans of Mohanjodaro and Harappa for making of beads. They did not appreciate that the increase in consciousness from wearing of beads is less than the decline of consciousness from the cutting of forests.

2. APPLICATION OF THE FRAMEWORK TO POWER SECTOR

We can use this framework to assess which source of energy would be good. This author has ranked the impact of three main sources of power—thermal, hydro and nuclear—on various parameters. The

TABLE 1

Sources of Power Reckoned on the Touchstone of Consciousness

Sl. No.	*Parameter*	*Impact*	*Thermal*	*Hydropower*	*Nuclear*
(1)	*(2)*	*(3)*	*(4)*	*(5)*	*(6)*
1.	Use of electricity	Students will be able to study etc.	+3	+3	+3
2.	Methane	Global warming kills life forms. Bad smell lowers consciousness. Methane emissions by thermal are assessed at 800 grams/KwH against 2154 for storage hydro plants in Brazil.	-2	-3	—
3.	Mining	Coal and uranium lying locked at lower level of consciousness is burnt and liberated into air. Large amount of coal is burnt.	+3	—	+2
4.	Storage of Nuclear waste	Accidental release of nuclear waste can kill life and lower consciousnes.	—	—	-3
5.	Quality of River Waters	Thermal and nuclear stations discharge hot water into rivers. Hydro deprives various living organisms of the kinetic energy of rivers; and lowers the quality of waters for bathing by pilgrims. Consciousness of flowing water is reduced by flowing it through canals and tunnels.	-1	-3	-1
6.	Sediments	Sediments are trapped in hydro reservoirs and deprives fish, flood-recession agriculture.	-	-2	-
7.	Forests	Forests are cut in mining; and are submerged in hydro reservoirs.	-2	-2	-2
8.	Earthquakes	Mining of coal leads to subsidence of cities like Jharia. Hydro reservoirs led to reservoir induced seismicity.	-2	-2	-1
9.	Displacement	Human beings are displaced in mining and hydro projects.	-2	-2	-1
10.	Health	SOx emissions from thermal, mosquitoes from hydro reservoirs and radiation from nuclear plants.	-1	-1	-1
11.	Biodiversity	Reduction of forest cover in thermal and hydro impacts negatively.	-2	-2	-1
12.	Aesthetic value	Cutting of forests and damming of rivers reduces their aesthetic value.	-1	-3	-1
	Total		-7	-17	-6

change in consciousness has been ranked on a scale of (-)3 to (+)3. The figures are best estimates by author. The learned reader is welcome to make alternative estimates. The importance lies in the framework rather than the figures.

The difference in this method is brought in focus when we compare it with the comparative assessment of the three sources of energy made by the Planning Commission in its Integrated Energy Policy Report, 2006. It has concluded that hydropower is the least negative source of energy.

Such difference in assessment arises because the various impact on consciousness enumerated above are not accounted for. The differences are explained in Table 2.

It will be seen that while thermal and nuclear energy raise the consciousness of dormant lying coal and uranium; hydropower reduces the consciousness of free-flowing waters. It is suggested that all development activities be reckoned on the touchstone of consciousness.

TABLE 2

Comparative Assessment of Sources of Energy by Planning Commission

Sl. No.	Option	SO_2 NOx TSPs	Solid Waste	Liquid effluent	Thermal impact on receiving waters	Flora, Fauna impacts	Forest loss	Involuntary Resettlement	Total Impact, Nos Y = Yes is Negative Impact
(1)	(2)	(3)	(4)	(5)	(6)	(7)	(8)	(9)	(10)
1.	Thermal power (including Mining)	Y	Y	Y	Y	Y	Y	Y	7
2.	Hydropower	N	N	N	N	Y	Y	Y	3
3.	Nuclear Power (including Mining)	N	Y	Y	Y	Y	Y	Y	6

Sl. No.	*Parameter*	*Planning Commission ignores impact of hydropower*	*Planning Commission ignores impact on consciousness*
(1)	*(2)*	*(3)*	*(4)*
1.	Use of electricity	—	—
2.	Methane	Methane emissions from hydro reservoirs ignored	—
3.	Mining	—	Increase in consciousness of coal and uranium is ignored.
4.	Storage of Nuclear waste	—	—
5.	Quality of River Waters	—	Decline in consciousness of flowing water is ignored.
6.	Sediments	Deprivation of sediments to downstream areas is ignored	—
7.	Forests	—	—
8.	Earthquakes	Impact of hydro on Reservoir Induced Seismicity is ignored	—
9.	Displacement	—	—
10.	Health	Ignored	—
11.	Biodiversity	—	Reduction in consciousness due to extinction of species is ignored.
12.	Aesthetic value	—	Reduction in aesthetic value of free flowing rivers is ignored.

Chapter 8

Ground Water Quality in Tiptur and Surrounding Areas

S.B. Basavaraddi, Ananthanag, B., Heena Kousar and E.T. Puttaiah

ABSTRACT

The Ground water quality in Tiptur town, Tumkur district, Karnataka. During Pre-Monsoon and Monsoon seasons in 2010 is discussed in this paper, water samples were collected from 50 bore-wells in and surrounding areas of Tiptur town. Physical-Chemical analysis of the collected samples were carried and for the parameters such as Ph, Total hardness, total dissolved solids, alkalinity, magnesium, chloride, sulfate, sodium, etc. From study it is revealed that in Tiptur town and surrounding areas, some of the sampling location water is fit for drinking and in few sampling locations not fit for drinking and irrigation. There is need to reduce pollutants by treating water through purifying centers.

Keywords: Physico-chemical parameters, water quality, chemical analysis.

INTRODUCTION

In recent years, the increasing threat to ground water quality due to human activities has become a matter of great concern, a vast majority of ground water quality problems present today are caused by contamination and by over-exploitation or by combination of both. Rapid urbanization and industrialization in India has resulted in steep increase of generation of both. Due to lack of adequate infrastructure and resources of the waste is not properly collected, traded and disposed, leading to the accumulation and in filtration causing ground water contamination. The problem is more severe in and around large cities due to various clusters of industries. In many of these areas ground water is only sources of drinking water. With this background water assessment studies, evaluation, of the quality ground water is an important as the quantity.

Estimation of physico-Chemical characteristics of water is essential for the suitability of water for various purposes like drinking, domestic, industrial and irrigation. The Ground water quality may also vary with seasonal changes and is primarily governed by the extant and composition of dissolved Solids.

STUDY AREA

Tiptur town is the area of the major towns of Tumkur district, famous for its Copra Business and Education. Tiptur is 75 Km from Tumkur district, 141 Km from Bangalore, Karnataka. Areas under study mainly depending on irrigation, especially Coconut Trees and other minor crops, for which quality of ground water has impact. The population of Tiptur has also impact on drinking water mainly supplied by tube-wells (bore-wells) in and surrounding of the town. Pollution free drinking water plays an important role in growth of town free from diseases. So the polluted water by sewages, waste disposals and other human activities may deplete the quality of ground water.

Materials and Methods

Water samples were colleted from bore-wells after pumping 10-15 minutes. In Tiptur town and surrounding areas about 10-12 Km distance. Samples were collected from 50 locations during Pre-Monsoon and Monsoon season the sampling locations are show in figure (1). Samples were analyzed to determine concentration of Ph, total dissolved solids, electrical conductivity, turbidity, total hardness, calcium, magnesium, alkalinity, iron, fluoride, chloride, sodium, potassium, nitrate, phosphate, etc. Using water analyzer Kit (Global make); Flame photometer, Visi-spectrophotometer (340-960 nm). The tests were carried out as per standard methods of APHA 1995 edition. The results were compared with IS standards (Table 1).

RESULTS AND DISCUSSION

The P^h values found well with in the permissible limits. The P^H value ranged from 7.08 to 7.73 and 4.9 to 9.2 during pre-Monsoon and Monsoon seasons respectively, except few location samples. TDS (Total dissolved solids) ranged from 213 to 1427 during Pre-Monsoon and 70 to 1610 mg/l during Monsoon It infers that Tiptur town showed high Values of TDS values like 1427 mg/l and 1610 mg/l respectively, most

TABLE 1

Ground Water Quality at Tiptur Town, Tumkur District, Karnataka

Sl. No.	Parameters	Pre-Mansoon				Mansoon				IS Limits 10500	
		Mini	Maxi	Mean	SD	Mini	Maxi	Mean	SD	Desirable	Max
(1)	(2)	(3)	(4)	(5)	(6)	(7)	(8)	(9)	(10)	(11)	(12)
1.	Ph	7.08	7.73	6.63	0.2653	4.9	9.2	6.96	0.7433	6.5-8.5	6.5-8.5
2.	TDS	213	1427	709.68	287.31	70	1610	772.64	351.48	500	1500
3.	EC	319	2139	1063.6	430.67	100	2420	1109.5	512.88	500	2000
4.	Turbidity	1	1.8	2.04	6.367	0.1	5	1.84	0.987	5	25
5.	TH	175	1180	485.6	223.97	200	920	496.6	171.2	300	600
6.	Ca	11	153	47.28	31.21	17.6	88	44.6	21.05	75	200
7.	Mg	0.4	223	89.66	57.06	27	204.5	94.5	40.65	30	100
8.	Alkalinity	230	720	504.4	407.26	180	730	503.58	115.12	200	
9.	Fe	0.27	2.3	0.70	0.76	0.24	1.71	0.76	0.29	0.3	1
10.	F	0.37	1.48	0.78	0.2588	0.4	1.45	0.79	0.219	0.6-1.5	1.2-1.5
11.	Cl	73	440	178.26	86.28	27	450	181	94.9	250	1000
12.	Na	38	187	105.24	34.83	0.9	289	117.88	50.6	200	
13.	K	0.2	118	17.3	25.79	0.3	115	16.04	24.75	0	
14.	NO 3	0.3	56	24.76	14.75	0.5	52	28.05	13.81	45	
15.	PO 4	0.01	0.1	0.055	0.137	0.01	0.09	0.036	0.0216		

TDS: Total dissolved solids in mg/l , EC: Electrical conductivity in μm. mohs/cm, Turbidity is in NTU, TH: Total hardness in mg/l, Ca: Calcium in mg/l, Mg:Magnesium in mg/l, Fe : Iron in mg/l, F: Floride in mg/l, Cl:Chloride in mg/l, Na: Sodium in mg/l, K: potassium in mg/l, NO_3: Nitrate in mg/l, PO_4:phosphate in mg/l.

of the sampling stations showed TDS well within the maximum limit. Hence, water is suitable for drinking and irrigation purposes in some areas.

Electrical Conductivity found high ranging between 310 to 2139 μ moh/cm in Pre-monsoon and from 100 to 2420 m.mohs/cm but in some of well EC values exceeded the permissible limits. If EC values exceeded 2000 m.mohs/cm then the water may not be suitable for irrigation as well as drinking purposes. EC values are useful to know about the extended pollution and too find out the presences of inorganic dissolved solids.

Turbidity values ranged from 1 to 1.8 in Pre-Monsoon and 1 to 5 in Monsoon.The values of turbidity well within the limits. The TH (total hardness) values ranged from 175 to 1180 mg/l in Pre-monsoon and 200 to 920 mg/l during post-monsoon. In many of sampling stations, few TH values of sampling locations exceeds the limit and few TH values are within the limits, in both the seasons.

The calcium ions ranged between 11 to 153 mg/l during pre-monsoon season and 17.9 to 88 mg/l during the pre-monsoon. The magnesium exceeded the limit in some places during pre-monsoon as well as in mansoon season ranged from 04 to 223 mg/l and 27 to 204.5 mg/l respectively. This indicates the hardness of water.

The alkalinity varied between 230 to 720 mg/l during Monsoon period respectively. The alkalinity values found increasing in Monsoon than in Pre -Monsoon, this may be due to movement of pollutants into the ground water during rainfall season. Large amount of alkalinity in water impacts bitter taste to water and excess alkalinity in water is harmful for irrigation which leads to soil damage and reduces crop yields.

Iron is the one of main constituents of rocks. It occur in ferric and ferrous states. In this study the iron is ranged between 0.27 to 2.3 mg/l during Pre-monsoon and 0.24 to 1.17 mg/l during Monsoon season. Most tubewells yield iron rich water on pumping after prolonged periods, excessive amount of iron imparts taste to water, which is detectable at very low concentration excessive presence has adverse effect on domestic uses. It also promotes growth of iron bacterial and causes health hazards. (Ministry of Rural Development 1994, Sawyer *et. al.* 1994; Todd, 1995)

The chloride concentration ranged from 73 to 440 mg/l during Pre-monsoon and 27 to 450 mg/l during Monsoon, chloride values in both seasons are well within the permissible limits. Water is suitable for construction otherwise high values may corrode the concrete.

The fluoride ranged between 0.37 to 1.48 in Pre-monsoon and 0.4 to 1.45 during monsoon. The study reveals, in some sampling location fluoride exceeds permissible limits. Fluoride ions in excess concentration in drinking water are know to damage teeth, skeleton and other organs (MRPR, 1994, Fluoride below 1.5 mg/l causes dental skeletal 14 over is and above 4 mg/l cams fluorosis (Sharma Kumar, 1998), fluoride low concentration need to be supplemented.

The Sodium raged from 38 to 187 gm/l during Pre-monsoon and 9 to 289 mg/l during monsoon sodium concentration makes water salt taste, it causes hyper tension and not fit irrigation, as it tends to deteriorate and 3 to 115 mg/l during Monsoon excess of sodium in drinking water causes nausea vomiting, etc. (Handa 1994).

The nitrate ranged from 3 to 56 mg/l during Pre Monsoon and 5 to 52 mg/l during monsoon. The presence of Nitrate in ground water is an indication of ground water pollution due to one or all other sources of Nitrates. However on its reduction it becomes health hazard and causes methenoglobinea (Gillii *et. al.*, 1984). In this study some of sample locations exceeded permissible limits and others are well within permissible limit.

The phosphate ranged between 0.01 to 1 mg/l during Pre monsoon and 0.01 to 0.9 during Monsoon. Phosphate is available in organic and inorganic forms, only inorganic compound is important for the growth. Phosphate controls corrosion in pipes and hence used in water supplies. Its high concentration causes eutrophication in water bodes.

CONCLUSION

The following conclusions were drawn on the basis of the chemical analysis of ground water of Tiptur town and surrounding areas:

- Electrical conductivity values were found greater in samples more than 1500μ moh/cm. In most of the locations ground water cannot be used for drinking as well as for irrigation.
- Total hardness ranged between 300 to 1500 mg/l except few sampling locations were hardness varied up to 2400 mg/l. This indicates the presence of calcium and magnesium in more concentration. The values of hardness were increased in monsoon season compared to pre-monsoon season.
- Turbidity values were within the maximum permissible limit in both the seasons in all locations. ·
- Calcium and Magnesium ions value were high in most of the

locations. This represents hardness of the water, and in some of the locations water is not fit for drinking.

- The values of chloride and fluoride were within the maximum permissible limit, except few locations.
- In general, the values of all studied parameters were high and above standards, except few during monsoon compared to pre-monsoon in 2010.

References

Guruprasad, B.J., Nature Environmental Pollution and Technology, 2(2), 173-78/2003.

M. Lenin Sundhar and M.K. Sasectharan, J.E.S. and ENG, Vol. 50, No. 3A187, July 2008, is specification for drinking water 10500, New Delhi.

Is standard 2296-1963(ND) 1974, Standard methods of examination of water and waste water APHA, 19th edition 1995, M.S. Nanjunda Swamy: Ground Water Quality of Jaglur Taluk, Chitradurga district, Karnataka, 2007.

Chapter 9

Sustainable Irrigation

A Study on Tank Irrigation in the Dry Zones of West Bengal

SEBAK KUMAR JANA

ABSTRACT

India is now facing a daunting challenge to provide food security to its burgeoning population. Rainfed areas in the country still accounts for about 60% of the cultivated area and these areas are home to majority of rural poor and marginal farmers. Repeated draughts and erratic rainfall have severally affected the livelihood of rural people in these areas. In this backdrop, there is an urgent need to explore the possibilities of sustainable forms of irrigation. Tanks considered as natural capital can play a vital role in this direction. Tank irrigation being based on surface water is an environmentally sustainable form of irrigation which can replenish the ground water level with other advantages like it is less costly and is more decentralized form of irrigation which can involve local communities. Inspite of these advantages, tank irrigation as a source of irrigation has a declining trend in India as well as in West Bengal for various reasons. The present paper attempts to look into following : (i) status and trends of tank irrigation in West Bengal at district level, (ii) the relation between tank irrigated area and rainfall for some selected districts in the dry zones of the state, (iii) impact of tank irrigation on foodgrains output, and (iv) policy suggestion for rejuvenating tank irrigation. The paper proposes to use panel data regression model for the analysis.

1. INTRODUCTION

The optimal management of available water resources has become a major issue world over. Global warming has made the monsoon more variable and unpredictable. Development of agriculture is essentially dependent on irrigation, not only in Rabi I Boro season but also in Khariff seasons in the event of drought or inadequate rainfall. The predicted increased variability of precipitation with longer drought periods would lead to an increase in irrigation requirement even if the total rainfall remains the same. Water storage to conserve flood water due to climate change will be an important agenda in the future (Palanasami *et al.*, 2010). Tanks can play a vital role to conserve water. Tank means a reservoir or place which is used for the storage of water. Tank irrigation, in certain parts of India particularly in the dry zones, provides a good alternative for irrigation development. Tanks can have a wider geological distribution than large-scale projects. Income distribution and employment generation effects are not limited to one area in case of tanks. Tank investments tend to be less capital intensive, have fewer negative environmental impacts and can better involve local communities in improvement and construction works. It is predicted that the impact of climate change is likely to be higher in the rain-fed regions where the tanks are major source of water storage and ground water recharge. Tank irrigation offers a good scope for adaptation to climate change through rain water harvesting. This warrants further intensive study on tank irrigation.

In spite of the advantages of irrigation development, development of tank irrigation deteriorated in almost every state in India including West Bengal. The share of tank irrigated area has declined from 13.2% in 1970-71 to 3.5% in 2003-04 (CWC). There are various causes behind this phenomenon. The main causes of degradation of traditional methods like tanks as has been observed in the literature are as follows (Agarwal, 2009, Dhawan, 1986, Shah, 2009, Narayanmoorthy and Deshpande, 2005).

1. The increasing demographic pressure on land has lead to encroachment of common lands and tank foreshore areas and feeder channels and conversion of tank beds to residential.
2. The introduction of drilling methods and energized pumping enable larger volumes of water to be lifted per unit of time which makes tank irrigation less attractive.
3. The construction of dams/reservoirs in the upper catchment results in a less quantity of water reaching the downstream tanks.
4. The deterioration of traditional local systems is also due to the weakening of local institutions for maintaining the existing facilities and managing them.

5. Due to government emphasis on more convenient systems of centralized storage arrangements in the form of water reservoirs and canals has lead to neglect of active community participation in maintenance of traditional systems.
6. Inadequate government expenditure for the construction/ renovation of tanks.

Though there is wide literature on irrigation in India, the study on tank irrigation in India particularly for the State of West Bengal is very limited. Sengupta (1985) emphasized that the knowledge of tradition is necessary to alter the distorted course of development in science and society under colonial rule. The book by Chiraneevulu (1992) discusses the problems of improvements of efficiency under conditions of tank irrigation in Andhra Pradesh and recommends for public policy for improvement of storage capacity. According to Sakthivadivel (2004) despite several state-sponsored tank rehabilitation programmes in many states, the number of tanks rehabilitated effectively is negligible as compared to the total number of tanks. The focus of the enquiry of Dhawan (1986) is to highlight regional differential in tubewell expansion, within the plains of the Ganges and the Indus rivers. Anbumozhi *et al.* (2007) observe that deferred maintenance of surface irrigation infrastructure over years has led to further deterioration of its physical service. According to Sreedhar (2007) while designing and implementing tank rehabilitation projects, it is important that these projects should be made an integral part of the poverty reduction programmes of the government because the decline in collective management of tanks would adversely affect the poorer sections and stricter rules of well irrigation should be implemented. The paper of Chakravorty *et al.* (1991) shows that in the presence of conveyance losses, efficient spatial allocation of irrigation water implies that the quantity of water applied should fall with distance from the water source. According to Palanasami *et al.* (1984), government should establish demonstration tanks in different agro-climatic zones with successful experiments on tank investments as well as new management strategies. Shah (2009) argues that Integrated Water Resources Management (IWRM) paradigm of direct demand management is not feasible in informal water economies. The rise of a class of intermediaries, i.e. water service providers between users and natural resources of water is a precondition to meaningful demand management. The central research question of Shah (2003) is to understand how democratic the water utilisation practices in tank irrigated areas are. Bardhan (2000) explores how does inequality among members of an integrating community affect the success of community undertaking of collective tasks like managing tank irrigation.

Tank Irrigation in West Bengal

West Bengal has a net cultivable area of of 57.51 lakh hectares which is about 64.8% of the total geographical area of the state. Agriculture employs around 58% of the rural workforce. The percentage of cultivated area covered by minor irrigation in West Bengal is 23.5% and the percentage of total irrigated area in the state is 35.6%. The state constitutes 7.8% of the geographical area of India and the percentage of irrigated area of the state is 9.4%. Per capita net irrigated area of the state is 0.03577 hectare which is below the national average of 0.05106 hec. As per the third minor irrigation census report of 2001, out of 9.182 lakh tanks in West Bengal, 1.617 lakh tanks are used for irrigation purposes. Most of these tanks were constructed in ancient times. After the abolition of Zamindari system the tanks did not get proper attention. Out of 30.48 lakh hectares of irrigated land in West Bengal in 2005-06, source-wise irrigated land are as follows: Government canal - 10.65 lakh hectares, Tubewell - 11.53 lakh hectares, tanks - 2.45 lakh hectares. District-wise share of different sources of irrigation in 2005-06 is presented in the Table 1. As the table reveals the dependence dry zones like Purulia, Bankura and Paschim Medinipur is higher than the state average.

Table 2 shows the district-wise area irrigated by Tank of West Bengal during the period 2001-02 to 2005-06 (in thousand hectares). It may be pointed out from table-that area irrigated by Tank in Bankura, Purulia and Paschim Medinipur are 20.8, 22.4 and 26.4 thousand hectares respectively in the year 2001-02. During the above-mentioned period, the tank-irrigated area decreased in Paschim Medinipur but increased in Bankura and Purulia.

2. OBJECTIVES, METHODOLOGY AND DATA SOURCE

We have selected three districts namely, Bankura, Purulia and Paschim Medinipur for our study as high proportions of these regions fall in the dry zones. Fluctuating rainfall with intermittent drought spell between two successive rainfalls make the crop, generally the Khariff crop, very vulnerable. Though the average annual rainfall in the districts is about 1500 mm, the districts suffer from vagaries of rainfall which seriously affect the crop yield. As the geology of the regions is not favourable to ground water irrigation, the option of tank irrigation should be explored further. The primary objective of the paper is to judge the importance to the farmer-households through willingness to pay of the households for renovation/reconstruction of the tanks and

TABLE 1

District-wise Share of Different Sources of Irrigation in West Bengal (2005-06)

	Govt. Canal	*Tank*	*DTW*	*STW*	*RLI*	*Others*	*Total*
(1)	*(2)*	*(3)*	*(4)*	*(5)*	*(6)*	*(7)*	*(8)*
Bankura	64	12.1	1.0	17.3	2.0	3.2	100
Birbhum	54.8	8.7	1.3	15.7	0.7	18.9	100
Burdwan	89.2		7.0	0.0	3.7	0.0	100
Coochbehar	1.6	5.5	20.9	47.2	13.4	11.4	100
Dakshin Dinajpur	—	36.9	22.3	20.2	20.6	0.0	100
Darjeeling	27.0	—	0.0	22.5	47.2	3.4	100
Hooghly	29.8		4.6	41.6	12.8	0.0	100
Howrah	68.8	17.4	5.4	3.9	4.5	0.0	100
Jalpaiguri	62.4	2.2	2.5	4.7	9.5	18.7	100
Malda	—	1.0	6.0	65.9	7.3	19.9	100
Murshidabad	16.1	3.8	5.4	0.2	5.4	69.1	100
Nadia	—	—	9.7	80.5	5.2	4.6	100
North 24-Parganas	4.8	9.7	4.9	70.6	7.6	2.4	100
Paschim Medinipur	21.7	8.1	10.6	37.2	5.2	17.2	100
Purba Medinipur	30.4	12.6	12.9	23.3	1.8	19.0	100
Purulia	42.0	39.3	0.0	0.0	1.8	16.9	100
South 24-Parganas	41.0	11.2	0.3	12.4	3.7	31.3	100
Uttar Dinajpur	1.8	4.1	6.7	82.8	4.7	0.0	100
West Bengal	34.9	8.0	6.4	31.5	5.7	13.5	100

Source : Govt. of W.B., District Statistical Handbook, 2006.

TABLE 2
District-wise Irrigated Area by Tanks of West Bengal in the Year 2001-02 to 2005-06 ('000 hec.)

Districts	*2001-02*	*2002-03*	*2003-04*	*2004-05*	*2005-06*
(1)	*(2)*	*(3)*	*(4)*	*(5)*	*(6)*
Bankura	20.8	24.9	26.7	32.4	33.1
Birbhum	40.8	40.5	25.4	25.4	25.3
Burdwan	—	—	—	—	—
Coochbehar	5.9	5.9	5.9	5.9	5.9
Dakshin Dinajpur	10.4	10.4	10.4	10.4	10.4
Darjeeling	—	—	—	—	—
Hooghly	35.2	33.3	33.4	37.5	37.5
Howrah	8.4	8.4	8.4	8.4	9.0
Jalpaiguri	2.3	2.3	2.3	2.1	2.1
Malda	1.3	1.3	1.3	1.3	1.3
Murshidabad	10.2	7.9	8.6	8.3	7.8
Nadia	—	—	—	—	—
North 24-Parganas	16.7	16.7	17.0	15.1	15.1
Paschim Medinipur	26.4	22.1	24.7	24.7	24.7
Purba Medinipur	40.0	26.3	26.3	26.3	26.3
Purulia	22.4	22.8	27.1	27.8	28.3
South 24-Parganas	34.3	35.7	40.0	34.7	12.3
Uttar Dinajpur	14.8	14.8	8.9	6.0	6.0

Source : Govt. of W.B., District Statistical Handbook.

their perception about the quality of tanks. For the willingness to pay analysis we have used Tobit Model and for the attitude study of the quality of the tanks we have used factor analysis. The data used for this paper has been collected both primary and secondary sources, though the work is mainly based on primary data. The primary data used in this study has been collected using the structured questionnaire meant for primary survey and the survey had been conducted during the period of 2008-09. The precondition for dry zones in the three districts of Purulia, Bankura and Paschim Medinipur did not allow us to adopt random sampling. We used purposive sampling procedure for the selection of blocks and villages. The blocks have been selected with the discussion of the officials of the departments of irrigation, Principal Agricultural Officer and other officials in the district. The villages have been selected in consultation with GP offices. But for the selection of households we have used random sampling procedure. Table 3 presents the number of districts, blocks, villages and households selected in West Bengal for our survey.

TABLE 3
Number of Districts, Blocks, Villages, Tanks and Households Selected in West Bengal

Districts	*No. of blocks selected*	*No. of GPs/ Municipalities selected*	*No. of villages selected*	*No. of tanks selected*	*No. of Households selected*
Bankura	2	5	26	52	155
Purulia	3	8	27	60	198
Paschim Medinipur	6	14	49	69	180
Total Sample	11	27	102	181	533

Source : Based on Primary (Field) Survey.

In section 4 we will discuss the willingness to pay for maintenance of old tanks or construction of new tanks. In section 5 we will discuss the household perception of tank quality.

3. SOCIO-ECONOMIC CHARACTERISTICS OF THE HOUSEHOLDS

As we have already stated we have collected data of 533 households. Some characteristics of the households are mentioned below:

(1) The percentage of female is 46%.
(2) Out of 533 households surveyed the percentage of households belonging to different land classes are as follows: marginal – 54.8%, small – 27.2% and medium – 18%.
(3) The average family size is 6.
(4) The caste-wise percentage of households are as follows: General – 38%, OBC – 32%, SC – 11%, ST – 19%.
(5) The average level of education of the head of the household is calculated as five.
(6) Out of 533 households surveyed 345 households have Kachha houses.
(7) Percentage of NREGS job card holder is 76% and average working days from NREGS per family is calculated as 27.
(8) The percentage of irrigated land of the sample households is calculated is 55.8%.

4. WILLINGNESS TO PAY (WTP) STUDY

We had asked each household about their willingness to pay (WTP) if the new tank/new bund is constructed or tank/bund is renovated and the water supply is ensured for them. The question was open-ended one. We have different responses about their willingness to pay. Some of the respondents are willing to pay zero amount or they avoided to answer the question. From their willingness to pay behavior, we have tried to measure the Total Economic Value (TEV) of the three districts under consideration.

Firstly, we have used the TOBIT regression model to identify the determinants of willingness to pay for improving the Bund condition. Then we have estimated the total willingness to pay. The persons who are not willing to pay or have not expressed anything about their willingness are assumed to have 0 WTP. The negetive value of WTP is not observable. The dependent variable is censored at the value 0. Thus in this analysis, the Tobit model can be written as

$$\begin{aligned} WTP &= WTP^*, \text{ if } WTP^* > 0 \\ &= 0, \text{ otherwise} \end{aligned}$$

It should be mentioned that method of estimation procedure depends upon the structure of the bidding game. If the bidding game is closed ended type, the relevant estimation procedure would be logit/probit. For open-ended question like ours, censored regression like Tobit is more appropriate. The estimated results are shown in the following table. We have tested the WTP model with the whole sample in three districts taken as a whole. We have tried with different variables that may affect household willingness to pay.

TABLE 4

Tobit Estimation Result of WTP

Method: ML-Censored Normal (TOBIT) (Quadratic hill climbing)

Variable	*Coefficient*	*Std. Error*	*z-Statistic*	*Prob.*
C	-115.5402	33.85175	-3.413123	0.0006
ALAND	58.67594	7.318268	8.017735	0.0000
EDUHEAD	10.24967	4.354270	2.353936	0.0186

The above results indicate that the decision of the individuals to contribute or not to contribute depends upon agricultural land (ALAND measured in) possessed by the households and education level of the head of the households (EDUHEAD). Both the factors have positive impact on their willingness to pay. In fact the potential causal factors in addition to these two include family size, age of the head of the household, male/female ratio, etc. But only two factors come out to be significant which has been presented in the table.

Estimation Total Willingness to Pay or Total Economic Value (TEV) of Tanks

Demand for a resource depends upon total economic value (TEV). TEV consists of both use value and non-use value. As water for the tank is used for irrigation, bathing and fishery purposes, it has certainly use value. Contingent valuation method (CVM) has been used for the valuation of use value (Chandrasekaran, 2009, Dutta, M. *et al.*, 2005) though CVM is generally used for valuation of non-use value. We have already stated our sample consists of 533 households. We have collected information about their annual willingness to pay. Total willingness to pay has been calculated by median WTP multiplied by the relevant population. Total willingness to pay reflects the demand for Bund/tank. In the following table we have reported the number of holdings for different size classes for the three districts that we have collected for our survey. There are 13.48 lakh holdings in total in the three districts.

TABLE 5

District-wise No. of Holdings for Different Land Classes

	Marginal	*Small*	*Semi-medium*	*Medium*	*Large*	*Total*
Bankura	239365	84960	36870	5827	20	367042
Paschim Med.	577469	82262	19001	1478	23	680233
Purulia	213085	63750	22455	2197	27	301514
Total	1029919	230972	78326	9502	70	1348789

Out of 13.48 lakh land holdings, we have tried to estimate the number of households those who are willing to pay on the basis of our sample observation. Out of 533 households surveyed we have observed that 121 households are not willing to pay for the renovation/ reconstruction of Bund. The distribution of 121 households by size classes are presented in the third row of the following table. In the next row we have estimated the number of households in the district not willing to pay given the proportion of households not willing to pay in the sample. Then we have estimated the number of households willing to pay in the three districts.

We have collected the data of willingness to pay for all the 533 households in the district. Out of these 121 households have 0 willingness to pay. So the rest of 412 households have some amount of

TABLE 6

Estimation of No. of Holdings Willing to Pay in the Three Districts

	Marginal	*Small*	*Semi-medium*	*Medium*	*Large*	*Total*
0 WTP (Nos. in sample)	46	28	30	11	6	121
0 WTP (Proportion in sample)	0.380	0.231	0.248	0.091	0.050	1.000
0 WTP (Population)	391539	53448	19420	864	3	465274
Estimated no. of holdings willing to pay	638380	177524	58906	8638	67	883515

willingness to pay. From these 412 households we have calculated the median willingness to pay for different categories of farmers, which has been reported in the third row of the following table. In the last row, we have reported the population wtp for the three districts. The total willingness to pay or TEV is calculated as 6.07 crores.

TABLE 7

Estimation of Total Willingness to Pay in the Land Classes

	Marginal	*Small*	*Semi-medium*	*Medium*	*Large*	*Total*
No. of holdings Willing to pay	638380	177524	58906	8638	67	
Median WTP (Rs.)	50	100	150	250	1000	
Total WTP	31918977.27	17752393	8835950	2159545	66528.93	60733395

5. FACTOR ANALYSIS : PRINCIPAL COMPONENTS ANALYSIS FOR THE QUALITY OF TANKS UNDER STUDY

Factor analysis is an interdependent technique when primary purpose is to define the underlying structure among variables in the analysis (Meyers *et al.*, 2006). Factor analysis enables us for analyzing the structure of the interrelationship among a large number of variables by defining sets of variables that are highly correlated known as factors. We have used principal component analysis for factor analysis. We have taken 13 factors to judge the quality of the tanks. The variables are taken as attribute variables and takes values 1 to 5 as judged by the farmers. We have included the following variables for tank quality:

(1) Adequacy of tank water - ADEWATER
(2) Control of water use - CONTWATER
(3) Water Entry and Exit System - ENTRYEXI
(4) Water Sharing System - WATRSHA
(5) Maintenance of Channels - MAINCH
(6) Maintenance of tanks - MAINTANK
(7) Lining between Channels and Land - LINING
(8) Certainty of getting water - CERTAINT
(9) Role of Panchyat in tank development - PANCHYAT
(10) Suez system - SUEZ
(11) Afforestation - FOREST
(12) Relation among the members - RELATION
(13) Fund availability - FUNDAV
(14) Maintenance Bund of the tank - MAINBUND
(15) Panchayat Tax - PANCHYAT

The scale values for the attributes are as follows: Very Bad -1, Bad - 2, Medium - 3, Good - 4, Very Good - 5. In the following Table 8, we report the average assessment of tank qualities on the basis of score given by households.

For factor analysis firstly, we have computed the unrotated factor matrix which contains the factor loading for each variable on each factor. The degree of correspondence between factor and variable is known as loading.

The KMO measure of sampling adequacy is 0.804, which is a rough indicator of how adequate the correlations are for factor analysis. As its value is greater than 0.7, it may be taken as adequate. A significant Barlett's test enables us to reject the null hypothesis of the lack of sufficient correlation among the variables.

KMO and Bartlett's Test

Kaiser-Meyer-Olkin Measure of Sampling Adequacy		.804
Bartlett's Test of Sphericity	Approx. Chi-Square	3234.434
	Df	91
	Sig.	.000

The communalities have presented in the Table 8 Communalities indicate the weights by which each degree to which each variable is participating or contributing to the components solutions. The initial communalities are 1 because each variable is fully involved in the

TABLE 8

Average Assessment of Tank Quality Factors

Tank Quality Variables	*Short name*	*Bankura*	*PM*	*Purulia*	*Total*
(1)	*(2)*	*(3)*	*(4)*	*(5)*	*(6)*
1. Adequacy of tank water	ADEWATER	Medium	Medium	Medium	Medium
2. Control of water use	CONTWAT	Medium	Medium	Medium	Medium
3. Water entry/exit System	ENTRYEXI	Medium	Medium	Medium	Medium
4. Water Sharing System	WAETRSHA	Medium	Medium	Medium	Medium
5. Maintainence of Channels	MAINCH	Very Bad	Very Bad	Very Bad	Very Bad
6. Maintainence of tanks (dissitation etc.)	MAINTANK	Very Bad	Bad	Very Bad	Very Bad
7. Channel/lining between tank and land	LINING	Very Bad	Bad	Very Bad	Very Bad
8. Certainty of getting water	CERTAINT	Medium	Bad	Medium	Bad
9. Role of Panchayat	PANCHYAT	Very Bad	Bad	Very Bad	Very Bad
10. Maintainence of pond bundh	MAINBUND	Very Bad	Very Bad	Very Bad	Very Bad
11. Suez System (lock gate)	SUEZ	Very Bad	Very Bad	Very Bad	Very Bad
12. Forestation	FOREST	Bad	Bad	Very Bad	Bad
13. Relation: owner-beneficiary	RELATION	Good	Medium	Medium	Medium
14. Fund availability	FUNDAV	Very Bad	Very Bad	Very Bad	Very Bad

solution. The extraction communalities indicate the degree to which each variable is captured by the component solution. As extraction communalities are greater than 0.4 for all the variables, there are no variables needed to be removed.

Communalities

	Initial	*Extraction*
ADEWATER	1.000	.600
CONTWAT	1.000	.677
ENTRYEXI	1.000	.550
WAETRSHA	1.000	.694
MAINCH	1.000	.795
MAINTANK	1.000	.636
LINING	1.000	.843
CERTAINT	1.000	.606
PANCHYAT	1.000	.808
MAINBUND	1.000	.721
SUEZ	1.000	.673
FOREST	1.000	.507
RELATION	1.000	.440
FUNDAV	1.000	.750

Extraction Method: Principal Component Analysis.

The following table summarizes the amount of variance accounted for each factor. Each row provides information on a component. The first component accounts for 32.8% variation, followed by second component (15.93%), third component (10.08%), fourth component (7.559%). The unrotated component matrix is shown in the following table. It shows the correlations between variables and components. Thus variable MAINCH is correlated 0.726 with component -1, -0.387 with component-4. In the component matrix, variables ordered by the correlation with each component.

Eigen value indicates the amount of variance of each factor. The eigen value of the first factor 4.6 indicates that the first factor has accounted for about 4.6 units of variance.

The scree plot is shown in figure. The scree plot confirms our previous observation that four components best describes our total variance explained in following Table 9.

The coefficient of rotated component matrix is called factor loading. Rotated component matrix is shown in the following table. Variables are ordered by correlations within each component starting

TABLE 9
Total Variance Explained

Component	*Initial Eigen values*			*Extraction Sums of Squared Loadings*			*Rotation Sums of Squared Loadings*		
	Total	*% of Variance*	*Cumulative %*	*Total*	*% of Variance*	*Cumulative %*	*Total*	*% of Variance*	*Cumulative %*
(1)	*(2)*	*(3)*	*(4)*	*(5)*	*(6)*	*(7)*	*(8)*	*(9)*	*(10)*
1	4.600	32.857	32.857	4.600	32.857	32.857	3.248	23.197	23.197
2	2.231	15.937	48.794	2.231	15.937	48.794	2.547	18.192	41.389
3	1.412	10.088	58.882	1.412	10.088	58.882	2.003	14.305	55.694
4	1.058	7.559	66.441	1.058	7.559	66.441	1.505	10.747	66.441
5	.864	6.169	72.610						
6	.734	5.245	77.855						
7	.594	4.241	82.096						
8	.584	4.172	86.268						
9	.500	3.572	89.840						
10	.427	3.050	92.890						
11	.349	2.494	95.384						
12	.314	2.242	97.626						
13	.199	1.418	99.044						
14	.134	.956	100.000						

Extraction Method: Principal Component Analysis.

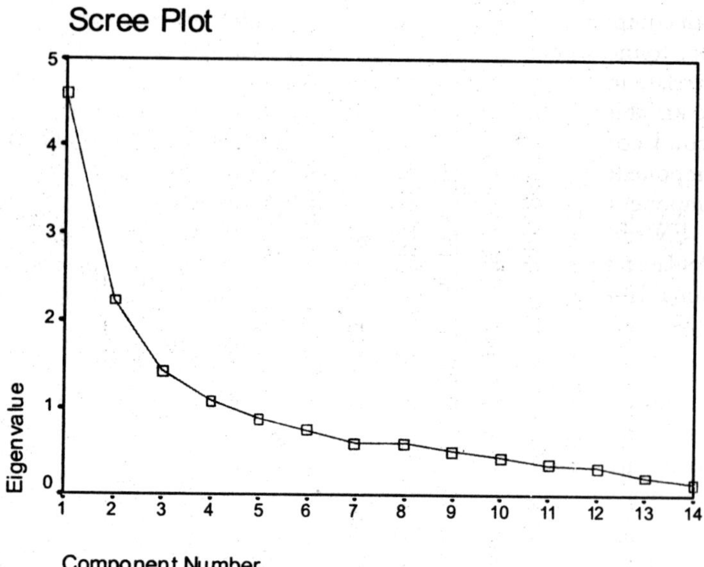

Rotated Component Matrix

	Component			
	1	*2*	*3*	*4*
PANCHYAT	.855	.145	-2.413E-03	-.237
FUNDAV	.810	.299	3.983E-02	-5.854E-02
MAINBUND	.793	.241	-.126	.137
MAINTANK	.614	.501	5.283E-03	-8.494E-02
LINING	.320	.852	6.746E-02	.100
MAINCH	.274	.840	.103	5.935E-02
SUEZ	.492	.542	-.138	.344
RELATION	-.102	-.423	.408	.291
WAETRSHA	-.118	2.939E-02	.811	.144
CONTWAT	.217	-.141	.774	.109
ENTRYEXI	-.181	.322	.638	-8.429E-02
ADEWATER	-2.843E-02	-6.663E-02	.244	.732
CERTAINT	-.436	.130	.109	.622
FOREST	.375	.278	-.223	.489

Extraction Method: Principal Component Analysis. Rotation Method: Varimax with Kaiser Normalization.
a Rotation converged in 11 iterations.

with component 1. It is evident from the table that the first four items were found to correlate to the first component. The next four items correlate high with the second component. Considering the nature of the variables, the first component may be called tank maintenance, the second component may be called channel maintenance, the third component may be called water sharing arrangement and the fourth component may be called certainty of getting water.

We have also tested the scale reliability. The Cronbeench's alpha is calculated as 0.85. As the value is greater than 0.7, it can be used as a seronable test of scale reliability. As the sample size is greater than 500, it is suitable for conducting factor analysis.

5. CONCLUSION

Sustainability of irrigation is an important agenda both from the point of view of farmers and farmers particularly in the context of climate change. Our study is based on the primary data collected from some dry zones in West Bengal. The factor analysis reveals that the maintenance of tanks is in bad shape in our study area. This is affecting the crop productivity and farming income in these areas. Local support is essential to ease the budgetary pressure on government for operation and maintenance. We have tried to determine the economic value of tanks through contingent valuation method. The study reveals that if proper institutional mechanism could be developed then farmers are willing to pay for the maintainenance of the tanks. The fees will bring an ownership feeling to the farmers so that the co-operation among the farmers for conservation get enhanced.

References

Agarwal, Anil *et al.* eds. (2009), "Making Water Everybody's Business", *Centre for Science and Environment*, New Delhi.

Anbumozhi, V., K. Matsumoto and E. Yamaji (2007), "Sustaining Agriculture through Modernization of Irrigation Tanks: An Opportunity and Challenge for Tamilnadu, India". *http://cigr-ejournal.tamu.edu/submissions/volume3/LW%2001%20002.pdf.*

Balasubramanian, K.R. and N. Selvaraj (2003) , "Poverty, Private Property and Common Pool Resource Management:The Case of Irrigation Tanks in South India", Working Paper Nos. 2-3, SANDEE.

Bardhan, Pranab (2000), "Irrigation and Cooperation: An Empirical Analysis of 48 Irrigation Communities in South India", *Economic Development and Cultural Change*, Vol. 48, No. 4, July, pp. 847-65.

Chandrasekaran, K. *et al.* (2009), "Farmers' Willingness to Pay for Irrigation Water: A Case of Tank Irrigation Systems in South India", Water, 1, 5-18, www.mdpi.com/journal/water.

Chakravorty, K.K. *et al.* eds. (2006), "Traditional Water Management Systems of India", Aryan Books International, New Delhi.

Chandrasekaran, K. *et al.* (2009), Farmers' Willingness to Pay for Irrigation Water: A Case of Tank Irrigation Systems in South India.

Chiranjaeevulu, P. (1992), "Tank Irrigation and Agricultural Development", Kanishka Publishing House, Delhi.

Dhawan, B.D. (1986), "Irrigation and Water Management in India", *Indian Journal of Agricultural Economics*, Vol. XLI, Oct.-Dec., pp. 271-81.

Government of West Bengal (1967), The Bengal Tanks Improvement Act, 1939 (As modified upto the 1st April, 1967).

Government of West Bengal (2003), "The Statistical Abstract", *Bureau of Applied Economics and Statistics.*

Government of West Bengal, *Economic Review*, Various Years.

Joy, K.J. *et al.* eds. (2007), "Water Conflicts in India", *Centre for World Solidarity.*

Palanasami, K. and K. William (1984), "Irrigation Tanks of South India: Management Strategies and Investment Alterηatives", *Indian Journal of Agricultural Economics*, Vol. XXXIX, April-June, pp. 214-23.

Palanasami K. and K.W. Easter (2000), *Tank Irrigation in the 21st Century*, Discovery Publishing House.

Rao, P. Narasimha (2007), "Irrigation Development—Issues and Challenges", Discovery Publishing House.

Sakhtivadivel, R. *et al.* (2004), "Rejuvenating Irrigation Tanks through Local Institutions", *Economic and Political Weekly*, July31.

Sengupta, Nirmal (1985), "Irrigation—Traditional *vs.* Modern", *Economic and Political Weekly*, November, pp. 1919-38.

Shah, Esha (2003), "Social Designs: Tank Irrigation Technology and Agrarian Transformation in Karnataka, South India", Oriented Longman Private Limited, New Delhi.

Shah, Mihir (2009), "Participatory Watershed Development: An Institutional Framework", in Anil Agarwal *et al.* eds. (2009), *Centre for Science and Environment*, New Delhi.

Sivasubramaniyan, K. (1994), "Decline of Tank Irrigation", *Economic and Political Weekly*, February 26.

Sivasubramaniyan, K. (2006), "Sustainable Development of Small Water Bodies in Tamil Nadu", *Economic and Political Weekly*, June 30.

Sreedhar, G. (2007), "Tank Irrigation Management in the Peril—Emerging Issues and Challenges", in Narsimha Rao, P. ed.(2007) "Irrigation Development—Issues and Challenges", Discovery Publishing House.

Vaidyanathan, A. (2006), "Inida's Water Resources, Contemporary Issues of Irrigation", Oxford India Paperbacks.

Chapter 10

Private Groundwater Market and Water Quality

A Case Study of Birbhum, West Bengal

SRIJIT CHOWDHURY

ABSTRACT

For more than 30 years, area under the summer paddy has witnessed a phenomenal growth rate in West Bengal. This phenomenal increase in area under boro paddy was achieved mainly due to increase in irrigation facilities because boro paddy—a summer crop with 100 per cent HYV, depends completely on irrigation. Historically, in West Bengal, primary source of irrigation was tanks and canals. But over time, groundwater irrigation expanded rapidly. Now the issue is the effect of largely unregulated private ground water irrigation on ground water quality in Birbhum district of West Bengal. This issue is significant because of the fact that in Birbhum district, 7 blocks out of 19 blocks are affected with high fluoride concentration in ground water. The primary objective of this paper is to test the hypothesis that increasing dependence on ground water leads to problem in ground water quality in Birbhum, West Bengal.

This paper consists of five sections. Paper starts with Introduction in Section 1, Section 2 deals with a Brief Review of Literature, Section 3 defines Methodology, Section 4 shows Results of Testing of Hypotheses and paper ends with Conclusion and Policy Prescriptions in Section 5.

SECTION 1

1.1 INTRODUCTION

In India, water becomes a scarce resource due to high growth in population along with related activities. According to a study of International Water Management Institute (IWMI), India is among the 50 most water poor countries. India is ranked 47th in Water Poverty Index. At present, out of total water utilized in the country, 84 per cent is used for irrigation. Nearly the entire utilizable water resource of the country would be required to be put to use by the year 2025. According to the Central Ground Water Board, a total decline of 4 m, i.e. 20 cm / year has been noted during 1981-2000 in several states of the country (*Singhal, 2003*).

After Green Revolution, growth in cultivation of boro paddy emerged as an engine of growth in West Bengal. Farmers prefer this summer paddy because of high yield rate compared to other principal paddy, aman paddy. For more than 30 years, area under the summer paddy has witnessed a phenomenal growth rate in West Bengal. In West Bengal, percentage of Gross cropped area under boro has increased from 4.52 in 1980-81 to 15.38 in 2000-01 and that of aman has dropped from 55.01 in 1980-81 to 39.92 in 2000-01. Percentage of boro area to total rice area has increased from 6.69 in 1980-81 to 25.79 in 2000-01. This phenomenal increase in area under boro paddy was achieved mainly due to increase in irrigation facilities because boro paddy—a summer crop with 100 per cent HYV, depends completely on irrigation. Historically, in West Bengal, primary source of irrigation was tanks and canals. But over time, groundwater irrigation expanded rapidly. In West Bengal, percentage of gross irrigated area under canal has gone down from 46.57 in 1960-61 to 42.22 in 1999-2000 and that of groundwater (wells, tube-wells, shallow and submersibles) has increased from 19.96 in 1960-61 to 47.13 in 1999-2000. In 2000-01, percentage of total boro area irrigated through groundwater in West Bengal was 77.11 (*Ray, S. and Ghosh, D., 2007*).

The emergence and expansion of private ground water markets has been a major driving force behind agricultural production and productivity growth in rice growing areas of eastern India and Bangladesh. Now the question is what is the effect of largely unregulated private ground water irrigation on ground water quality in Birbhum district of West Bengal. This question is significant because of the fact that in Birbhum district, 7 blocks out of 19 blocks are affected with high fluoride concentration in ground water.

1.2. STUDY AREA AND ITS RATIONALE

Rationale for Selection of Birbhum

Birbhum district lies in the south-western part of the state of West Bengal. The district lies between 23°32'30" and 24°35'0" north latitude and 87°05'25" and 88°01'40" east longitude. The district is bounded by Murshidabad in the east and south-east, river Ajoy in the south, Burdwan in the south-west and state Jharkhand in the north-west and west. The district is spread over an area of 4545 sq.km, with a triangular shape, where apex point towards the north. The district has three administration sub-divisions, (i) Suri Sadar, (ii) Rampurhat, (iii) Bolpur, and 19 administrative blocks.

Geomorphological Features

The rocks of Chhotanagpur Plateau, have extended into the western portion of the district, comprising Rajnagar, Khayrasol, Dubrajpur and Md. Bazar blocks and parts of Suri-1, Rampurhat-1, Nalhati-1 and Muraroi-1 blocks. These blocks are situated at a higher elevations, which gradually slope down till flat plains appear towards further east of this district. The land slope is < 10 meter/km in the eastern part of the district and it is 10-20 m/km in the central part, while land slope is of the order of 20-80 m/km in the western part of the district. The part of the district along the north of Ajoy River and to the south of Labhpur and Bolpur block is absolutely flat.

The rivers and rivulets of the district mainly originate from Chhotanagpur hills, which enter into the western part and pass through the eastern portions. The river Ajoy divides the districts Burdwan and Birbhum. The river Mayurakshi, Bansloy, Hinglo, Kopai, Bakreswar, Brahmani, Dwaraka, etc., pass through the different blocks of the district. Majority of these rivers flows from west to east and their chief characteristic is that they flow with tremendous velocity in the monsoon months but are almost dry in the months of winter.

Climate Conditions

The prevailing temperature and rainfall precipitation in the district indicates sub-tropical to sub-humid climate. The average rainfall (annual) of the district to be 1605 mm. The relative humidity ranges from 25% in the pre-monsoon period to 77% in the post-monsoon period. The temperature ranges from 7°C in the winter to 40°C in summer.

Hydrogeology

- **Major Water Bearing Formation**

  (1) Weathered mantle within 12 mbgl and fractures in

Archaean granites, Gondwanas and Rajmahal traps in the depth span of 70 mbgl.

(2) Shallow (within 50 mbgl) and deeper aquifers (within 400 mbgl) in Alluvium.

- **Pre–Monsoon Depth to Water Level During 2006:**
  In general 5-10 mbgl and as deep as 16.35 mbgl.
- **Post-Monsoon Depth to Water Level during 2006:**
  In general 5 mbgl and as deep as 14.75 mbgl.
- **Long-term Water Level Trend in 10 Years From 1996-2006 in M/Year:**
  During pre-monsoon, rise of water level to the tune of 0.003-0.74 m and fall to the tune of 0.0095-0.399 m.
  During post-monsoon, rise of water level to the tune of 0.0067-0.32 m and fall to the tune of 0.01-0.522 m.

Groundwater Resources

The dynamic groundwater resources of Birbhum district has been estimated by CGWB and SWID, Govt. of West Bengal.

- Total Groundwater Resources : 166715 Ham
- Net Annual Groundwater Availability : 152612 Ham
- Net Annual Groundwater Availability (For Future Irrigation Development) : 111252 Ham
- Stage Of Groundwater Development : 25.79
- All Blocks are Categorized under “Safe” Except 4 Blocks, Namely, Nanoor, Nalhati-2, Muraroi-2 And Rampurhat-2 are under “Semi-Critical Category”.

Birbhum District at a Glance

A. General

(1) Location : 23°32’30” : 24°35’0” N latitude
87°05’25” : 88°01’40” E longitude

(2) Area : 4545 Sq. Km.

(3) Sub-divisions : (i) Suri Sadar
(ii) Rampurhat
(iii) Bolpur

(4) Blocks : 19

(5) District Headquaters : Suri

(6) No. of Villages : 2472

(7) No. of Towns : 6

(8) Population

		(2001 Census)	: 3015422
	(9)	SC and ST Population (%)	: 37
	(10)	Density of Population	: 663/Sq.km.
	(11)	Male : Female	: 1000 : 950
	(12)	Literacy	: 61.48
	(13)	Avg. Height Above Mean Sea Level	: 66.8 m

B. Climate And Rainfall

(1)	Average Annual Rainfall	: 1605 Mm.
(2)	Max. Temperature	: 40°C.
(3)	Min. Temperature	: 7°C.
(4)	Relative Humidity	: 45 To 91.

C. Land Use (Sq. Km.) as on 2004-05

(1)	Forest Area	: 158.50
(2)	Net Swon Area	: 3206.10
(3)	Gross Irrigated Area	: 3165.40
(4)	Non-Agricultural Area	: 917.70
(5)	Others	: 228.90

D. Area Under Principal Crops (Sq. Km.) as on 2004-05

(1)	Total Cereals	: 4173.00
(2)	Total Pulses	: 157.00
(3)	Total Oil Seeds	: 425.00
(4)	Total Fibre	: 6.00
(5)	Total Misc. Crops	: 133.00

E. Groundwater Resources (2004) In Mcm

(1)	Net Annual Groundwater Availability	: 1526.12
(2)	Existing Groundwater Draft for all Uses	: 393.59
(3)	Projected Demand (For Domestic and Industrial Uses After 25 Years)	: 64.73
(4)	Stage of Ground Water Development	: 25.79

F. Ground Water Quality

- Sporadic Occurrence of High Fluoride (Above 1.5 Mg/L and as High as 17.9 Mg/L) Reported From 7 Blocks of the District
- Type of Water:
  Mainly Calcium Bicarbonate Type, at Places Sodium Chloride Type.
  Source : D.I.B., Birbhum, (2007).

TABLE 1

Area (Hectare) Irrigated by different Sources in the Blocks of Birbhum during 2000-01

Block	*Area (hectare) Irrigated by different sources in the Blocks of Birbhum during 2000-01*							
	Canal	*Tank*	*R.L.I.*	*D.T.W.*	*S.T.W.*	*O.D.W*	*Others*	*Total*
(1)	*(2)*	*(3)*	*(4)*	*(5)*	*(6)*	*(7)*	*(8)*	*(9)*
Bolpur- Sriniketan	16940	1660	182	137	3170	13	4267	26369
Dubrajpur	6455	1875	82	712	1460	54	4524	15162
Illambazar	6035	1975	24	1412	4227	2	5131	18806
Khoyrasol	2485	2195	135	0	1254	146	5176	11391
Labpur	13280	2004	168	420	4777	108	3429	24186
Md. Bazar	5925	1285	195	0	291	50	3441	11187
Mayureswar-1	15850	1207	112	101	1241	25	2413	20949
Mayureswar-2	5591	857	16	40	2408	2	1403	10317
Muraroi-1	7796	348	250	20	1858	2	951	11225
Muraroi-2	10884	402	86	150	4064	4	4703	20293
Nalhati-1	8736	917	180	100	2722	27	2430	15112
Nalhati-2	7717	540	7	120	3090	4	1032	12510
Nanoor	18280	1785	146	0	7040	3	2970	30224
Rajnagar	1590	1217	0	0	10	138	2452	5407
Rampurhat-1	12116	1575	58	100	929	8	2153	16939
Rampurhat-2	14437	1158	8	56	2528	2	1689	19878
Sainthia	16484	1982	72	352	3915	2	2931	25738
Suri-1	2410	1158	305	0	255	48	2044	6220
Suri-2	10670	1135	58	0	442	2	1395	13702

Source : D.S.H., Birbhum (2006).

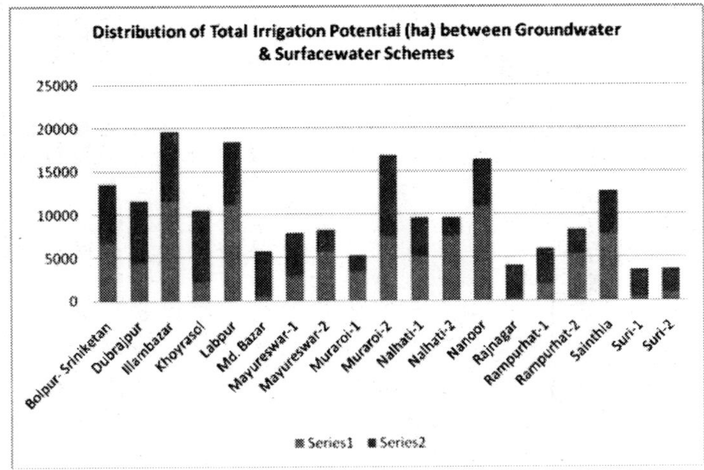

Fig. 1

Series 1: Irrigation Potential of Groundwater Scheme (ha.)
Series 2: Irrigation Potential of Surface-water Scheme (ha.)
Source : GOI (2001).

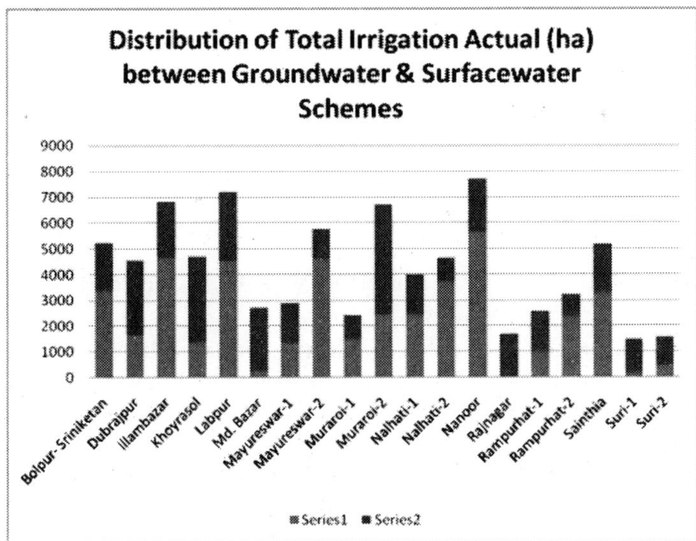

Fig. 2

Series 1: Actual Area Irrigated by Groundwater Scheme (ha.)
Series 2: Actual Area Irrigated by Surface-water Scheme (ha.)
Source : GOI (2001).

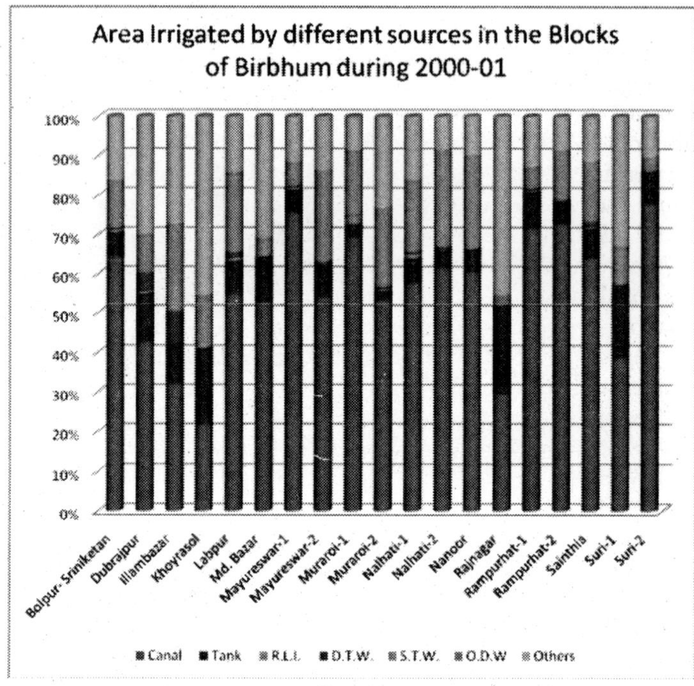

Fig. 3

Source : D.S.H., Birbhum (2006).

SECTION 2

BRIEF REVIEW OF LITERATURE

Scenario of Depth of Water levels in West Bengal in January, 2007

[*Source*: G.Y.B., West Bengal, (2006-07)]

Based on the data collected during January, 2007 by Central Ground Water Board (CGWB), measurement from 814 wells out of 877 existing wells, four grouping were made viz. 0 – 2, 2 – 5, 5 – 10 and 10 – 20 mbgl.

In the Northern part of river Ganga comprising Malda, Uttar and Dakshin Dinajpur, Darjeeling, Jalpaiguri districts are characterized by shallow water level except in 'Bhabar' zone in the northern fringes of Jalpaiguri and Darjeeling and in 'Barind' tract in Malda district. This deeper water level is attributed to the geomorphological condition of the area. The major part of these districts are mainly experienced with the depth to water level ranging between 2.00 – 5.00 mbgl, however,

5.00-10.00 mbgl has also been recorded. In the southern part of the state, the depth to water level ranges mainly from 2.00-5.00 mbgl and 5.00-10.00 mbgl. The minimum and maximum values of water level is 0.66 (Birbhum) and 20.10 (Barddhaman) respectively. The maximum number of wells (67.24%) ranging between 2.00-5.00 mbgl is recorded in Nadia district. The Piezometric level in confined aquifer in South 24 Parganas and coastal part of Medinipur district ranges mainly between 2.00-5.00 mbgl (48.75%) and 5.00-10.00 mbgl (31.25%).

According to *Rudra, K.* (2009), West Bengal covers 2.7 per cent of the national territory and renders home to 8 per cent of the Indian population. The State is endowed with 7.5 per cent of the water resource of the country and that is becoming increasingly scarce with the uncontrolled growth of population, expansion of irrigation network and developmental needs.

The decades of 1950 and 1960 witnessed large scale dam-building either beyond the western border of the State or just along the boundary and even today we have no sustainable system to conserve the rain-water that falls within the geographical territory of West Bengal. The massive capital-intensive engineering interventions into the fluvial system during the post-independence era was guided by a reductionist engineering logic of remaking the Nature for meeting increasing and indiscriminate needs of population and economy. The engineers denied the basic ecological tenets and laws governing the fluvial regime and consequently the projects have not fared well in delivering due benefits and have instead been subject to serious and prolonged controversy over grim social, economic and environmental repercussions.

Since 1970 there was the beginning of over-exploitation of the ground water often beyond the naturally replenishable limit. This was directly related to the introduction of high-yielding but water-intensive seeds that replaced the traditional ones. Now more than 0.60 millions of shallow and more than 5000 deep tube wells are operating in the agricultural fields of the State.

It is learnt from the experience of large dam projects that the crucial gap between storage and actually utilised water must be reduced by shifting the place of conservation within or near the agricultural field. The only option to achieve this target is to have decentralised rainwater harvesting which can be attained by community participation and with the aid of low capital investment ventures. Such bottom-up options would be farmer-centered, eco-friendly and cost-effective. The major shift of paradigm should be reduction of over-dependence on ground water and that is to be utilised within the rechargeable limit. Finally, the community should continue to be the custodian of water.

In India, water becomes a scarce resource due to high growth in population along with related activities. According to a study of International Water Management Institute (IWMI), India is among the 50 most water poor countries. India is ranked 47th in Water Poverty Index, which is based on resource, access, use, capacity and environment. In India, the total utilizable water potential is 1140 km^3, out of which 690 km^3 is from surface water and 450 km^3 from groundwater. Out of total water utilized in the country, 84% is used for irrigation, about 4.4% for drinking and municipal use, 4% for industry, 3.6% for energy development and remaining 4% for other purposes. Nearly the entire utilizable water resource of the country would be required to be put to use by the year 2025 (Singhal, 2003). This is mainly on account of increased demand of water for irrigation required to grow more foodgrains for the increasing population, which is estimated to reach about 1.25 billion by the year 2025 (Singhal and Gupta, 1999).

The use of ground water for irrigation has increased from 6.5 km^3 in 1951 to over 50 km^3 in 1997. The number of wells has also considerably increased from merely 4 million in 1951 to nearly 17 million in 1997. According to the Central Ground Water Board, a total decline of 4 m i.e. 20 cm/year has been noted during 1981-2000 in several states of the country (Singhal, 2003).

Fluoride contamination in the groundwater is a natural phenomenon, influenced basically by the local and regional geography as well as hydro geological conditions. The main sources of fluoride in the groundwater are the fluoride bearing minerals in the rocks and sediments. The weathering and leaching of rocks are expected as major governing factor. Although fluoride is an essential element for all living beings from the health point of view because it helps the normal mineralization of bones and formation of dental enamel, but excess intake of fluoride replaces the calcium component in teeth and bones and hinder building of calogen leading to damages of the teeth and bone.

Fluoride contamination in drinking water is a worldwide problem. 29 countries are reported to be affected with fluorosis and the fluoride related disease. It is estimated that in India, 20 states covering more than 65 million people including 6 million children are affected with fluoride. Fluoride ion (F) concentrations in India's groundwater vary widely from 0.01 mg/l to 48 mg/l.

Gupta et al. (2006) examined the concentration of fluoride ion (F) in groundwater samples from different sites in Nalhati-1 block of the Birbhum district, West Bengal by SPADNS colorimetric analysis. Most of the F levels were within permissible limits, whereas a significantly higher concentration of 1.95 mg/l was found in artesian well samples of Nasipur village. The F in groundwater appears to be controlled by the distribution of Ca^{2+} and to some extent SO_4^{2-}, ionic strength, and the presence of complex ions. From correlation coefficient analysis, F was found to be inversely related to Ca^{2+} and positively related to Na^+. Basaltic rock and water interaction with it may be responsible for the increased F in the artesian well samples from the study area.

Rapid Assessment of Fluoride Contamination in Groundwater of West Bengal

As per the decision taken by the Fluoride Committee, a Rapid Assessment of fluoride contamination in groundwater of West Bengal was conducted in 2006-07 to have a firsthand knowledge of the fluoride problem so as to draw up a plan of action for mitigation of the problem. Depending on the hydro-geological condition and vulnerability of fluoride contamination of groundwater, 101 blocks were selected spread over 12 districts. The sample size constitutes 10% of the estimated number of tube-wells in the study area. However, to cover the entire area uniformly it was decided to collect one sample from each gram sansad. As per Bureau of India Standard (BIS) of drinking water (IS: 105001), acceptable limit for presence of fluoride is 1.0 mg/l and cause for rejection is 1.05 mg/l, accordingly, the test results have been arranged in three categories viz, samples having fluoride <1.0 mg/l, $>1.0<1.5$ mg/l and >1.5 mg/l. Summary results for Birbhum district are given in Table 2.

Central Pollution Control Board [*CPCB (2006)*] carried out a study to assess the level of contamination of groundwater in the fluoride-contaminated areas of Birbhum district of West Bengal. The groundwater of nine blocks of Birbhum District namely, Nalhati-1, Rampurhat-1, Mohamadbazar, Khoyrasole, Rajnagar, Sainthia, Suri-1 and Suri-2, Mayureswar-1 and Dubrajpur were considered for monitoring. In this present study, 46 tube-wells and one dug well in affected villages of four blocks, identified by PHED were selected for the monitoring.

75% tubewells in Rampurhat-1, 46% in Khoyrasole, 20 % in Suri-2 have been found contaminated with fluoride but Nalhati-1, no such contamination has been observed. The highest fluoride concentration of 14.48 mg/l and pH 8.01 has been found in Khoyrasole block. It has been observed that the groundwater below 100 ft. depth was found safe and within the permissible limit whereas fluoride concentration was detected very high in depth higher than 180 ft.

TABLE 2

Blocks of Birbhum Tested for Fluoride Contamination

Sl. No.	Name of Block	No. of samples tested	No. of samples having F conc. in mg/l			Max. conc. In mg/l	Remarks
			F < 1.0	1.0<F<1.5	F>1.5		
(1)	(2)	(3)	(4)	(5)	(6)	(7)	(8)
1.	Dubrajpur	120	115	5	0	1.37	Not Affected
2.	Khayrasole	114	104	5	5	15.90	Affected
3.	Md. Bazar	102	102	0	0	0.965	Not Affected
4.	Muraroi-1	100	100	0	0	0.704	Not Affected
5.	Muraroi-2	121	121	0	0	0.563	Not Affected
6.	Nalhati-1	127	124	2	1	1.52	Affected
7.	Rajnagar	54	44	9	1	1.52	Affected
8.	Rampurhat-1	126	118	3	5	17.90	Affected
9.	Suri-1	79	75	4	0	1.30	Not Affected
10.	Suri-2	59	57	1	1	2.56	Affected
11.	Mayureswar-1	115	113	1	1	1.54	Affected
12.	Mayureswar-2	90	90	0	0	0.588	Not Affected
13.	Sainthia	142	50	91	1	1.52	Affected
14.	Lavpur	142	142	0	0	0.666	Not Affected
15.	Nanoor	159	159	0	0	0.833	Not Affected
16.	Bolpur-Sriniketan	139	139	0	0	0.892	Not Affected
17.	Illambazar	116	116	0	0	0.875	Not Affected
	Total	1905	1769	121	15		

Source : Office of PHED (2008), Kolkata.

TABLE 3
No. of Tubewells covered in Different Blocks under the Study

Name of the Block	*No. of Users (Approximately)*	*No. of tubewells monitored*
Rampurhat-1	2000	12
Suri-2	6000	16
Nalhati	9000	8
Khoyrasole	7500	11

Source : Central Pollution Control Board.

Groundwater Management Strategy and Recommendation by CGWB (D.I.B, (Birbhum) 2007)

Groundwater Development

In Birbhum district, about half of the district area falls under hard rock terrain and the remaining area is underlain by alluvium. A general strategy, as applicable, in the district is summarized below:

- ➢ In hard rock terrain, constituted by hard Crystalline rocks/ Gondwana sedimentaries/Rajmahal Traps, the following groundwater abstraction structures are feasible:
  - Large dia dug wells within 12 mbgl can only meet domestic needs.
  - Bore wells/dug-*cum*-bore wells within 70 mbgl, may yield to the tune of 60-150 lpm and at places as high as 330 lpm.
  - Considering limited potentialities, attempts are to be made to augment the groundwater resources, especially for the weathered zones, by rainwater harvesting.
  - Special care is to be taken on the concentration of fluoride in groundwater for drinking purpose and fluoride contaminated water can be used only after proper de-fluoridation.
- ➢ In alluvial terrain, thickness of alluvium gradually increases from west to east. In the transitional area, the thickness of alluvium is limited within the depth range of 50 to 70 m, and it is very thick in within the depth of 400 m in the western part of the district. Here, groundwater may be developed through different abstraction structures, considering the availability of potential and potable aquifers, thickness of potable aquifers, stage of groundwater development, etc.
  - Shallow tubewells are generally constructed within the

depth ranges of about 10 to 50 mbgl. Average yield of tube wells varies from 18 to 22.7 m^3/hr.

- In general, medium to heavy duty tube wells within 200 mbgl are feasible in the eastern part, average yield of which is to the tune of 102.19 m^3/hr to 276.75 m^3/hr.
- In the alluvial part of the district, four blocks namely Nalhati-2, Nanoor, Muraroi-2 and Rampurhat-2 are in 'semi-critical' category. These blocks need special attention for augmenting the groundwater resources.
- Fluoride contaminated groundwater has been reported from the tube-wells within 60 m depth in the blocks of Nalhati-1, Rampurhat-1 and Mayureswar-1. The deeper aquifers, beyond 60 mbgl, may be exploited for drinking purposes.

SECTION 3

METHODOLOGY

This study tries to test the hypothesis that Increasing dependence on groundwater leads to problem in ground water quality in Birbhum, West Bengal.

STEP-1

Selection of blocks of Birbhum district by the availability and non-availability of public irrigation (canal, RLI) in adequate amounts, by soil and hard rock features, by intensity of use of inputs mainly tractor and sh./sm. There are 19 blocks in this district. The study covers 6 (six) blocks at the time of the first survey and 9 (nine) blocks in the second survey (that is more than 50 per cent of total blocks).

STEP-2

Selection of mouzas within each block selected in Step1 through stratified sampling. There are 2472 mouzas in Birbhum district. Approximately ten per cent of total mouzas of the district is selected from the blocks selected in Step 1.

STEP-3

A survey of mouzas selected in Step 2.

SURVEY-1 (2002)

The first survey on 285 mouzas of the following 6 (six) blocks was done in 2002. The blocks are Muraroi-2 (7GP), Rampurhat-1 (7GP),

Rampurhat-2 (8GP), Mayureswar-1 (8GP), Bolpur (7GP) and Nanoor (5GP).

SURVEY-2 (2008)

The second survey on 249 mouzas of the following 9 (nine) blocks was done in 2008. The blocks are Muraroi-2, Rampurhat-1, Rampurhat-2, Mayureswar-1, Bolpur, Nanoor, Suri-1, Khoyrasol and Rajnagar. In addition to earlier blocks three new blocks (Suri-1, Khoyrasol and Rajnagar) are added to capture the features of the western side of the Birbhum district.

First, five gram panchayats (GP) from each block are chosen on the basis of cropping intensity. Two GPs are of high cropping intensity, Two GPs are of medium cropping intensity and one GP is of low cropping intensity.

Finally, except Rajnagar block, six mouzas from each GP are chosen on the basis of cropping intensity. Two mouzas are of high cropping intensity, Two mouzas are of medium cropping intensity and two mouzas are of low cropping intensity. In Rajnagar block, two mouzas from each GP are chosen on the basis of cropping intensity.

Out of 249 mouzas of Survey-2, 84 mouzas belong to Survey, 1.i.e. a repeat survey has been done in these 84 mouzas in 2008. The distribution of mouzas is as follows:

19 mouzas from Muraroi-2, 15 mouzas from Rampurhat-1, 15 mouzas from Rampurhat-2, 8 mouzas from Mayureswar-1, 12 mouzas from Bolpur, and 15 mouzas from Nanoor.

A repeat survey of these 84 mouzas of above mentioned six blocks helps to test the following hypotheses. The hypothesis is as follows:

Increasing dependence on groundwater leads to problem in groundwater quality in the area.

Binary Logit Model has been used to test the hypothesis.

STEP-4

Report.

SECTION 4

RESULTS OF TESTING OF HYPOTHESES AND ANALYSIS

Hypothesis

Increasing dependence on groundwater leads to problem in groundwater quality in the area.

Logistic Regression

Case Processing Summary

Unweighted Cases[a]		N	Percent
Selected Cases	Included in Analysis	84	100.0
	Missing Cases	0	.0
	Total	84	100.0
Unselected Cases		0	.0
Total		84	100.0

a. If weight is in effect, see classification table for the total number of cases.

Dependent Variable Encoding

Original Value	Internal Value
no change in water level	0
fall in water level	1

Block 0: Beginning Block

Classification Table[a,b]

			Predicted		
			water level		
	Observed		no change in water level	fall in water level	Percentage Correct
Step 0	water level	no change in water level	0	40	.0
		fall in water level	0	44	100.0
	Overall Percentage				52.4

a. Constant is included in the model.

b. The cut value is .500

Variables in the Equation

		B	S.E.	Wald	df	Sig.	Exp(B)
Step 0	Constant	.095	.218	.190	1	.663	1.100

Variables not in the Equation

			Score	df	Sig.
Step 0	Variables	DGWP	14.096	1	.000
		DCHSM	1.171	1	.279
		DCHSH	.022	1	.883
	Overall Statistics		15.446	3	.001

Block 1: Method = Enter

Omnibus Tests of Model Coefficients

		Chi-square	df	Sig.
Step 1	Step	16.988	3	.001
	Block	16.988	3	.001
	Model	16.988	3	.001

Model Summary

Step	-2 Log likelihood	Cox & Snell R Square	Nagelkerke R Square
1	99.270	.183	.244

Classification Table[a]

Observed			Predicted: water level: no change in water level	Predicted: water level: fall in water level	Percentage Correct
Step 1	water level	no change in water level	15	25	37.5
		fall in water level	2	42	95.5
	Overall Percentage				67.9

a. The cut value is .500

Logistic Model for Water Level

Variables in the Equation

		B	S.E.	Wald	df	Sig.	Exp(B)
Step 1[a]	DGWP	2.723	.839	10.546	1	.001	15.227
	DCHSM	.291	.508	.327	1	.567	1.337
	DCHSH	-.653	.547	1.425	1	.233	.520
	Constant	-1.927	.771	6.240	1	.012	.146

a. Variable(s) entered on step 1: DGWP, DCHSM, DCHSH.

Analysis

This analysis is based on a comparative study during the period 2002 to 2007. It incorporated 6 blocks namely, Muraroi-2, Rampurhat-1, Rampurhat-2, Mayureswar-1, Bolpur-Sriniketan and Nanoor covering 84 mouzas across Birbhum district, West Bengal. The first survey was done in 2002 and last one is done in 2008. Over the period, this study tries to examine the effect of dependence on private groundwater on groundwater quality.

Many studies show that the Birbhum district is affected with Fluoride contamination in groundwater. From the data given by PHED (Table 2) department, it has been observed that among all (17) the blocks surveyed, 7 blocks are affected. These 7 blocks includes Rampurhat-1 and Mayureswar-1. Thus out of 6 blocks this study includes, only two blocks are affected with fluoride contamination.

A study from Central Pollution Control Board (*CPCB*, 2006) shows that 75% tubewells in Rampurhat-1, 46% in Khoyrasole, 20% in Suri-2 have been found contaminated with fluoride but Nalhati-1, no such contamination has been observed. The highest fluoride concentration of 14.48 mg/l and pH 8.01 has been found in Khoyrasole block. It has been observed that the groundwater below 100 ft. depth was found safe and within the permissible limit whereas fluoride concentration was detected very high in depth higher than 180 ft.

Our analysis shows that presence of private ground water increases the probability of lowering the ground water level. The corresponding coefficient (DGWP) is positive and statistically significant at 0.01% level. The other variable such as increase in number of submersibles does have a positive effect on lowering the water level though the coefficient is not statistically significant. Thus the null hypothesis that increasing dependence on ground water leads to problem in ground water quality in the area is rejected on the basis that out of 6 blocks that this study includes, only two blocks are affected with fluoride contamination. Moreover, this study does not find a single mouza where water level goes down below 180 ft.

SECTION 5

CONCLUSION

From the data given by PHED department (Table 2), it has been observed that among all (17) the blocks surveyed, 7 blocks are affected. These 7 blocks includes Rampurhat-1 and Mayureswar-1. Thus out of 6 blocks this study includes, only two blocks are affected with fluoride contamination. A study from Central Pollution Control Board (*CPCB*, 2006) shows that the groundwater below 100 ft. depth was found safe and within the permissible limit whereas fluoride concentration was detected very high in depth higher than 180 ft. Fluoride contamination in the groundwater is a natural phenomenon, influenced basically by the local and regional geography as well as hydro geological conditions. The main sources of fluoride in the groundwater are the fluoride bearing minerals in the rocks and sediments. The weathering and leaching of rocks are expected as major governing factor.

Logistic Regression analysis shows that presence of private ground water increases the probability of lowering the groundwater level. Thus heavy dependence on groundwater reduces the water level that may be one of the factors that are responsible for fluoride contamination in the groundwater.

Policy Prescriptions

Improvement of Groundwater Quality

Many studies show that the Birbhum district is affected with Fluoride contamination in groundwater. From the data given by PHED (Table 2) department, it has been observed that among all (17) the blocks surveyed, 7 blocks are affected. These 7 blocks includes Rampurhat-1 and Mayureswar-1. Thus, out of 6 blocks of this study, only two blocks (Rampurhat-1 and Mayureswar-1) are affected with fluoride contamination. Thus heavy dependence on groundwater reduces the water level that may be one of the factors that are responsible for fluoride contamination in the groundwater.

Fluoride contamination in the groundwater is a natural phenomenon, influenced basically by the local and regional geography as well as hydrogeological conditions. The main sources of fluoride in the groundwater are the fluoride bearing minerals in the rocks and sediments. The weathering and leaching of rocks are expected as major governing factor.

This analysis suggests the following courses of actions:

(1) Extension of Surface irrigation through building of Dams and Canals, River Lift Irrigation, Rainwater Harvesting, etc. so that dependence on groundwater comes down.

(2) Artificial recharge of aquifers through the use of modern methods such as remote sensing, geographical information systems (GIS) and information technology.

(3) Application of available defluoridation techniques.

References

Alagh, Y.K. (2004): *State of the Indian Farmer—A Millennium Study: An Overview*, New Delhi, Academic Press.

Bardhan, P.K. (1984): *Land, Labour and Rural Poverty*, Columbia University Press, New York.

Basu, K. and Bell, C. (1991): Fragmented Duopoly—Theory and applications to backward agriculture, *Journal of Development Economics*, Vol. 36, pp. 145-65.

Behara, B. and Mishra, P. (2007): 'Acceleration of Agricultural Growth in India: Suggestive Policy Framework', *Economic and Political Weekly*, Vol. 42, No. 42, October 20, 2007, pp. 4268-71.

B.D.G.: Birbhum (1996): *Bengal District Gazetteers: Birbhum*, West Bengal District Gazetteers, Government of West Bengal, Calcutta.

Boyce, James K. (1987): *'Agrarian Impasse in Bengal: Institutional Constraints to Technological Change'*, Oxford University Press.

Chatterjee, B. (2008): 'Growth and Instability of Agricultural Production in West Bengal', Chatterjee, Biswajit (ed): *Growth, Distribution and Public Policy—A Study of West Bengal.*

Chatterjee, B. and Bhattacharyya, R. (2008): 'Productivity Growth and Use of Modern Farm Inputs in West Bengal', Chatterjee, Biswajit (ed): *Growth, Distribution and Public Policy- A Study of West Bengal.*

CPCB (2006): 'Human Health and Environment Health Impact of PAH in Kolkata Study', *http//www.cpcb.nic.in/Highlights/2006/Human Health And Environment%5B1%5D.pdf*

D.I.B, Birbhum, (2007): *District Information Booklet* (Birbhum), Central Ground Water Board.

D.S.H., Birbhum (2006): *District Statistical Handbook, Birbhum*, Bureau of Applied Economics & Statistics, Government of West Bengal.

Fujita, K. and Hossain, F. (1995): Role of groundwater market in agricultural development and income distribution: A case study in a northwest Bangladesh village. *Developing Economies*, 33(4), pp. 442-63.

Gazdar, H. and Sengupta, S. (1999): 'Agricultural Growth and Recent Trends in Well-Being in Rural West Bengal', Rogaly, B., B. Hariss–White, Bose, S. (eds) *SONAR BANGLA?: Agrarian Growth and Agrarian Change in West Bengal and Bangladesh*, Sage Publications.

GOI (2001): *Report of the 3rd Minor Irrigation Cencus*, Ministry of Water Resources Development, New Delhi, Government of India.

GOI (2007): Inaugural Address of Manmohan Singh, Prime Minister of India, 53rd Meeting of National Development Council, Government of India, New Delhi, May 29.

Gupta, S., Banerjee, S., Saha, R., Datta, J. and Mondal, N. (2006): 'Fluoride Geochemistry of Groundwater in Nalhati-1 Block of the Birbhum District, West Bengal, India', *Research Report*, Fluoride, 39(4), 318-20.

G.Y.B., West Bengal (2006-07): *Groundwater Year Book of West Bengal*: 2006-07, Central Ground Water Board.

Kumar, M.D., Patel, A., Ravindranath, R, and Singh, O.P. (2008): 'Chasing a Mirage: Water Harvesting and Artificial Recharge in Naturally Water-Scarce Regions', *Economic and Political Weekly*, August 30, 2008, pp. 61-71.

Narasimhan, T.N. (2008) : 'Groundwater Management and Ownership', *Economic and Political Weekly*, Vol. XLIII, No. 7, pp. 21-27.

Narayanamoorthy, A. (2007): 'Deceleration in Agricultural Growth: Technology Fatigue or Policy Fatigue?', *Economic and Political Weekly*, Vol. 42, No. 25, June 23.

Ray, S. and Ghosh, D. (2007): 'Modern Agriculture and the Ecologically Handicapped.

'Fading Glory of Boro Paddy Cultivation in West Bengal', *Economic and Political Weekly*, Vol. XLII, No. 26, 2534-42.

Roy, K.C. (1989): 'Optimisation of Unconfined Shallow Aquifer Water Storage for Irrigation', *Ph.D. Dissertation*, Utah State University.

Rudra, K. (2009): http://gangajal.org.in/blog/2009/01/the-status-of-water-in-west-bengal-a-brief-by-kalyan-rudra/

Singhal, B.B.S. and Gupta, R.P. (1999): 'Applied Hydrogeology of Fractured Rocks', Kluwer Academic Publishers, Dordrecht, Netherlands; p. 400.

Singhal, B.B.S. (2003): 'Some Socio-economic aspects of groundwater development in India', *RMZ-Meterials and Geoenvironment*, Vol. 50, No. 1, pp. 345-48.

Vaidyanathan, A. (2007): 'Irrigation' in Kaushik Basu (ed.), *The Oxford Companion to Economics in India*, Oxford University Press.

World Development Report (2008).

Chapter 11

Farm Forestry in India

An Overview

S. PUTTASWAMAIAH

ABSTRACT

Recognising the increasing demand for fuelwood, fodder, and timber the National Commission on Agriculture advocated the Farm Forestry Programme. The programme aimed at bringing fragile and uncultivated lands owned by farmers under tree cover to meet the wood requirements. Implementation of the Farm Forestry programme attracted large number of farmers and also several criticisms against planting of certain tree species. Considering the requirement of promoting tree cultivation activities in the present circumstances of environmental degradation, this paper makes an overview of the Farm Forestry programme in terms of development, farmers' participation, and economics of tree cultivation.

SECTION I

Social Forestry programme with its components like Farm Forestry, Extension Forestry and others was initiated in India since the eighties to meet certain basic needs like fuelwood, fodder, fruits and small timber, as well as regenerate and improve tree cover on degraded forest and common lands. Thereby, seeking to reduce pressure on surviving natural forests, as well as improve the natural resource base of ecologically fragile regions, which are depleting fast due to economic and demographic factors. Farm Forestry programme was intended to induce farmers, especially in ecologically fragile and economically disadvantaged regions such as arid, semi-arid and hill regions of the country to take up tree growing activities. This would help farmers to make better and optimum use of their lands, as well as earn income by meeting the needs of rural and urban markets for fuelwood, bamboo, pulpwood, small timber, etc. Although the terms Social Forestry and Farm Forestry have been ambiguously used, the two are strictly speaking not the same. While the Farm Forestry has been promoted largely on commercial considerations and with profit motive in view, the same is not the case with Social Forestry which has broader social objectives in view such as improving tree cover on degraded forestlands and village commons, making productive use of the country's wastelands, promoting soil and water conservation and improving the landscape. Promotion of the Farm Forestry programme attracted large number of farmers for tree growing activity and they participated by selecting commercially viable tree species which could meet the cash and other requirements. Considering these points the present paper attempts at an overview of the Farm Forestry programme by examining the development of the Farm Forestry programme in India in later part of Section I; farmers' participation; relative economics of trees grown under the Farm Forestry programme; marketing aspects in Section II and controversies raised against adoption of some tree species in Section III and Section IV makes a concluding remark.

Development of Farm Forestry Programme

The National Commission on Agriculture (NCA, 1976) advocated taking up Social Forestry Programme in India by involving people. Farm Forestry programme was implemented as a component of the broad Social Forestry Programme, to grow trees on privately owned agricultural lands, on bunds and wastelands by farmers. The NCA defined the programme as "the practice of forestry in all its aspects on farm or village lands, generally integrated with other farm operations. It is a programme of planting trees on bunds and boundaries of the fields of the farmers and to be taken up by the farmers themselves (GOI, 1976)". Thus, the principal concern was to integrate tree cultivation into the land-use pattern and produce fuel wood, fodder, etc., to meet rural requirements. With this programme the government aimed at motivating farmers to afforestation activities and make the locals understand the necessity of and threat faced by forests due to extensive exploitation of forest resources. The Farm Forestry programme got added impetus after the announcement of the National Forest Policy (NFP) in 1988. The NFP providing thrust to the programme, advocated that as far as possible forest-based industries should meet their raw materials by establishing a direct relationship with farmers. Although this is quite contradictory to the recommendations of the NCA, the NFP recognised the motivation behind planting trees on farmlands.

Since late 1970s, the Farm Forestry programme has been widely practiced by farmers in states like Gujarath, Karnataka, Tamil Nadu, Haryana, Uttar Pradesh and West Bengal (Saxena, 1994). The government encouraged farmers by distributing seedlings, saplings and giving guidelines for planting. This attracted large participation as the number of seedlings distributed during the initial period of the Farm Forestry programme crossed the target in some states as shown in Table 1. In Orissa 67 million seedlings were distributed against the target of 5 million. Farmers in Gujarat planted over 713 million trees by surpassing the target fixed at over 311 million trees (Sharma *et al.*, 1995). Similarly, West Bengal brought an area of over 62 thousand hectares under plants crossing the target of 52 thousand hectares (NCAER), which indicates the extent of tree growing activity adopted by farmers. The intention of farmers was to meet multiple requirements like fuel wood, fodder and income from their farmlands, by selecting commercially viable tree species like eucalyptus, poplar, casurina, etc. (GOI, 1987). The area under trees on farm grew at a rate of 53 per cent per annum between 1975 and 1984 in Haryana.

TABLE 1

Progress of Farm Forestry Programme in the Initial Phase (early 1980s)

States	*Seedlings (million)*	
	Target	*Achievement*
Gujarat (Phase I)	150	375
Orissa	5	67
Bihar	200	206.5
Tamilnadu	40.08	101.4

Source : NCAER.

Poplar was popularly adopted by Haryana farmers, considering its contributions to the rural economy in terms of employment, availability of fuelwood and its service as a windbreak (Sodhi and Ansari, 1996). In Uttar Pradesh among the afforested broad-leaved species, eucalyptus occupied 8.2 thousand hectares out of total 53.9 thousand hectares constituting 15 percent of the planted area up to 1978-79 (Mathur *et al.*, 1984). Farmers in Karnataka also actively participated by bringing large area under tree cultivation. The area under eucalyptus, which was about 48 thousand hectares during 1987-88 increased to 66 thousand hectares in 1996-97 (GoK, 1997). Further, for motivating people a new scheme called 'tree patta' was launched during 1993-94 in rural and urban areas of Karnataka. Under this the participant i.e., pattadar was entrusted with protection of trees and at the time of harvesting the produce would be shared in the ratio of 75:25 between the pattadar and government. The programme achieved success as over 9000 pattalands were distributed against the target of 5000 during 1994-95 (GoK, 1994). The Farm Forestry programme was also encouraged from external agencies, which brought large area under tree cover. As shown in Table 2 the total area covered under the external aid was 1241 thousand hectares. Among the states higher afforestation activity has been reported from Gujarat (20.75 per cent), Uttar Pradesh (11.84 per cent), Karnataka (10.82 per cent) and Andhra Pradesh (9.74 per cent).

SECTION II

PARTICIPATION BY FARMERS : SIZE NEUTRAL

Tree growing activity under the Farm Forestry programme attracted all farmers irrespective of size of holdings, and is size-neutral in its reach to farmers (Aziz, 1995). Farmers participated in the activity considering many factors: (i) natural and technical like uncertainty of rainfall, non-availability of land for field crops, distant location of field from the residence, etc., (ii) economic reasons such as low human and animal labour requirement, higher returns from tree crops, non availability of family labour for field crops, availability of seedlings at lower price; (iii) domestic needs like fuel, small timber and fodder (GOI 1987; Bisalaiah, 1995; Sharma *et al.*, 1995; Saxena, 1995). Further, other factors like plantation of trees by neighbour farmers, to avoid leaving land unused, etc., (Puttaswamaiah, 2001) also have influenced farmers to opt for tree crops. It is significant to note that economic returns are the determining factors, which motivated large number of farmers. Information on the level of participation by farmers in different states in the Farm Forestry programme, in Table 3, shows that in Karnataka and Gujarat small farmers participation is higher compared to that of medium and large farmers (Bisalaiah 1995; Sharma *et al.*, 1995). The Indian Institute of Public Opinion also reported that in Karnataka participation by marginal and small farmers was high, where out of a sample of 3014 tree growers 44 per cent were marginal and small farmers (quoted in GoK, 1997). The marginal farmers who got land under the pattaland scheme in West Bengal also largely cultivated trees. The small farmers have allotted relatively a larger share of their total land to tree cultivation. For instance, in Karnataka, the small farmers who accounted for 24 per cent of the total land had put 37 per cent of it under trees, while large farmers who accounted for 53 per cent of the total land reported 41 per cent to be under trees (Bisalaiah, 1995). All this indicates

TABLE 2

Area Brought Under the Externally aided Farm Forestry Projects in India State-wise

(Area in Thousand Hectares)

States	*Area*	*Percent*
Gujarat	230	20.75
Bihar	72	6.49
Uttar Pradesh	147	11.84
Andhra Pradesh	108	9.74
Karnataka	120	10.82
Rajasthan	91	8.21
Himachal Pradesh	67	6.04
Tamil Nadu	103	9.29
West Bengal	52	4.69
Kerala	69	6.22
Orissa	89	8.03
Maharashtra	44	3.97
Haryana	30	2.71
Jammu and Kashmir	19	1.71
All India	1241	100

Source : Forestry Statistics India, 1995 (adopted from Compendium of Environmental Statistics, GoI, 1997).

that irrespective of size classes of holdings farmers participated in tree growing activity on farmland.

Economics of Tree Cultivation under the Farm Forestry Programme

Farmers' participation in any programme, particularly one that seeks to induce them to grow new or non-traditional crops on farmlands depends upon its returns. In the Farm Forestry programme also this factor played a major role as revealed by farmers' preference towards commercially viable tree species, like eucalyptus, poplar, etc., for higher economic benefits along with meeting fuelwood and fodder requirements. Trees cultivated on farm land under the Farm Forestry programme are economically profitable as revealed in Table 4. In Karnataka eucalyptus cultivation has generated a return of Rs. 1009 per acre per annum (undiscounted) to a cultivation cost of Rs. 92 per acre per annum (undiscounted). Eucalyptus plantation in a village of Midnapur district of West Bengal showed a return of over Rs. 41 thousand per acre, which gave a discounted (15 per cent discount rate)

TABLE 3

Level of Farmers' Participation in Farm Forestry Programme

Author	*State Referred*	*Districts Covered*	*Year of Reference*	*Categories of Farmers*			*Total Sample*
				Small	*Medium*	*Large*	
(1)	*(2)*	*(3)*	*(4)*	*(5)*	*(6)*	*(7)*	*(8)*
Aziz (1995)	Karnataka	Kolar	1989-90	21 (25.92)	39 (48.14)	21 (25.9)	81
Bisalaiah (1995)	Karnataka	Kolar	1990	48 (58)	21 (25)	14 (17)	83
Sharma *et al.* (1995)	Gujarat	Kheda, Panchamahal Junaghad	1990	40 (40.40)	28 (28.28)	31 (31.3)	99
Sodhi and Ansari (1996)	Haryana	Ambala Kurukshetra Karnal	Na	Na	Na	Na	126
Singh and Bhattacharjee (1995)	West Bangal	Midnapur	1988 –89	Land less 10 (15)	Marginal Small & Medium 51 (75)	7 (10)	68

Note : Na = Not available.
Figures in brackets are percentages to the total sample.

Source : Compiled from the Sources quoted in the Table.

TABLE 4
Economics of Tree Cultivation under the Farm Forestry Programme

Author and Tree species	*Reference Year*	*State*	*Districts Covered*	*Cost (Rs.)*	*Return (Rs.)*
(1)	*(2)*	*(3)*	*(4)*	*(5)*	*(6)*
Eucalyptus					
Aziz (1995)	1989-90	Karnataka	Kolar	92 per acre	1009 per acre
Singh and Bhattacharjee (1995)	1981-82 to 1987-88	West Bengal	Midnapur	7192 per acre	41674 per acre
Sharma *et al.*(1995)	NA	Gujarat	Kheda	NA	Net 4331 per acre
		Panchamahals	NA	Net 2545 per acre	
			Junaghad	NA	Net 5247 per acre
Saxena (1994)		Uttar Pradesh	Muzafarnagar	Rs. 1.02 per plant	NA
			Nainital		
			Allahabad		
Mathur *et al.* (1984)	NA	Uttar Pradesh	Mohanwali	1263	Net 12079
			Barahpur	1148	Net 4459
Bamboo					
Ashuthosh *et al.* (1996)	NA	NA	NA	NA	NA
3000/Ha, 9000/Ha, 19000/Ha, 31000/Ha.					
Poplar Sodhi and Ansary (1996)	NA	Haryana			
			Ambala	NA	NA
			Kurukshetra	NA	NA
			Karnal	NA	NA

(*Contd.*)

TABLE 4 (*Contd.*)

Author and Tree species	*NPV*	*B:C Ratio*	*Sample Size*
(1)	*(7)*	*(8)*	*(9)*
Eucalyptus			
Aziz (1995)	NA	NA	81
Singh and Bhattacharjee (1995)	NA	NA	68
Sharma *et al.*(1995)	NA	NA	8
	NA	NA	8
	NA	NA	2
Saxena (1994)	NA	Small Farmers 2..9 Large Farmers 1.6	
Mathur *et al.* (1984)	4598 (12% Dic. Rate)	1:5.08	NA
	1290(12% Dic. Rate)	1:2.3	NA
Bamboo			
Ashuthosh *et al.* (1996)	Size of poles / Land Group*[1]: 3000/Ha, 9000/Ha, 19000/Ha, 31000/Ha.		
	3*4 metres 17792 13995 5175 -7403		NA
	4*5 metres 13673 11394 6099 -1451		NA
	5*6 metres 11099 9581 6055 1027		NA
Poplar			
Sodhi and Ansary (1996)	At 15% discount rate		
	Small / Medium / Large	Small / Medium / Large	Total 126
	29 / 29.5 / 33	6 / 7 / 8	NA
	25 / 30.9 / 25.7	7 / 8 / 9	NA
	33 / 37 / 39	6 / 8 / 8	NA

Note : NA = Not Available, *Land group based on the agricultural value of land, 1 at 8% Discount rate.

Source : Compiled from the Sources quoted in the Table Rs. 7 thousand per acre to a return of Rs. 41 thousand per acre in West Bengal (Singh and Bhattacharjee, 1995). The analysis reveals that trees cultivated on farmlands under the Farm Forestry programme are profitable to farmers.

net benefit of Rs. 12,563 per acre. Asuthosh *et al.* (1996) found the NPV of bamboo cultivation to be higher on a low value agricultural land than medium and high value agricultural land. Similarly, Mathur *et al.* (1984) showed eucalyptus cultivation to be more economical and profitable in Uttar Pradesh. An economic analysis of poplar trees in three regions of Haryana showed that the poplar trees (age 4 years) are economically viable to all categories of farmers (Sodhi and Ansary, 1996). Further, there is a difference between the cultivation cost of trees and annual crops where the cultivation cost of trees is less than that of annual crops, as the number of operations involved in tree cultivation are less and the amount of labour and other material inputs used are low as compared to annual crops. For instance, in Karnataka eucalyptus growing involves a cost of Rs. 92 per acre per annum for a return of Rs. 1009 per acre per annum (Aziz, 1995). The total cultivation cost is over

Marketing Aspects

Expected profit is one of the significant determinants of farmers' participation in tree cultivation activity. Profitability in turn depends upon access to market, prevailing demand and supply conditions and the price of the product in the market. Further, adoption of new technologies in agriculture depends upon the development level of a region. A region with commercialised agriculture has more potential for farmers' participation in adopting modern technologies than a non-commercialised region. This is also true in the case of Farm Forestry programme, which intends to motivate farmers to take up tree cultivation activity. For instance, a dry region in Uttar Pradesh experienced low participation and low investment in tree cropping activity by farmers as compared to the agriculturally advanced regions (Saxena 1995). Contrarily in West Bengal and Karnataka tree cropping is practised well on dry lands due to market advantages (Singh and Bhattacharjee, 1995; Aziz, 1995). This indicates that tree growing is more profitable and successful given stable and remunerative market conditions (Saxena, 1995; Aziz, 1995).

Farmers have differential access to market viz. (i) they can sell their tree plantations through village petty traders (ii) they can contact middlemen/brokers in big towns to sell their plantation (Aziz, 1995). Usually trees are sold through middlemen or intermediaries. Saxena (1994) observes that between the producers and retailers there were about three to four layers of intermediaries, showing no direct contact between tree growers and consumers. Even though farmers contacted the intermediaries to sell their products, most of the farmers resorted to pre-harvest selling than post-harvest selling. For instance, study by

Chatha *et al.* (quoted in Saxena, 1994) found that 77 per cent out of 53 farmers who planted eucalyptus on farm bunds and 56 per cent out of 23 farmers who grow eucalyptus in block plantation resorted to pre-harvest selling. Farmers tend to pre-harvest disposing because of the problems involved like high labour input, technical skill for coppicing, transportation of products, finding consumers, etc., in harvesting and selling of tree production. It is known that the wood market is not well developed in India and the tree growers have to face many problems to dispose off their tree products. Factors like concentration of market activities among few traders, existence of brokers, etc., dissuade farmers from participating in tree growing and marketing activities (Saxena, 1995; Sharma *et al.*, 1995; Singh and Bhattacharjee, 1995). The price of Farm Forestry products like eucalyptus was remunerative in the beginning of the programme and hence it attracted large number of farmers. But, in the course of time due to oligopsnonistic (dominated by few buyers) nature in the market prices declined (Singh and Bhattacharjee, 1995) in some states. Further, it was observed that large farmers are in an advantageous position in the market as compared to small and medium farmers, and this adversely affected the prices of products. The price received by farmers for eucalyptus trees in different states and at different points of time is presented in Table 5. The price received by small and medium farmers (Rs. 784.92 and Rs. 945.59 per acre of eucalyptus respectively) is less than the average price received by all farmers as a whole (Rs. 993.80) and large farmers (Rs. 1292 per acre of eucalyptus) in Karnataka. In Uttar Pradesh also farmers with a holding of more than 2.5 hectares received a price of Rs. 45 per eucalyptus tree as against only Rs. 25 per tree for the farmers with a holding of less than 2.5 hectares. Apart from the difference in price across size classes of holdings, the price fluctuation too adversely affected the returns, as some of the studies show the price of eucalyptus products experienced wide fluctuations over time. In West Bengal the price for a eucalyptus pole of size 16 feet long 3 inches diameter was Rs. 14 in 1985, and declined to Rs. 5 in 1989 (Singh and Bhattacharjee, 1995). Similarly, in Uttar Pradesh the price of eucalyptus reduced from Rs. 40-42 per quintal in 1987-88 to Rs. 33-35 per quintal in 1989-90 (Saxena, 1995). The prices declined owing to several factors like increased aggregate supply of eucalyptus poles causing glut in the wood market; lockout of many paper mills decreasing the demand for eucalyptus products; unethical practices of middlemen and agents to keep lower price for the product; supply of low quality wood; etc. Additionally, lack of adequate information about the wood market also played a major role in adversely affecting the market price. An Evaluation Study of Social

TABLE 5

Prices of Eucalyptus Products in Different States

Author	*Ref. Year*	*State*	*Unit*	*Price (Rs.)*				
				1985	*1986*	*1987*	*1988*	*1989*
(1)	*(2)*	*(3)*	*(4)*	*(5)*	*(6)*	*(7)*	*(8)*	*(9)*
Sharma *et al.* (1995)	1989	Gujarath	One pole	39	35	35	33	34
Singh and Bhattacharjee (1995)	1989	West Bengal						
			Per pole of girth	*1985*	*1986*	*1987*	*1988*	*1989*
			3 Inches	14	13	8	5.6	5
			4 Inches	26.47	22.4	14.6	11.3	10.2
			5 Inches	55.35	48	31.7	28.35	24.1
			6 Inches	92.1	96.2	58.2	52.1	41.3
Bisalaiah (1995)	1988-89 to 1989-90	Karnataka		*Small farmers*	*Medium farmers*	*Large farmers*	*All*	
			One ton	340.68	463.16	409.5	386.29	
Aziz (1995)	1989-90	Karnataka	Eucalyptus	*Categories of Farmers*				
			Per acre	*< 5 acres*	*5 to 10*	*> 10*	*All*	
				784.9	945.5	1292.2	993.8	
Saxena (1995)	1990	Uttar Pradesh	Per tree	*> 2.5 Ha.*	*< 2.5 Ha.*	*Average*		
				45	25	39		

Source : Compiled from the sources quoted in the Table.

Forestry Project in Karnataka reports that only a few farmers were aware of market conditions and the price they could expect for their product (ODA, 1989). But, it is significant to note that in Karnataka the price of eucalyptus has not fluctuated much as in other states because of the presence of paper mills, which are the major source of demand for eucalyptus products (Aziz, 1995). This indicates that stable and enough demand can sustain the tree growing activity on farmlands. Hence, measures like establishing wood markets with easy access to growers and consumers, providing transport facilities, developing the wood market by finding new diversified uses of trees, market information dissemination to farmers, etc., (Bisalaiah, 1995; Sharma *et al.* 1995) need to be taken up to provide development support to tree growers.

SECTION III

CRITICISMS AGAINST EUCALYPTUS TREE CULTIVATION UNDER THE FARM FORESTRY PROGRAMME

Large-scale tree cultivation, particularly eucalyptus under the Farm Forestry programme raised several criticisms. Most of the criticisms were directed on environmental effects of the widely accepted specie, and other effects like change in labour use, displacement of food crops, etc., due to large scale tree plantation. Some of the allegations and their validity against the scientific findings as observed by studies are summarised in Table 6. The critics argued that mass plantation of eucalyptus increases water run off and soil loss; reduces ground water table; depletes soil productivity; prevents under growth and decreases productivity of nearby crops, etc. (Krishnamurthy, 1984; Shiva *et al.* 1984). But, this argument is found to be untrue as tree cover reduces water runoff and soil loss. In fact, eucalyptus trees have no measurable effect on soil loss, instead help to reduce the soil loss on degraded lands. Surface run-off and soil loss would be reduced and water yield can be increased by 10 per cent of the total flow by cultivating eucalyptus (Centre for Industrial Information, 1989).

Another criticism levelled against eucalyptus trees was that it draws down the groundwater table by consuming more water (Krishnamurthy, 1984; Shiva and Bandhyopadyay, 1984). The controversy that eucalyptus lowers the water table by increasing its demand on underground water reserves at times of stress thereby drying up pools (Shiva and Bandopadyay, 1989) has to be considered by the relative ability of the root system of different species to tap the groundwater resources and the rate at which those ground water resources could be recharged (Centre for Industrial Information, 1989).

TABLE 6

Environmental Aspects in Eucalyptus Cultivation under the Farm Forestry Programme

Nature of Problem/ Criticism	*Author, Year of Reference*	*Study Area*	*Method of Study*	*Sample Size*	*Remark of the Study*
(1)	*(2)*	*(3)*	*(4)*	*(5)*	*(6)*
Water runoff and Soil Loss	1. Krishnamurthy (1984)	Karnataka	Forest Land	NA	Increases the water runoff and soil loss
	2. Shiva and Bandhyopadhyay (1984)	Karnataka	NA	NA	
Ground water level	1. Krishnamurthy (1984)				
	2. Shiva and Bandhyopadhyay (1984)				Decreases ground water
more	3. Rajan (1982-83)	Uttar Pradesh	Forest Research Laboratory	NA	Efficient water user and& produces biomass.
	4. Centre for Industrial Information (1989)	Uttar Pradesh	Forest Research Laboratory Dehra Dun	NA	Eucalyptus roots do not grow more than 3.5 metres, so it does not draw more water
Effect on Soil Productivity	1. Krishnamurthy (1984)	Karnataka	Forest Land	NA	Decreases soil Productivity
	2. Shiva and Bandhyopadhyay	Karnataka	Farm Land	NA	

	(1984)				
	3. Rajan (1987)	Uttar Pradesh		NA	Beneficial effects on soil
	4. Centre for Industrial Information (1989)				Beneficial effects on soil
	5. Kushalappa (1985)	Karnataka	Scientific		Contributes to the soil nutrients
Effect on Neighbour Crop	Saxena (1994)	Uttar Pradesh	Farmers' Experience	28	Observed loss in the production of nearby crops

Note : NA – Not Availalbe.
Source : Compiled from the Sources quoted in the Table.

A study of the root system of eucalyptus hybrid of 21 years old showed that the taproot was of 3.2 meter length in Marasandra Eucalyptus Plantation of Hosakote taluk in Bangalore Rural district. Since the root of eucalyptus does not go deep, the tree does not draw more water and hence drying up of the water pools is ruled out (Rajan, 1987). Further, it has been proved that eucalyptus is an efficient user of water. An experiment carried out at Forest Research Laboratory of Uttar Pradesh Forest Department, shows that among several species tested eucalyptus is the most efficient water user (Rajan, 1987). To produce one gram of biomass consumes least quantity of water i.e., 0.48 litre, whereas pongamia pinnata utilises 0.88 litre. Further critics pointed out that the land used for eucalyptus subsequently turns out to be unfit for cultivation of other crops due to depletion of soil nutrients by eucalyptus (Krishnamurthy, 1984). But, a study in Dehra Dun Forest Division of Uttar Pradesh noted that eucalyptus hybrid produces nutrients through litter fall (Rajan, 1989). The study found out the quantity of each of the nutrients that are contributed to the soil separately by leaf, twig and bark litters in Kg. per hectare per annum in 5, 7 and 10 years old eucalyptus plantation. In each of these plantations the highest quantity of nutrients returned to soil is Calcium (Ca) i.e., 40.2, 42.8 and 73.2 Kg/hectare in ascending order of age, then follow the nutrients of Nitrogen (N), Potash (K), Magnesium (Mg) and Phosphorous (P) in which case contribution of 10 years old plantation is roughly double those returned in 5 years old plantation. Considering these facts it can be said that eucalyptus trees add to soil nutrients instead of depleting them.

It is also argued that nothing grows underneath eucalyptus trees. This criticism has been levelled without any proper examination (Rajan, 1989). Eucalyptus generally has narrow crown that cast very little shade even in stands consisting of several thousands of trees per hectare. If there is paucity of undergrowth reasons for its absence are more likely to be excessive grazing, fires, besides the previous vegetation on the site and surface soil conditions (Centre for Industrial Information 1989) than any adverse effects of eucalyptus itself on the site. Contrary to the above criticism in Sulikere Reserve Forest of Karnataka, Kushalappa (1985) found thick undergrowth in the eucalyptus plantation. There were as many as 20 species in the plantation. The eucalyptus plantation had allowed 65 different species to grow underneath than the sal plantation which had 37 species in Uttar Pradesh Forest Research Centre (Rajan, 1989). Based on these facts, the allegation that undergrowth is not possible in eucalyptus plantation found to be untrue. It was also argued that eucalyptus reduces the productivity and production of nearby crops. Saxena's (1994) study covering a sample of 28 farmers in western Uttar Pradesh noted a loss in crop production along the line of plantation (Saxena, 1994). The loss was more in rabi than in kharif crop and it was only after the second year of the plantation of the tree these losses became more conspicuous. However, these observations are based on a limited sample i.e., 28 observations. As evident from Table 7 except the World Bank (1989) study in Uttar Pradesh that observes negligible crop loss, all other studies have reported crop loss on farmland adjoining eucalyptus plantation. For instance, Khyber (1992) finds decline in wheat yield ranging from 4 per cent in the first year to 61 per cent in the ninth year.

TABLE 7

Crop Losses due to Bund Plantation of Eucalyptus

Author	*State*	*Space in meter*	*Density per Ha.*	*Crop Loss*
Wilson and Trivedi (1987)	Gujarat	1.5	—	Mustard and castor crops were destroyed
Ahmed (1989)	Haryana	1.8	250	8.2% of total crop in the first year to 48.8% in the tenth year
World Bank (1989)	Uttar Pradesh	—	200	Negligible
Malik and Sharma (1990)	Haryana	1.5	—	Mustard and wheat crop less by 41%, losses up to 10 meter of the tree line
Khyber (1992)	Uttar Pradesh	—	100	Decline in wheat yield up to 5 meter ranging from 4% in the first year to 61% in ninth year
Saxena (1994)	Uttar Pradesh	0.3 to 3.0 meter	21 to 800	Poor crop in 2-10 meter from the tree line, losses from 5 to 25% of the total crop

Source : Saxena (1994).

Other Criticisms Against Eucalyptus Cultivation

Eucalyptus cultivation has also been criticised on the grounds that it displaces the area under food crops, benefits only large farmers and displaces labour from agricultural activities, etc. Rao (1988) observed a decline in the total food crop area after 1987-88 due to Social Forestry activities in Khammam district of Andhra Pradesh. But, it is difficult to accept the above report as Deshpande and Chandrashekar (1984) and Aziz (1995) note the contrary in Kolar district of Karnataka. They state that the increase in area under eucalyptus is not at the cost of cereals but by garden crops like mulberry and vegetables. Besides, increase in the area of eucalyptus is not from the cultivable land but from the barren, uncultivable land and cultivable wasteland and when there was a decrease in the area of cereals it was mainly because of bad weather and not due to eucalyptus cultivation.

The argument that large-scale tree cultivation displaces labour has been examined by some studies. After the advent of the Social Forestry programme in 1984-85 eucalyptus cultivation in an acre produced only 104 mandays over 6 years, while the annual crops cultivation in an acre generated 228 mandays of job opportunities thus increasing it by 124 mandays than eucalyptus cultivation (Rao, 1988). But, this is not based on the actual farm level data as the author has collected data from official records. Saxena and Srivastava (1995) found that in Muzafarnagar and Nainital of Uttar Pradesh eucalyptus cultivation of 6 years required 398 mandays per hectare for a high density plantation and 296 mandays per hectare for a low density plantation. Eucalyptus absorbs more labour on per hectare basis as compared to the other annual crops. But, annualization of labour use shows poor labour absorption by eucalyptus. NCAER has presented the demand for labour in growing *ragi* (i.e., finder millet) an annual/seasonal crop and eucalyptus as given by the Karnataka Forest Department. The study reveals that over a 5-year cycle, mandays of employment created is 250 per hectare in ragi cultivation, whereas in eucalyptus plantations it is 525. Over a 10-year cycle mandays of employment provided by *ragi* is 500, while corresponding employment in raising eucalyptus plantations is 800. Therefore, the aggregate employment generated in eucalyptus plantations over its life cycle is greater than that in the cultivation of ragi. The Evaluation Report on Social Forestry Programme (GOI, 1987), adopting a before and after the project approach, indicated that the average person days of employment on own work with Farm Forestry increased from 312 mandays in 1982-83 (the year before adoption) to 363 person days in 1983-84. The average employment opportunities created by Farm Forestry is 15 person days taking all states together. Among the states the Farm Forestry programme has generated more employment in West Bengal, while in Karnataka the increased labour absorption is less (only 2 per cent). In total, adoption of Farm Forestry has increased the utilisation of labour over the previous period.

Another criticism levelled is that the programme of Farm

Forestry, which was introduced to induce farmers to take up tree growing as a commercial activity has attracted only large farmers, with small and medium farmers left out. It is claimed that this has increased the social tension (Shiva and Bandhyopadhyay, 1984). But, studies have shown that all farmers including small, medium and large are participating in tree growing activity. If the size of the land brought under eucalyptus is small in the case of small farmers, it is because of the small size of their holdings and other resource constraints like lack of income, constraints in shifting cropping pattern and farmers dependency on land for food crops, etc. Large farmers have ready access to these resources and they can put more area under eucalyptus. According to Aziz (1995) eucalyptus cultivation is size-neutral; to the small and medium farmers it serves as a cushion against the not too dependable monsoon that largely determines the performance of annual crops. Therefore, the criticism that only large farmers are practising Farm Forestry is not supported with empirical evidence. Many of the criticisms levelled against eucalyptus under the Farm Forestry programme are found to be not true. Studies have shown that eucalyptus is an efficient water user by producing high biomass, it does not deplete the soil nutrients, etc. But, the growth of eucalyptus may affect neighbouring crops as evidenced by some studies.

SECTION IV

CONCLUSION

Though the Farm Forestry programme was introduced to produce fuelwood and fodder on farmlands, and to reduce the pressure on forests, farmers grew trees for commercial purpose. The initial success of the programme indicates that farmers expect more income from the changed land use. As the available literature shows tree cultivation under the Farm Forestry programme is economically viable in states like Gujarat, Karnataka, West Bengal, Uttar Pradesh. But, in North-western states and western Uttar Pradesh farmers incurred loss from tree cultivation in later stage of the farm forestry programme. Many states experienced participation by all size classes of holding i.e., small, medium and large farmers and received more income from tree cultivation. But, this success did not last for too long in some states due to reasons like high profit expectation by farmers, over production, inadequate and unfavourable marketing systems. These factors created a glut in the market that brought down the price of tree products thus driving farmers away from tree cultivation in some western states. In fact, farmers in many states practised it as a substitute for annual crops, which attracted attention. Considering the success of the Farm Forestry programme it indicates that a change in the land-use pattern from annual crops to tree crops is profitable and is more suitable particularly in dry and semi-arid regions, which suffer from several environmental constraints. However, mixed plantations consisting of fuelwood, fodder and fruit trees are considered to be some socially desirable than mono - cultivation like eucalyptus, since it can meet the diverse needs of rural communities for fuelwood, fodder and fruits.

REFERENCES

Asuthosh, S., Paresh Vishwakarma and Sharad Nema (1996) "Financial Analysis of Bamboo Based Agroforestry Models", *Indian Forester*, Vol. 122, No. 6, July.

Aziz, Abdul (1995) "Economics of Alternative Uses of Marginal Lands by Class of Farmers and Their Market Links", in *Farm Forestry in South Asia*, ed. by Saxena, N.C. and Ballabh, V., Sage Publication, New Delhi.

Bisalaiah, S. (1995) "Decision Making on Farm Forestry: Role of Socio-Economic and Institutional Factors", in *Farm Forestry in South Asia*, ed. by Saxena, N.C. and Ballabh, V., Sage Publication, New Delhi.

Centre for Industrial Information (1989) Eucalyptus: Planting Techniques, Disputes and Commercial Profitability, Centre for Industrial Information, New Delhi.

Deshapande, R.S. and Chandrashekar, H. (1984) "Is Eucalyptus Farming Really Uneconomic?" Workshop on Eucalyptus Plantation, Indian Statistical Institute, Bangalore

Government of India (1976) Report of the National Commission on Agriculture. New Delhi: Ministry of Agriculture and Cooperation, Government of India.

Government of India (1997) Compendium of Environmental Statistics, Central Statistical Orgnisation, Ministry of Planning and Programme Implementation, New Delhi.

Government of Karnataka (1994-95) Economic Survey, 1994-95, Bangalore.

Government of Karnataka (1997) Karnataka Social Forestry Project Phase II: Project Report—Karnataka Forest Department, 1990-95.

Sodhi, Herminder and M.Y. Ansari (1996) "Economics of Farm Forestry in Haryana—An Economically Viable and Ecologically Sustainable System", *Indian Forester*, Vol. 122, No. 6, July.

Krishnamurthy, B.V. (1984) Ecological Destruction through Government Policies, Workshop on Eucalyptus Plantation, Indian Statistical Institute, Bangalore.

Kushalappa (1985) "Productivity of Mysore Gum (Eucalyptus Hybrid) under different Eco-system in Karnataka", Ph.D Thesis, University of Mysore, Mysore.

Mathur, R.S., S.R. Sagar and M.Y. Ansari (1984) "Economics of Eucalyptus Plantations, With Special Reference to Uttar Pradesh", *Indian Forester*, Vol. 110, Feb., No. 2.

NCAER (Year not available) An Evaluation of Social Forestry Programmes in Selected States in India, New Delhi.

ODA (1989) Evaluation of the Karnataka Social Forestry Project, London: Overseas Development Assistance.

Puttaswamaiah, S. (2001) Economics of Farm Forestry in Karnataka, Ph.D. Thesis, Bangalore University, Bangalore.

Rajan, B.K.C. (1987) Versatile Eucalyptus, Diana Publications, Bangalore.

Rao, Krishna (1988) "Social Costs of Social Forestry", *Economic and Political Weekly*, December 10, Vol. 23, No. 50.

Saxena, N.C. (1994) India's Eucalyptus Craze, The God That Failed, Sage Publications, New Delhi.

Saxena, N.C. (1995) "Liberalisation at the State Level: A Study of Laws Affecting Forest Product Markets", *The Administrator*, Vol. XL, January-March.

Saxena, N.C. and Srivastava (1995) Tree Growing as an Option for Supervision-constrained Farmers in North-west India", in *Farm Forestry in South Asia*, ed. by Saxena N.C. and Ballabh V., Sage Publication, New Delhi.

Sharma, Kailash, Viswa Ballabh, and Ajay Pandey (1995) "An Analysis of Farm forestry in Gujarat", in *Farm Forestry in South Asia*, ed. by Saxena, N.C. and Ballabh, V., Sage Publication, New Delhi.

Shiva, Vandana, Somashekara, Reddy, S.T. and Bandhyopadhyay, J. (1984) "The Ecology of Eucalyptus and Farm Forestry Policy in Rainfed Areas", Workshop on Eucalyptus Plantation, Indian Statistical Institute, Bangalore.

Singh, Katar and Saumindra Bhattacharjee (1995) "The Economics of Eucalyptus Plantations on Degraded Lands, West Bengal", in *Farm Forestry in South Asia*, ed. by Saxena, N.C. and Ballabh, V., Sage Publication, New Delhi.

Chapter 12

Commons and Individuals

Is Forest Rights Act Changing Debates Around Forest Commons?

SANJOY PATNAIK

ABSTRACT

This paper tries to re-visit the debates around commons established since the 1960s specifically focusing on their governance and control interface between state and community. The paper endeavours to analyse different forms of protections for individual and common rights and offers a discussion on how security of commons are linked to secured individual land rights. It also presents a comparative analysis of commons debate of the 60s that focused mostly on breaking commons for individual rights and benefits; and the probable focus, especially after the Forest Rights Act, on creation of commons by squeezing individual domains. Therefore, tries to draw and establish a link between individual and common domains and creation of commons and sustaining state control. Thereby initiating a debate on concepts around 'fuzzy zones', 'vanishing frontiers', 'new commons' focusing on critical questions like, how are they formed, why are they formed and who benefits, who legitimizes them and in whose interest, is common good a move to defy common rights, how the new commons serve as a threat to the traditional common, etc.

I. BACKGROUND

Commonly understood as resources accessible to the entire community without any specific individual rights, Common Pool Resources (CPR) and debates around it have been greatly influenced by changing state policies and programs. In rural India, some of the most important village commons include community forests, pasture or 'wasteland', river banks, river beds, ponds and tanks and even forest department land. In total, commons roughly account for around 20 percent of India's total land area. This resource provides a wide range of physical products (e.g. food, fuel, fodder), income and employment benefits (e.g. supplementary crops or livestock, drought period sustenance, off-season activities), and broader social and ecological benefits (e.g. groundwater recharge, drainage, renewable resources, maintenance of a favourable microclimate).

Common property discourse in India has, since the 1960s, primarily focused around individualization and encroachment. While individualization of commons is widely perceived as an investment and a settled issue, encroachment by nature has mostly been tentative, temporary and runs the risk of attracting enforcement of relevant laws. Though perceived to be essentially the same, there is a subtle difference between the two. While all individualizations, mostly state consented and long drawn, can be seen as essentially encroachments, all encroachments may not necessarily be individualizations strictly from an accumulation stand point.

Beyond the semantics of 'individualisation' and 'encroachment', is the debate on erosion of commons as a result of undefined, unclear, undemarcated and inaccessible individual rights. Examples of such individual-right-induced erosion are both numerous and varied, suggesting the state's inability to respond to rights settlement as well as pro-activism to programatise and break commons to meet individual needs within poverty and livelihoods promotion programs.[1] It is now a foregone conclusion that no commons is secured if individual rights are unclear and unsettled.

The current discussion will be dealt with in two sections. While the first will focus on state induced or guided erosion processes in either of the ways mentioned. The second will deal with camouflaging continued control over commons through creating provisions for benefits to local communities by denying individual rights and in a sense creating a buffer between the 'individual' and the 'commons'.

The debate around 'buffer commons' or 'new commons' or 'fuzzy zones' or 'vanishing frontiers' reached new heights with the coming up of the Forest Rights Act, 2006 and its elaborate provisions on Community Forest Resources. The Forest Rights Act opened up new areas of interpretations of the 'individualization of commons' debate. 'Community forest resource'—a common resource to protect where communities had traditional access—'and community forest rights'—a statutory right to various commons which have been pre-existing—are much neglected subjects in the implementation of FRA. The new menace of recognizing less forestland than what claimants have applied for under individual claims has brought in a different angle to the earlier 'commons' debate. Since forest land use would continue legally in the remaining claimed forestland (earlier under individual use), the smart thing to do has been to design community-based farm forestry programmes to pull out and establish a 'common' out of an erstwhile individual land to continue 'negotiated' control.[2] This newly formed artificial 'common' will serve as a buffer between individual rights and the traditional commons. A possible fall out of this will be that traditional commons will continue to be restrictive with continued institutional uncertainty for communities and shift the control and livelihoods debates to the newly formed commons.[3]

SECTION I

II. TRENDS OF EROSION OF COMMONS OVER THE YEARS

One way in which the rural poor and other socially excluded groups compensate for their lack of access to and control over privately owned arable land is through encroaching upon common and public land. Various studies have revealed that a huge amount of common land is under the possession of landless and socially excluded groups. Commons, therefore, significantly influence poverty and food security in rural India. As mentioned, (i) commons are critical to livelihoods of the rural poor and other socially excluded groups, including women and tribal populations; (ii) they complement private land and other asset holdings; (iii) threats to the extent and quality of the commons harm the poor relatively more compared to better-off and more powerful groups; (iv) of these threats, various forms of institutional changes in recent decades have undermined the local capacity to manage these resources through customary arrangements without replacing them with effective alternative arrangements (state, private, or civil society). In such an environment of institutional uncertainty, non-poor groups with a stronger ability to influence rural institutions in their own favour have managed to encroach on commons with impunity, while the landless, who may be legally entitled to acquire occupancy rights over a plot of cultivable 'wasteland', are unable to realize their legal claim in practice.

The depletion of village commons has been brought about by various processes operating in parallel. The failure of land ceiling laws to bring about any significant redistribution of privately owned ceiling-surplus land in practice has led many states to resort to redistribution of some public land ('wastelands') among landless households. Such *de jure* privatization of commons has not always led to *de facto* control over land by the landless. In Odisha, for example, the very act of pressing a claim to such land is regarded as illegal, making the establishment of rights practically impossible to 'regularize' (Mearns and Sinha, 1998). In many cases, the land that has been redistributed is of low quality and generates low and uncertain crop yields. Many poor beneficiaries also do not have access to the complementary resources (labour, capital, draft animals) required to make more productive use of such land.

Erosion of Commons due to Unclear Individual Rights

In order to improve the economy of the weaker sections of the society and to boost agricultural production in the State, ceiling surplus land has been allotted to the landless persons for agricultural purpose since mid-70s. Though the land ceiling law was imposed to acquire surplus land from the big landlords and redistribute it among the landless, the poor and landless, especially the tribals, have not been able to benefit much from the ceiling surplus distribution. Either the land has not been transferred to the allottee or the allottee does not have physical possession of the allotted land[4] leading to encroachment of commons more often than not.

A number of studies undertaken to assess ceiling surplus and homestead land distribution in Odisha reveal startling evidences of households having patta without possession and possession without patta. Ceiling surplus and government wasteland distributed since the mid 70s have led to various anomalies viz.; land has been allotted but patta not been issued; land has been allotted but is yet to be taken possession of; allotted land is being cultivated by previous owner, land quality: is not suitable for cultivation.[5] In most of these cases, due to denied access to settled farm land, a good amount of common land has been encroached upon for farming and homestead as well.

Similarly, limited access to secured homestead land rights has also forced families, especially tribals in the scheduled areas and scheduled caste in coastal Odisha, to occupy *gochur* (grazing land). The move to relocate these families by providing homestead land has not been very successful as 8 out of every 10 families have received Indira Awaas Yojna (Central Rural Housing Scheme) allocation by producing the patta for the homestead land given elsewhere and have built their houses on gochur instead. There are thousands of examples where people have been given homestead land under homestead land distribution schemes (Vasundhara program in Odisha) that are neither identified nor demarcated. As a result, these homestead land patta holders have gone to common lands (mostly gochur-grazing land) and have constructed IAY houses.

In some cases, IAY houses have been constructed on nearby forest lands. A major reason for the reluctance to relocate has been that most of these homestead land provided are neither identified nor demarcated.[6]

Land Settlement Led Erosion

There are some interesting cases where settlement has changed the land category (kissam) impacting individual rights and causing serious damage to village commons. Such change of land category has created a number of complications; a good number of displaced households do not have access to land, though records say they have been given land compensation, change of land category not only bars access but also does not record rights,

It would be interesting to analyse two specific examples of Settlement changing the kissam of the land and causing land alienation and encroachment into nearby commons in the process. The first of these is the Rengali Irrigation Project Affected People in undivided Sambalpur district getting agriculture land in village forest. In the second, both homestead and agriculture land in most part of coastal Odisha under the erstwhile Burdwan Estate were termed as Village Forest by the 1980 settlement making thousands of Stitiban (settled) patta holders landless overnight.

Double Jeopardy: Legal Jugglery

During the early 1980s, the Rengali Irrigation Project Affected People from Sambalpur and Angul districts, were given land in the Khelei panchayat of Deogarh district—50 decimal each as Gharabari (Homestead land) in Khelei village and 6 acres in the nearby Siarmalia and Jhirpani village for agriculture. They started living in Khelei in 1983 and got the provisional patta from the zonal office in the same year. They, however, got the original patta from Deogarh Tahsildar in 1993. The pattas specify homestead under Ryoti and agriculture land under Godaeka (barren land) kissam. The provisional patta (K form) mentions that 'the villagers will not have any right over trees other than fruit bearing ones, which would be auctioned by the Forest Department.'

However, when these Rengali outstees went to their agricultural field to demarcate and start cultivation, it turned out to be a 'village forest' that was under protection by the adjacent villagers (Jadumani Youth Club of Siarmalia village), who claimed that the area in question is a village forest area which they have been protecting since 20 years. Even the local forest officials corroborated that it was indeed a 'village forest', which can not be given for agricultural purpose. Interestingly, these 6 oustees carry land passbooks distributed to them recently, which corroborates outsees' version and does not recognise the agricultural land as a village forest.

The confusion over the land type arose because is, the land earmarked for these families, previously used by the Forest Department for plantation, later came under cultivable waste (Abada Jogya Anabadi) whose protection responsibility was given to the local villagers, who had rights over the fruits, but not the trees. The land was used more as a common, though not a legal one. Settlement during the 80s turned this cultivable waste into village forest, which by that time was already allotted to these oustees under the rehabilitation plan. Therefore, though these families were given land compensation for displacement, but they didn't have access to those lands for reasons mentioned above. Pushed to the wall, these families slowly started encroaching on nearby commons and forestland.

The problem with recognition of forestland under FRA has been as follows; Sec. 8 of Ch. 3, says the forest rights recognized and vested under this Act would include the right of land to FDSTs/OTFDs who can establish that they were displaced from their dwelling without land compensation due to State development interventions. This is double-jeopardy for these displaced families, one they cannot prove that they were not given land compensation as record says they were, and secondly, being OTFDs they cannot prove forestland under occupation for three generations.[7]

The coastal and north-eastern portions of Odisha were administered by the Bengal Presidency during British India where land revenue was collected through the erstwhile Burdwan Estate. Till Estates Abolition in 1952, people in undivided districts of Balasore and Cuttack were carrying revenue records of Burdwan Estate including titles (pattas) for both homestead and agricultural land. Post-independence and 1952, these Stitiban (settled) pattas were operational till 1980 when the Settlement took place in these districts and changed these homestead and farm lands to 'village forests'. For about three decades, since early 80s till 2010, these families (numbering more than 100,000) do not have patta (title) for the land they reside in and cultivate. The old Burdwan Pattas are no more recognized, though people continue to reside and cultivate in these newly formed village forests. The area does not have a single leaf of grass only houses and farm fields, but continues to be called village forest. People have brought this to the notice of the district administration but the revenue department washed off its hands saying since village forest means *forest kisam*, it would attract provisions of Forest Conservation Act. Therefore, nothing much can be done. Now the situation is either they quietly reside and continue occupation of the village forest without patta or relocate to a revenue area. Now since relocation is almost impossible as both socio-cultural and economic investments are done in this area, residing without patta looks as the only feasible option till not asked to vacate.

Being non-tribals, these families are unable to gain any significant benefits from the Forest Rights Act. Besides since there are instances of proof of residency before independence people do not want to take this option as proving 75 years of residency would become almost impossible. Moreover, since the benefits would be differential across the community, it might mean that some might have to relocate, so they are not seriously considering claims under the FRA.

As against the previous examples of people encroaching on commons due to lack of access to land distributed by Government, this is an example of creation of commons that has hindered secured access to individual rights forcing families to encroach on and occupy nearby commons. In both cases, unsecured individual land rights have been responsible for gradual erosion of commons.

SECTION II

III. FOREST RIGHTS ACT[8] AND COMMONS

After almost two years of its introduction, the one common factor in the forestland recording and recognition process across the country has been tribals getting much less than what they had applied for. Assessments and studies in states reveal that on average tribals have received around 1/3rd of the forestland that they had claimed. Mostly these allotted plots are considerably small than what they had applied for and vary between 75 decimals to one acre—a figure that matches with most assessments, though not many have questioned why. There are examples of villages with unbelievable similarity in the amount of land that have been recognized to each claimant. This is one area where even RTI (Right to Information) has failed to provide any remedy. For instance, one Sub-Divisional Level Committee (SDLC) in Daspalla, Odisha was asked to furnish detailed information on what the Forest Rights Committee (FRC) had filed and the verification process. The concerned SDLC candidly confessed in writing that it doesn't have this information. This, however, raises a number of very pertinent questions; in the absence of the above information how to ascertain the basis on which a certain amount of forestland is being recognized; how does then one verify claims and recognition; is there a link between the said uniformity and lack of information, is there a need (if yes, then how) to verify recognition against claims. It is very difficult to get clear and specific answers to these questions. The standard official answer to the less land allocation issue is tribals do not have a sense of measurement, though the basis of such an answer is debatable.

The post patta distribution issue has been; what happens when people get less than what they have been occupying or cultivating? Theoretically, less land would mean reduced yield/production raising some questions around household food security. Secondly, what happens to that portion of land that the claimant was cultivating but which had not been recognized under FRA? Ideally, forest land use continues.

Though legally speaking, forest land use would continue in the area not recognized in the individual land patta, it is almost impossible for the Government, especially the Forest Department, to map out those areas and initiate any conservation process, partly because of its agricultural land use history. Government was smart enough to recognize this paradox and started gearing up to see what measures could be taken to ensure; firstly, that the forest area regains the green cover through plantation/afforestation program, secondly, in order to facilitate protection and management, some institutional processes to be set up with clear benefit sharing between local communities and the Forest Department; thirdly and most importantly, work with an objective to develop these areas in a manner that would slowly grow up to reduce anthropogenic pressures on the traditional commons—local/ nearby forests, Community Forest Resources, now with FRA. The idea of creating a buffer common could be argued for and against, though reaching any specific conclusion at this point would be grossly premature.

The discussions on commons over the years have been mostly around management, institutions, encroachment, etc. But it is expected with legal frameworks like FRA and PESA (Panchayat Extension to Schedule Areas Act), the commons debate and most importantly the nature of state control may undergo some fundamental changes. In other words, the strong statutory provisions for individual and common rights protection with extremely decentralized control and management mechanism is bound to challenge the *Eminent Domain* and other state control structures. Therefore, there would be a by-design creation of new or artificial common almost delivering the same functions and services as that of the traditional common. Working to create new commons could well be a design that maintains and continues the control status quo in the traditional common and shift the attention to the new commons. Besides, through a long drawn process, such substitution and shift of functions and services of a traditional common with that of a new one would be followed by a shift of the control debate to the new common, leaving the traditional one to different stakeholders.

An assessment of the implementation of Forest Rights Act was undertaken in January-February 2010 in 28 villages across four schedule V districts in Odisha. The districts are Gajapati, Koraput, Kondhamal and Kalahandi. Out of total forest land *patta* distributed (in OTELP field area) to the claimants as per Forest Rights Act, 2006, around 10 per cent households were selected for the study. About 395 households were selected and surveyed using purposive sampling.

Total Forestland Recognised and Recorded (*in acre*)

District	*Land recognized*	*Land allotted to different households*						*Total HHs*
		0-1	*1-2*	*2-3*	*3-4*	*4-5*	*More than 5*	
Kalahandi	75.27	55	18	2	2	1	0	78
Gajapati	122.15	54	26	12	5	4	0	101
Kandhamal	30.70	59	0	0	0	0	0	59
Koraput	240.94	34	40	30	10	4	0	118
Total	469.06	202	84	44	17	9	0	356

Land Applied *vs.* Land Recognised (*in acres*)

District	*No. of Claimants*	*Area Claimed*	*Average area claimed*	*No. of claims recognised*	*Total Area*	*Recognized Average area recognised*
Kalahandi	78	185.73	2.38	78	75.27	0.97
Gajapati	140	936.61	6.69	101	122.15	1.21
Kandhamal	59	30.81	0.52	59	30.7	0.52
Koraput	118	356.91	3.02	118	240.94	2.04
Total	**395**	**1510.06**	**3.82**	**356**	**469.06**	**1.32**

Study conducted in January-February 2010.

Out of the total 1510 acres of forestland applied by 395 households, only 469 acres have been recognised for 356 households. While there is a 10% less in the total number of households that has been recognised rights over forestland, about 70% of the forest area claimed has not been recognised. Out of 185.73 acres of land claimed by 78 families in Kalahandi, only 75.27 acres of land have been recognised. In Gajapati, out of 936 acres of claimed lands, it is only 122 acres. In Kandhamal, all the land claimed has been recognised. In Koraput, out of 365 acres of land only 240 it is quite clear from the data that most of the land recognised are under the first two categories, i.e. up to 1 acre and 1-2 acres.

Odisha Tribal Empowerment and Livelihoods Programme (OTELP), a bilateral livelihoods promotion project, in collaboration with Rural Development Institute carried out the study in Odisha.

The information on application and recognition of land is based on discussions with the claimants and the Forest Rights Committee (FRC). Records couldn't be collected from the SDLC as the applications submitted by claimants were unavailable for cross verification.

IV. CHANGING PARADIGM OF FOREST COMMONS

As a prominent forest rights, "Community Forest Resource" and the institutional governance around it has offered a fairly new trend to the ongoing commons debate. The Forest Rights Act defines CFR as rights over customary common forest land within the traditional and customary boundaries of the village or seasonal use of landscape in case of pastoral communities within forests of all types including protected areas. The criticality of forest commons within the CFR framework has been; an excellent blending of customary practices and existing laws in force to define areas of control, move to democratise use rights by differentiating between users, and most importantly challenging the existing notions of 'rights-prohibited zones' by clearly specifying its operational area. While the local communities will have the right to protect, regenerate or conserve or manage any community forest resource that they have been traditionally protecting, they will also have to ensure that the decisions taken in the Gram Sabha to regulate access to community forest resources and stop any activity that adversely affects forests and environment.

The efforts of limiting, to the extent of denying, state's role in forest commons management in the CFR framework is evident from the institutional governance and processes that are put in place by bringing in Gram Sabha centrestage. Some prominent processes are;

Gram Sabha to initiate the process of determination of community forest resource, decentralized conflict resolution – Forest Rights Committee and Gram Sabhas to settle conflicting claims on traditional and customary boundaries, Gram Sabha to constitute committees for protection of forests and wildlife from among its members (Section 5 of FRA), etc.

Besides, such decentralising trends have continued with the provisions under proposed amendment to PESA. The proposed amendment says, 'Gram Sabha to make regulations to impose conditions for protecting environment and forest', definition of 'community resources' to include land, water, forest, minerals and other resources located in the territorial domain of the community, Gram Sabha recommendations 'binding' on all authorities unless otherwise decided by the State Government for reasons to be recorded in writing, local communities will have legal rights over customary common forest land with right to define and demarcate traditional boundaries and rights over all forests. Moreover, communities will have the rights to decide what constitutes a traditional boundary and Gram Sabha is the ultimate authority in terms of regulating protection of forests through imposition of conditions, and communities to define and decide forest use practice.

The above mentioned statutory framework on one hand would influence commons governance by defining in many ways the decentralized management processes, on the other it is expected to encourage changed state action to ensure that set regulations on commons use (or misuse) is not relinquished. While legal provisions like FRA and PESA have made elaborate provisions for enhanced community role in the management of commons, actions or inactions like delaying or deliberate minimal importance for community claims, creating buffer commons, vaguely defining (or not defining) community resources, creating separate interpretations of provisions, are some standard methods used by state to "balance" centralized control and decentralization. Besides, one set state response to such decentralizing resource management provisions is silence, inaction and indifference. For instance, if PESA provided for the ownership framework for minor forest produce (MFP) and FRA went one step ahead to define it, most states (with the exception of Odisha) still continue to control its management and MFP still operates in a state controlled and monopolistic arrangement in the entire country.

Moreover, forest commons, community rights and the legal framework for recognizing local/indigenous rights within the REDD+ framework is going to be extremely critical in days to come within the climate change discourse. The governance of forest commons will be put to an acid test with the need to blend green and livelihoods rights on one hand and individual and common rights on the other. Besides, forest commons with growing green demands and with a highly Gram Sabha/ local community driven management process will probably be one of the most prominent subjects for debate within the natural resource management regime in India.

Notes and References

1. There are numerous examples of common property resources being individualized under 20 point program, one prominent is portions of village pond going to individuals for beneficiary pond program. Besides, panchayat's role in auctioning village ponds for fishing rights is also a violation of CPR norms.
2. Letter No. 38708 – II-NREGS-43/09/PR dated 5th Dec., 09 - Land Development, Horticulture Plantation and Farm Pond in the land of the beneficiaries under Forest Rights Act under NREGS, Govt of Odisha.
3. The case studies, data and figures cited in this paper mostly pertain to studies undertaken in Odisha.
4. Patnaik, Sanjoy, 'Status Report: Land Rights and Ownership Status in Orissa', August 2008, UNDP and Ministry of Rural Development, Government of India.
5. A number of studies undertaken to assess ceiling surplus and homestead land distribution in Odisha by NIRD, Hyderabad, RDI, Odisha, OTELP and RCDC, a local NGO in Odisha, reveal startling evidences of households having patta without possession and possession without patta. Ceiling surplus and government wasteland are being distributed since mid-70s where land has been allotted but patta not issued, land allotted but yet to take possession, allotted land cultivated by previous owner, land quality: not suitable for cultivation.

   About 57% of households do not have access to farm land and 40% do not have patta for current house sites.
6. In a sample of 856 households in 45 villages across 6 district in Odisha, allocated land of about 82% are neither identified nor demarcated (RDI:2009).
7. Patnaik, Sanjoy, 'Land Rights and Ownership Status in Orissa, 2008', UNDP and Ministry of Rural Development, Government of India.
8. Schedule Tribe and Other Traditional Forest Dwellers' (Recognition of Forest Rights) Act, 2006 is a landmark legislation that recognizes, records and vests forest rights to forest dwelling schedule tribes and other traditional forest dwellers basing on a cut-off date of 13th December 2005 for the tribals and three generations and 75 years (whichever is earlier) for OTEFDs.

References

Report of the Sub-Committee on Land, Planning Commission (2006).

Report of the Expert Group on Prevention of Alienation of Tribal Land and its Restoration , Ministry of Rural Development (2006).

Report of the PESA Enquiry Committee, Ministry of Panchyat (2006).

Economic Survey, 2006-07, Directorate of Economics and Statistics, Government of Orissa.

Orissa Development Report, Planning Commission, Government of India.

The Scheduled Tribes and Other Traditional Forest Dwellers (Recognition of Forest Rights) Act, 2006.

The Scheduled Tribes and Other Forest Dwellers (Recognition of Forest Rights) Rules, 2007.

Government of Orissa (1994), 'A Decade of Forestry in Orissa, 1981-90', Principal Chief Conservator of Forests, Orissa, Bhubaneswar.

Fukuyama, Francis: State Building: Governance and World Order in the Twenty-first Century, Profile Books, London, 2004.

Baseline Report on Restoration of Land to the Allotees of Ceiling Surplus and Government Wasteland, Study undertaken by RCDC with support from NIRD, Hyderabad in Koraput district of Orissa, November 2006.

Hereinafter, Forests, Down to Earth, CSE (New Delhi), June 15, 2007, pp. 36-44.

Chopra, Kanchan, *et al.* l, 'Participatory Development: People and Common Property Resources,' Sage Publications, New Delhi, 1990.

Hardin, Garret, John Baden, 'Managing Commons', W.H. Freeman and Company, San Francisco, 1977.

Karan Manveer Singh, Land use and food security, Open Page, *The Hindu*, 8th October 2006.

Kumar, Kundan, *et al.* l, A Socio-Economic and Legal Study of Scheduled Tribes' Land in Orissa, 2005.

Patnaik, Sanjoy, 'Decades of Forest Governance: Alienating Locals from Forests,' Community Forestry, January 2006.

Patnaik, Sanjoy, Status Report: Land Rights and Ownership in Orissa', August 2008, UNDP and Ministry of Rural Development, Government of India.

Prasad, Archan, Hunger and Democracy: *Political Economy of Food in Adivasi Societies,* People's Democracy, Vol. XXV, No. 44, November 04, 2001.

Prasad, Archana, 'Survival at Stake,' *Frontline* (Chennai), 12 January 2007, pp. 4-10.

Report of the Working Group on Land Relations for Formulation of 11th Five Year Plan, Planning Commission, Govt of India, New Delhi, July 2006.

P. Viegas, Encroached and Enslaved: Alienation of Tribal Lands and its Dynamics, New Delhi, Indian Social Institute, 1991.

Keohane, Robert O, Elinor Ostrom, 'Local Commons and Global Interdependence: Heterogeneity and Cooperation in Two Domains,' Sage Publications, London, 1992.

Saxena, N.C, "Policies for Tribal Development: Analysis and Suggestions", Occasional Paper, 2004.

Singh, Kartar, 'Managing Common Pool Resources: Principles and Case Studies', Oxford University Press, Delhi, 1994.

Singh, S.K., 'A Series on Self-Governance for Tribals: Tribal Lands and Indebtedness,' National Institute of Rural Development, Ministry of Rural Development, Govt. of India, Hyderabad, 20C5, Vol. 1, p. 226.

Chapter 13

Environment and Forestry Planning in Arunachal Pradesh

Facts and Policy Implications

A. MITRA AND KAJU NATH

ABSTRACT

Environmental Conservation and sustainable development are closely inter-linked such that one cannot be achieved at the expense of other. The preservation of natural resources is one of the important components of environmental conservation in the hilly areas. Forests are the most important natural resources in the hilly area and they play an important role in the economic life of the indigenous people of the hilly areas. At the same time forest represent a complex economic resource. In addition to their timber value, they are a valuable source of biodiversity, as a carbon store and in reducing the severity of floods. At the same time forest may be regarded as capital stocks of such a nature that they yield two distinct flows of goods fundamentally different in nature-flow of private goods such as timber, fuelwood, other minor forest produce and a flow of public goods like maintenance of environment, causation of rainfall, prevention of soil erosion, etc. However, there is a conflict between these two types of goods which the forests yield. Hence, in this paper an attempt has been made for an in-depth study of environment and forestry planning in context of Arunachal Pradesh which consists of only 2.5% of India's land mass but it contains 16% of total timber growing stock of the country.

INTRODUCTION

Environmental Conservation and sustainable development are

closely inter-linked such that one cannot be achieved at the expense of other. The preservation of natural resources is one of the important components of environmental conservation in the hilly areas. Forests are the most important natural resources in the hilly area and they play an important role in the economic life of the indigenous people of the hilly areas. At the same time forest represent a complex economic resource. In addition to their timber value, they are a valuable source of biodiversity, as a carbon store and in reducing the severity of floods. At the same time forest may be regarded as capital stocks of such a nature that they yield two distinct flows of goods fundamentally different in nature-flow of private goods such as timber, fuelwood, other minor forest produce and a flow of public goods like maintenance of environment, causation of rainfall, prevention of soil erosion, etc. However, there is a conflict between these two types of goods which the forests yield. Hence, a compromise has to be brought between the demand for private goods and that for public goods from forest for achieving sustainable development in Arunachal Pradesh and here lies the importance of effective forestry planning in a hill state like Arunachal Pradesh.

The present paper makes an in-depth study about the need for an effective approach on forestry planning. The paper is divided into three sections. The first section concentrates on ownership and management of forest in the state. The second section devotes to the assessment of forestry planning of the state during the last five year plan. The third section deals with an alternative approach of forestry planning of the state and finally the conclusion follows.

SECTION I

There are some unique features of forest situated in Arunachal Pradesh. Around 61.5% of the total geographical area is covered under forest which is highest among the seven states of North East India. As per the estimates of the forest survey of India, 2003 based on satellite imagery, forest area consists of around 81.25% of the total area of the state, out of which 78.67% consists of dense forest (density 40% and above) and 21.33% consist of open forest (10% to 40%). An analysis reveals that in around 2.5% of India's land mass, the state of Arunachal Pradesh contains nearly 16% of total timber growing stock of the country (the highest among the individual states) and more than 30% fauna of India (Basic Statistics of North-Eastern Region, 2006).

In Arunachal Pradesh, the ownership of land as well as the forest land and the individual right to use it are governed by local traditions and custom of the tribes. Under the prevailing land tenure system, there are three types of land ownership namely : (a) community land (b) clan land, and (c) individual land. Regarding the forest land, almost all the tribes have the community forest which is controlled by the village council. In some areas, clan ownership is recognized in the forest areas falling within the village jurisdiction. That is why if we look at the data on the basis of legal status of forest in the state, it is found that around 62.17% of total forest is under community ownership which is reported as Unclassified State Forest (USF) (Table 1). However, at present there is a growing tendency of individual ownership of forest which is a recent phenomenon in the state. For example, in the Apa Tani plateau of Lower Subansiri district, the forest has become increasingly privatised.

TABLE 1
Classification of Forest in Arunachal Pradesh in 2000-01 (Legal Status)

Sl. No.	*Legal Status of Forests*	*Area (in Sq. Km.)*	*% of Total Forest Area*	*% of Total Geographical Area*
1.	Reserved Forests	20,319.16	39.42	24.26
2.	Anchal Forests	256.08	0.50	0.31
3.	Protected Forests	7.79	0.02	0.0093
4.	Unclassified State Forests (USF)	30,956.39	60.06	36.96
	Grand Total	51,539.42	100.00	61.54

Source : Compiled from Arunachal Pradesh State Biodiversity Strategy and Action Plan, Department of Environment and Forest, Government of Arunachal Pradesh, 2002.

The reserve forest constitutes 20319.16 sq. km, i.e. nearly 39.42% of the total forest area of the state which is much less than the all India average of around 53%. However, there is a, steady increase in reserve forest as compares to 1950, when there was only 526 sq. km of reserve forest in the state. Although the forest department is enthusiastic of such a steady expansion of reserve forest but possibly it has had its impact on shifting cultivators who are forced to shift to a more dense forest area.

On the other hand, Anchal Reserve Forest covered only 256.08 square kilometers (only 0.5% of the total forest area). Such forests are managed by the forest department with the provision for sharing the net revenue in the ratio of 50:50 (share of village: share of Government).

MAP 1

SECTION II

In Section II, an attempt is made to assess the forestry planning of the state. Regarding forestry planning in Arunachal Pradesh, we find that the state government had not formulated a forest policy of its own and therefore. The forestry planning in Arunachal Pradesh slowly evolved out of the forestry planning enunciated at the national level. For example, when the National Council of Applied Economic Research was requested in 1964 to conduct a techno-economic survey, it recommended three suggestions regarding development afforest namely: (i) a massive programme of industrialization based on forest resources: (ii) emphasis on the immediate development of interior areas: and (iii) strengthening of the machinery implementation i.e., the Forest Department (Techno-Economic Survey, 1967). On the basis of recommendations therein, we find the introduction of Arunachal Forests in 1975-76, the establishment of Arunachal Pradesh Forest Corporation in 1978-79, encouragement for the mushroom growth of saw mills and as a result the forest-based industries were steadily initiated in this state. At the same time, the programmes of forest resources-based industrialization in the state has largely been synonymous with the massive supply of forest-based raw materials and semi-finished products to the other parts of the country. For example, according to the official estimate, Arunachal Pradesh contributes nearly 50% of timber supply made from the North-Eastern region of India to other parts of India, (Government of Arunachal Pradesh, 1995),

With this background, it will be quite interesting to study the detail of the financial outlay for forestry development during the planning period which is given in Table 2. Table 2 show that there was no doubt steady increase in plan allocations in the forestry sector from Rs. 23.15 lakhs in the First Five Year Plan to around Rs. 7700 lakhs during the Draft Eleventh Five Year Plan. But there was a drastic reduction in percentage of forestry allocation to total allocation from 7.71% in First Five Year Plan to only 1.98% in Draft Eleventh Five Year Plan. At the same time, the planned allocation to the forestry sector is not in keeping with the revenue generated. For example, the forest revenue contributes nearly 20% of the State Plan fund during the Seventh and Eighth Five Year Plan. However, the allocation during the

TABLE 2

Allocation under Forestry Sector and Revenue from Forest during Five Year Plan Periods

(in Rs. lakh)

Plan Period	*Allocation to Forestry*	*Revenue from Forests*
1st Plan	23.15 (7.71)	37.00
2nd Plan	32.20 (6.32)	101.57
3rd Plan	51.00 (7.13)	236.67
4th Plan	158.00 (8.78)	707.05
5th Plan	348.00 (5.50)	1368.70
6th Plan	1397.24 (6.26)	3487.82
7th Plan	2741.80 (4.99)	7004.55
8th Plan	4710.00 (4.07)	N.A.
9th Plan	6018.12 (2.42)	N.A.
10th Plan	7022.40 (1.81)	N.A.
11th Plan (Draft)	7700.00 (1.98)	N.A.

Note : The figures in the bracket indicate the percentage of allocation to forestry sector to total allocation.

Sources : (1) Report of Planning Commission, Government of India, New Delhi.
(2) Forest Statistics, Arunachal Pradesh, 1980.

Eighth Five Year Plan to forestry sector was only 4.07%. The figures show that the amount allocated for the forestry sector was inadequate and it was due to inadequate budget allocation the plantations and development programmes of the forest department suffered a major setback. Hence, fund allocations should be increased substantially so that more investment in the forestry sector could be made to increase the productivity of forest.

At the same time, the state government policy towards the development of forestry suffers from the conflicting choice between production-oriented approach and protection-oriented approach to forestry. The approaches seem to be mutually contradictory. The Sixth and Seventh Five Year Plan in Arunachal Pradesh paradoxically had based its strategy on this contradictory footing (Mitra, 1998). Thus, while it was stated, strategy during the Seventh Five Year Plan that in order to protect ecological balance, it is necessary to take up "afforestation", the strategy on creating artificial regeneration method to increase the area under economically valuable species for meeting the industrial and economic needs of the state (Draft Eighth Five Year Plan, 1987) undermined the very purpose of the strategy, as emphasis was laid more on the schemes like economic plantation of industrial and commercial species (i.e., Production Sector) have got much higher priority than Forest Conservation and Development (i.e., Forest Protection Sector). The production sector got more priority during the planning period because of a strong explicit demand backed by purchasing poor. As a result, the state suffers from deforestation although the government statistics claim that around 61.5% of total area is covered under forest for the last two decades.

SECTION III

The experience of Arunachal Pradesh shows that the private goods which the forest yields got more priority particularly as the initial stage. This may be due to the fact that the public goods which the forest yields are of such a nature that the benefits of their presence or the harms caused by their absence are felt slowly in the long-run. At the same time it should be noted that the forest have externalities but essentially it is a resource (Ganguly, 1986). Though, as forest must be protected to conserve soil, environment, wildlife and bio-diversity at the same time the demand for resources yielded by forest is expected to grow along with the grow1h of population and economic development in order to satisfy direct human wants and requirement of the forest-based industries and construction activities. Therefore, an effective planning for forestry development in Arunachal Pradesh should take into account the following consideration :

(a) Control of Indiscriminate Destruction of Forest

It is found that accessible natural forests particularly in the foothills of Arunachal Pradesh are under great pressure to a large scale due to extraction of timber and illegal felling of trees. Although the demand for wood for local consumption is relatively low due to low population of the state but in view of the increasing demand for industrial timber within the state and other parts of the country, the forests in the state are under great pressure. As per the official estimate, the state contributes to nearly 50% of the timber supply made from the North Eastern region of India. At the sometime, it is observed that there is large scale illegal felling of trees, which is many times more than the official permits issued for felling of trees. The tree permit system in unclassified state forest (U.S.F.) which was introduced to enable the local people to earn their livelihood in logging and extraction of timber with a view to generate income has led to the emergence of a 'neo-rich' class in the traditional tribal society to make easy and quick money in collaboration with private forest contractors. There is a growing social and political pressure to over exploit the forest and the protection of forest are becoming increasingly difficult.

Moreover, the forest located particularly along the inter-state border with Assam and Nagaland are mainly prone to illegal fellings of and smuggling of timber. Some illicit fellings are also reported in the forest adjacent to the tributaries of the Brahamaputra river and the logs are thrown into the river which are collected in the downstream of Assam plains for sale to saw and veneer mills in different parts of the country. The state forest department finds itself ill-equipped to fight such timber poachers and smugglers and contain this menace due to the limited resources available with it for protection of forest. For example, only Rs. 300.00 lakhs were allotted for forest protection during the Tenth Five Year Plan. Therefore, there is an urgent need to protect the rich forest resources of the state by firmly dealing with such organized gangs indulging in illicit tree felling and also by controlling the indiscriminate destruction of forest in unclassified State Forest.

(b) Control of Shifting Cultivation

Shifting cultivation ('jhum') is one of the factors adversely affecting the forest conservation efforts in North-Eastern states in general, and Arunachal Pradesh in particular. Jhumming is mainly practiced in the USF areas in the district of Tirap, East and West Siang, Lower and Upper Subhansiri, Papumpare and East Kameng. Population-wise around 54,000 families are practising shifting cultivation in the state (North-Eastern Council, 2006). Thus, the dominance of shifting cultivation in the whole economy is quite evident. At the same time, it is a well-known fact that much of the forest land in the state is lost due to shifting cultivation. For example, according to Forest Survey of India, 2003, there was a net decrease of forest area in Arunachal Pradesh by around 26 square kilometers and out of these 78.32 per cent is lost due to the shifting cultivation. Forest gets denuded when old jhum land is left uncared for and new land is taken for jhum. The danger of ecological imbalance due to deforestation looks large when the claim for such new land soars up with the growth of population. Hence, in order to save forest and to control environmental degradation, most social scientists and environmentalist agree that the shifting cultivation has to be reduced. Efforts have to be made for better utilisation and scientific management of jhumland regeneration of forest in abandoned jhumland and a greater stress on horticultural crops, which lies in the border line of forestry and agriculture. However, the question is whether these alternative measures are sustainable from the point of view of hill agro eco-system or from the socio-economic framework of tribal societies of the state. Hence, it is necessary to have a vigorous study regarding the recent trend of shifting cultivation in various districts of the state which must address socio-anthropogenic-economic issues and impact of various developmental measures and strategic corrections needed to redesign the developmental programmes benefiting the jhumies.

(c) Emphasis on the Role of Non-Governmental Organisation

One of the important components of effective forestry planning is involving the people in the protection and regeneration of forest because forest can never be protected unless the people are made to feel that they have a say in the matter. This is why there was a major change in forest policy in 1988 which envisaged people's involvement in the development and protection of forest. The policy stressed that it is one of the essentials of an effective forestry planning that the forest communities should be motivated to identify themselves with the development and protection of forests from which they derive benefit. In the pretext of change in forest policy in 1988, the idea of Joint Forestry Management (JFM) came and it was adopted by various states of the country. However, the JFM model will require suitable modifications to be workable in the hill state like Arunachal Pradesh where 62.17% of the total forest areas are mainly under the traditional ownership of local village committees. At the same time, there is enough scope to shift focus from government forests by bringing more virgin USF to JFM network. The concept of Apna Van has received enthusiastic and wide acceptance among some of the local tribes in Arunachal Pradesh. So, the success stories of Apna Van in people's sector can be replicated and promoted. In this respect the NGOs can also play an important role in forest conservation and preservation among the people of Arunachal Pradesh. The tribes of Arunachal Pradesh had their own system of forest conservation and management but due to modernization and introduction of monetary economy, much of it has been lost. Hence, it is the NGOs who can make the local people more aware about the age-old tradition of the preservation of forest resources and help them to conserve the forest resources of the state. It should be noted that unless social consciousness about forest promotion and conservation do not grow or revive among the people, the government's efforts will be of no use. It is the NGOs, who can motivate people at the grassroot level.

CONCLUSION

We find from the experience of Arunachal Pradesh that rapid annual growth of population (average annual exponential growth rate of population of Arunachal Pradesh is 2.39% which is much higher than the all India average of 1.93% in 1991-2001) and a strong desire for the improvement in the standard of living in this state (where the majority of population are poor) have resulted in a lack of sustainability in the use of forest resources At the same time during the planning period, the private goods which the forest yield got more priority in the name of modern development. In fact the development of the state through industrialization docs not mean the mushrooming growth of the forest-based industries and over exploitation of forest resources. The state can also be developed through setting up of fruit processing industries and through promotion of nature based tourism in which the state has enough potentiality, provided certain infrastructural facilities are built up. Forest used to contribute a major source of total revenue from the local source to the state exchequer, in which around 85% of its total budgetary expenditure comes from the centre in the form of grants-in-aid. The situation has worsened with the royalty of forest products declining due to the Supreme Court's interim order banning of felling of trees in mid nineties. Hence, alternative internal resources and employment opportunities have to be created within the state itself and the promotion of nature-based tourism appears to be the best way in this respect which is considered the least ecologically disturbing industry in hill regions. In fact, under the present circumstance a vigorous study is required to make the people aware of how additional income and employment can be generated locally due to the existence of forest resources by promoting ecotourism which may also help to conserve the forest resources. In fact, the flow on academic thinking on forestry has been always directed towards issues like timber demand, survival of forest based industries or sustainability and biodiversity. However, very few studies have been stressed on the valuation of recreational aspects of forests in the context of India and it is urgently required, particularly in the context of Arunachal Pradesh in order to conserve the rich forest resources of the State.

References

Dutta Ray, B. (ed.) 1987: *North-East India: 2000 AD.*, New Delhi.

Elwin, V. (1957): *A Philosophy for NEFA*, published on behalf of advisor to the Governor of Assam, Shillong.

Ganguly, J.B. (1986): 'Planning for Forestry Development in the North-Eastern Region. An Approach' in Das Gupta *et al. Forestry Development in North East India*, (Guwahati: Omsons Publications).

Government of Arunachal Pradesh (1990): Arunachal Pradesh Forest Statistics, Forest Department, Itanagar.

Government of Arunachal Pradesh (2002): Arunachal Pradesh State Biodiversity Strategy and Action Plan, Forest Department, Itanagar.

———, (1992): Draft Eighth Five Year Plan (1992-95): Planning and Development Department, Itanagar.

———, (1993): Arunachal Pradesh Forests, Environment and Forest Department, Itanagar.

———, (1995): Stale Forestry Action Programme Report (Draft), Environment and Forest Department, Itanagar.

Government of India (1993): The State of Forest Report, Forest Survey of India, Dehra Dun.

Majumder, D.N. (ed.) (1990): *Shifting Cultivation in North-East India*, (New Delhi: Omsons Publications).

Mitra, A. (1998): 'An Evolution of Forestry Planning in Arunachal Pradesh' in Dutta Ray *et. al.* (ed.) *Forest Resource in North-Eastern India*, pp. 259-67, (New Delhi: Omsons Publications).

Mitra, A. (1998): "Environment and Sustainable Development in the Hilly Regions of North-East India : A Study in Arunachal Pradesh", *International Journal of Social Economics*, Vol. 25, Nos. 2-4.

National Council of Applied Economics Research (1967): *Techno Economic Survey of NEFA*, New Delhi.

North-Eastern Council (2006): *Basic Statistics of North-Eastern Region*, (Shillong: NEC).

Chapter 14

Unlocking Sustainable Livelihood Opportunities for Forest Dwellers of Nagarhole National Park

Where is the Way?

C. NANJUNDAIAH

ABSTRACT

Forests play a profound role in protecting environment and meeting the livelihood needs of 500 million people all over the world (WRI, 2002). However, deforestation at an unprecedented rate affects environment and economy locally as well as globally in recent years. The annual loss of forest cover globally is currently about 1 per cent and this exceeds its regeneration rate (Roba, 2000). This has important policy implications for forest conservation as well as access to livelihood needs. Sustainable development (SD) is positively correlated with practices of sustainability in forest conservation and management. However, there is a tradeoff between environmental conservation and meeting the livelihood needs. The best strategic imperative thought out by policy-makers for conserving forest resources has been creation of Protected Areas across the country and to provide alternative livelihood needs to forest dependents. Despite rigorous implementation of several forest and wildlife conservation policies in India, deforestation and threat to wildlife continue to exist. These policies however, have failed to yield desired result and paradoxically there is a greater extent of deforestation mainly owing to growing conflict between local communities and foresters. Against this backdrop the present paper makes an attempt to realize sustainable development objectives that are embedded in conserving forest resources and meeting sustainable livelihood needs, through studying the interface between forests, environment and local community in Nagarahole National Park in the Western Ghat region of Karnataka.

Keywords: Poverty; Environment; Forest Resources; Livelihood needs; Sustainable Development; Local Communities; Forest Conservation; Peoples' participation; National Park; Forest and Wildlife policies.

INTRODUCTION

Degradation of environmental resource-base beyond its carrying capacity in recent decades is a cause for concern because of its adverse consequences (Dasgupta and Maler 1997). Tropical deforestation, for instance, influences the climate and contributes to a loss of biodiversity, continues to be the critical environmental issue. Moreover, the environmental problems that countries of the world are facing today have posed the greatest challenge in the way of achieving inter-generational and intra-generational equity in development (World Development Report, 1992). Forest conservation is therefore, a *sine qua non* for continuously supplying vital ecosystem services for lasting human well-being. Concern is rising all over the world about adverse consequences of tropical deforestation and a considerable effort both in research and policy circles has been undertaken in recent years towards environmental stewardship. The World Commission on Environment and Development (WCED), also known as the Brundtland Commission, in its report entitled 'Our Common Future' (1987), has earnestly called the world community for making development sustainable one by "meeting the needs of present generation without compromising the needs of future generations". The Earth Summit held at Rio-de-Janeiro in 1992 has further made a fervent message to the world community that "without better environmental stewardship development will be undermined and without accelerated development in poor countries, environmental policies will fail". The Earth Summit also made forest conservation and management a major focus (Roba, 2000).

Tropical forests are the vital segment of earth's ecosystems and they harbour about two-thirds of the known world's biological diversity. Senseless destruction of forests at an unprecedented rate both in quality and quantity has resulted not only in environmental degradation but also threatened the livelihood security of millions of people particularly in developing countries. The root and proximate causes of tropical deforestation are however, traced to the market failure, policy failure, institutional failure, population pressure, agricultural expansion, grazing, timber logging, forest fires, exploitation of timber and non-timber products and other socio-economic reasons (Pearce and Moran, 1994, Dasgupta, 1997 and Perrings, 2000). However, most prominently, governments' policy interventions in recent decades for internalizing negative externalities through the creation of national parks, sanctuaries and reserve forests aimed at preserving biodiversity, have excluded local communities without giving them a stake in conservation and management of forests so as to meet their bonafide usufructs. These inappropriate forest conservation efforts seem to be caused widespread depletion of forests in developing countries. The local communities are squarely blamed for the deforestation in official circle by indicating poverty-environment nexus. A thorough research therefore, can comprehend linkages whether conservation policies in protected areas impose significant human costs by excluding people or generate important local benefits by preserving access to the resource through people's participation. Besides this, the paper probes in to what new policy initiatives for conservation of forests need to be undertaken to address tropical deforestation and livelihood needs. In the light of these issues, the present study examines the interface between forest conservation and meeting livelihood opportunities in the context of Nagarhole National Park of Western Ghat region of Karnataka.

2. FOREST RESOURCES AND SUSTAIANBLE DEVELOPMENT

Forest resources play an important role as a source of achieving environmentally sustainable development. Sustainable use of forest guarantees the use of forest resources at a rate that does not lead to the long-term decline of forest ecosystem, thereby maintaining its potential to meet the needs and aspirations of present and future generations. Forests offer multiple benefits in ensuring sustainable development by generating income and employment to millions of forest dependent communities in developing countries. Conservation of forests from further degradation for meeting environmental prerequisite and sustainable use of these resources for meeting bonafide needs of millions of local community can be construed as the two faces of the same coin. Sustainable forest management practices need to be reoriented with the changing concept of forest resources, which may not be treated as renewable resources any more and it is more appropriate to construe as "conditionally renewable resources". If these resources are over-exploited they may be irreversible and their renewable capacity is challenged. Forests lose not only their tangible and intangible services but also their resilience or carrying capacity if they are exploited beyond their regeneration rate (Markandya and Dale, 2001). Forests, therefore, need to be exploited strictly on the basis of their carrying capacity and sustained yield principle. Sustainable development is possible when consumption of forest involves not only sustainable use but also reinvestment of rents from the exploitation of forest resources (Pearce, 1998).

Pearce and Markandya (1992) opine that "fundamental to an understanding of sustainable development is the fact that the economy is not separate from the environment in which we live. There is interdependence both because the way we manage the economy impacts on the environment and because environmental quality impacts on the performance of the economy". Sustainable development places emphasis on providing for the needs of the least advantaged in society (intra-generational equity) and on a fair treatment of future generations (intergenerational equity).

Forests cover 1/5 of India's land area and they are considered as lifeblood of 375 millions of people who are leading biomass based subsistence existence. Conservation programmes implemented by government institutions to conserve forest resources must involve local community because they are the victims of both deforestation and alienation. The realization of sustainable development would require a careful policy implementation in economic decision taken by assessing social cost-benefit of the project. The increasing divergence between social and private cost needs to be abridged through policy initiatives. For better conservation of forests, participatory style of management is preferred to top-down management.

It is not rational on the part of the government to expect from the local community to shape their behaviour on the onerous task of forest conservation while they do not have any rights and benefits from forest resources either directly or indirectly. Consequently, the achievement of sustainable development remains impossible if the stakeholders, historical users and culturally dependents are not given adequate stake in forest use and chance to participate in forest conservation and management programmes. Conservation of forest resources along with meeting indispensable human needs sustainably is therefore, considered as rational approach towards realizing sustainable development.

3. FOREST RESOURCES AND LOCAL COMMUNITY

Local communities are those indigenous forest-dwellers or group of stakeholders or a single homogenous entity who have historical rights over forest resource use in a particular area. They possess traditional environmental knowledge and experience, which are highly invaluable in managing and conserving both forests and wildlife (Kothari *et al.*,

1998). Local communities are dwelling in forest areas in many developing countries, and they have been managing and conserving resources for generations. They are leading subsistence livelihood by extracting various forest products sustainably for ages.

India's forests are not only inhabited by wilderness but also by indigenous forest-dwellers for centuries. "Today, in India there are about 100 million forest-dwellers in the country living in and around forest areas and another 275 million for whom forests have continued to be an important source of their livelihood and means of survival" (Lynch, 1992 and Saxena, 2000). Local communities in India besides, getting fuel wood, fodder, wood for house construction, also earn about one-third of their income from the sale of MFP (Raymond, 1997). These evidences have validated the importance of forests for meeting livelihood needs of local communities all over India.

Depletion of environmental resource base in recent decades owing to various factors has affected local communities' essential livelihood needs and they have to spend invariably more time in collecting forest produce. In a number of regions in India as much as 40-50 per cent of working hours of villagers are devoted to fodder and fuel wood collection, animal care and grazing (Raymond, 1997). "Commercial cutting of Indian's forests and conversion of forests to agriculture have left the traditional system of management of local forests in shambles. This has brought shortage of fuel wood and building materials to many of the 275 million rural Indians who draw on local forest resources" (Gadgil and Guha, 1992).

Appropriation of forestlands by the government for in situ conservation has further deprived local communities' dependence on forests. "The number and area of protected areas throughout the world have been growing quite rapidly in recent years, by more than a factor of five since 1950. Presently, more than 400 million ha. are under protection, more than one half of the areas are in National Parks and a bit under 50 per cent are located in developing economies" (Brown, 1997). Government of India has enacted and implemented Wildlife (Protection) Act, 1972 to protect biodiversity and has established protected areas throughout India, their number are rising from 131 in 1975 to 496 at present. National Parks and Sanctuaries now cover some 4.3 per cent of the country's geographical area. The protected area notification to create National Parks and Sanctuaries has the harshest impact on local communities especially where these policy initiatives have made no provisions to incorporate those living within the forest. Local communities are systematically excluded from forest management and use of forest resources. These policy decisions have completely ignored the social cost incurred by the local communities. Obviously, the success of the conservation programs is hindered by growing conflicts between people and law enforcers.

As noted by Raymond (1997) there is no need to fear about local communities' exploitation of forests beyond their carrying capacity because, "most of traditional societies use some form of control to prolong the use of needed resources, and they have intimate knowledge of their physical environment and resources in which they live-flora and fauna. The life of these forest-dwellers is adapted to their respective environment. The traditional environmental knowledge of these societies engenders care for the forest resources and this care results in economic systems that are sustainable". Gadgil (2000) points out that the non-sustainable, centralized and bureaucratic approach to the management of forests, parks, sanctuaries and protected areas that excludes the local communities is at the root of widespread deforestation in the country.

The governments need to understand the fact that "individuals who do not benefit directly or indirectly from the continued viability of wildlife cannot be expected to shape their behaviour in a manner that preserves wildlife, particularly when the preservation options have private opportunity costs" (Brown, 1997). It is evident that the present wildlife and forest policies have failed to achieve twin objectives viz., protection of environment through conservation of forests and ensuring livelihood security of local community. The new institutional arrangements can best serve the local communities' participation in conserving forests.

International Convention on Biological Diversity (ICBD) is legally binding signatory countries to adopt the required policy and legal changes in realizing local communities' participation in biodiversity conservation. The convention decisively states that "each signatory country will, subject to its national legislation, respect, preserve and maintain knowledge, innovations and practices of indigenous and local communities embodying traditional life styles relevant for the conservation and sustainable use of biological diversity and promote their wider application with the approval and involvement of the holders of such knowledge, innovations and practices and encourage the equitable sharing of the benefits arising from the utilization of such knowledge, innovations and practices" (Kothari *et al.*, 1996). From the above, it is evident that the role of local communities in forest conservation and environmental protection is increasingly recognized worldwide. It may be inferred that that livelihood needs of local community must to be integrated with forest conservation.

4. THEORITICAL FRAMEWORK OF THE STUDY

Many research studies (Pearce and Moran, 1994, Perrings, 2000, Dasgupta, 1997 and Repetto, 1997) have established the fact that the extent of forest degradation in a region is influenced by the underlying and proximate factors. Deforestation is however, closely associated with human behaviour and activities hence, it can be checked to a large extent provided, these interfaces are clearly comprehended. In fact, local communities would make decisions about use and abuse of forestlands and forest resources by taking into account the underlying causes. For instance, wrong forest policies, which distance or disregard the role of local communities in forest conservation are accountable for fast depletion of forests in most of the developing countries. In fact, there are strong linkages among the root and proximate causes, the state of forests and environment and the local communities' behavior towards forest appropriation and conservation in the in the case study area of NNP.

The theoretical framework illustrates generally the interrelationships among forests, environment and local community. It depicts that how the root causes such as existing market, policy decisions, institutional provisions, existing property rights and population size and proximate causes including cropping pattern, alternative use of forest land for agricultural and plantation crops, collection of NTFPs, demand for timber, grazing and forest fire in the study area influence local communities' behavior towards forest resource use and abuse or practices of sustainability or unsustainability. Households' behavior is intrinsically shaped around opportunity cost of resource use, which is eventually resulted in state of forest. Finally, the economic decision is ended up in sustainability or unsustainability in the study area, NNP (Figure 1).

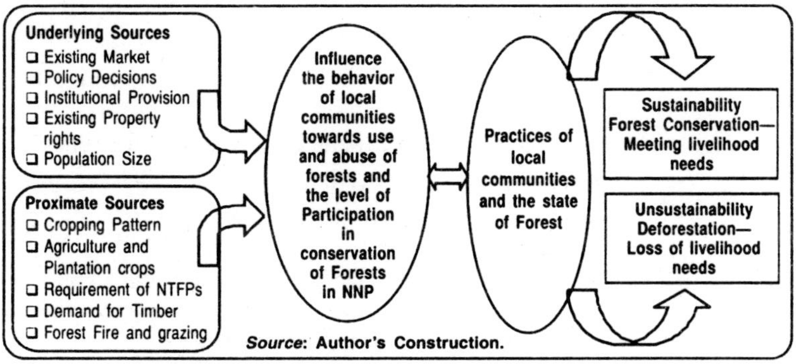

Underlying Sources
- Existing Market
- Policy Decisions
- Institutional Provision
- Existing Property rights
- Population Size

Proximate Sources
- Cropping Pattern
- Agriculture and Plantation crops
- Requirement of NTFPs
- Demand for Timber
- Forest Fire and grazing

Influence the behavior of local communities towards use and abuse of forests and the level of Participation in conservation of Forests in NNP

Practices of local communities and the state of Forest

Sustainability Forest Conservation— Meeting livelihood needs

Unsustainability Deforestation— Loss of livelihood needs

Source: Author's Construction.

Figure 1 : Theoretical Framework of the Study Depicting the Interface Among Forests, Environment and Local Community

The framework further depicts that how forest policies and deforestation undermine and affect different social groups and how local communities respond when their livelihoods are affected by environmental change. The crucial emphasis of the study is on how forests are integrated in the livelihood system of local community, and how the process of deresponsibilisation or disempowerment of local community by the centralized decision leads to loss of control over resources. The study further analyses how the present decision-making process can lead to the breakdown of forest resource management system and environmental degradation and how the new emerging conditions affect the local communities' access to forest resources, how local communities' perception differs with the new situation, what solutions can motivate local communities' willingness to participate in the process of conservation and what are the costs and benefits associated with the conservation of forests? These and other important interfaces are analyzed in the study.

5.1. Objectives

The main objectives of the study are:

1. To study the importance of forests for meeting bonafide needs of local community and sustaining environment.
2. To estimate what extent sample households are dependent on forest for income.
3. To evaluate the status of livelihood of local community on account of deforestation and environmental degradation.
4. To explore the perception and attitude of local communities towards their participation and conservation of forest resources and achieving sustainable development.
5. To review the existing forest policies in conservation of forests
6. Finally, to prepare a blueprint for forest conservation with peoples' involvement in the study area.

5.2. Hypotheses

The study has identified the following hypotheses. They are:

1. Local communities are more dependent on forest resources for meeting their livelihood needs in relation to other sources of income.
2. There is a positive linkage between people's willingness to participate in forest conservation and access to forest resources for meeting their livelihood needs.

5.3. Methodological Framework

To answer the questions identified in the research proposal, two sorts of data viz., secondary and primary were collected through various sources. The primary sources of data were collected by carrying out field surveys through questionnaires administered to village population. Village level data were collected from the records of Revenue Department and records of Forest Department.

The following detailed information was collected from the study area, NNP.

(A) The basic socio-economic data.

(B) The data on local communities' dependence on a wide range of NTFPs and TFPs for meeting livelihood needs and employment and income generation.

(C) The perceptions of local communities on importance of forests, causes for deforestation, environmental implications and declaration of NP and its effects on livelihood needs of local communities.

(D) Peoples' perceptions and their willingness to participate in conservation and management of forests and their reactions towards forest and wildlife policies.

(E) The role of local communities in conservation and management of the National Park.

5.4. Sample Design

NNP in Virajpet taluk was chosen for carrying out the field survey. The survey was carried out by giving equal representation to people living in and around the Nagarahole National Park. An appropriate number of households i.e. 10 per cent of the total households (175) were selected for the survey. For the study, a random sampling technique is used for the selection of households from hadis of NNP. For this purpose 4 hadis from inside the NP (INP)—Gadde hadi, Gonigadde hadi, Siddapura hadi and Murkal hadi and 3 hadis from outside the NP (ONP)—Bommadu hadi, Chandanakere hadi and Karekandi hadi were selected.

5.5. The Tools for Analysis

The collected data was computerized in a database program such as Microsoft Excel; the results were used to test hypotheses by adopting Chi-square test. The research study is principally descriptive and analytical in nature. In addition, standard deviation, co-efficient of variation, percentages and averages were used to analyze different issues that emerged in the course of the study.

6.1. A Profile of the Study Area: Nagarahole National Park

The study area—Nagarhole National Park (NNP)—was christened

as Rajiv Gandhi National Park (RGNP) in 1992. It is situated in Coorg and Mysore districts of Southern Karnataka. The park covering an area of 643 sq. km. is a part of the Bandipura Tiger Reserve (BTR) and it falls in Nilgiri Biosphere Reserve (NBR) comprising adjoining forests of Kerala and Tamil Nadu. Conservation of this biodiversity 'hotspot' has received a world-wide attention including the World Bank's Global Environmental Facility (GEF) to launch India Eco-Development Project in 1996, by recognizing its rich and unique biodiversity such as (fauna) large carnivores like tiger, panther, herbivores like wild Asian elephants and (flora) dense moist deciduous forests.

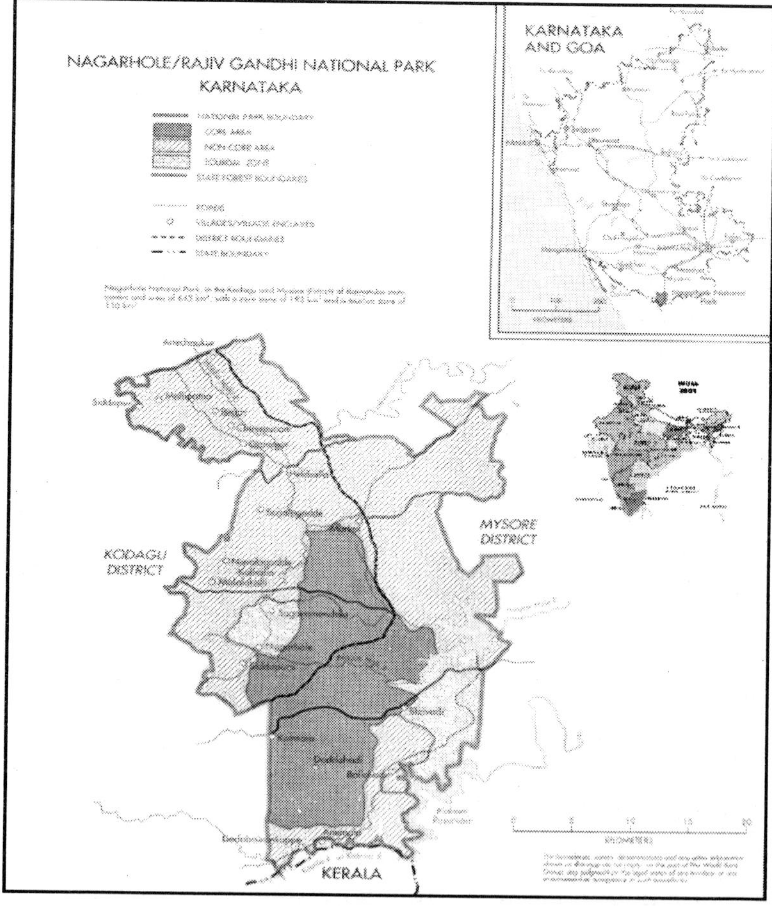

Figure 2 : The Location of the Study Area: Nagarahole National Park

NNP consists of many aboriginal tribal settlements or hadis. In order to give due representation to households living inside and outside (fringe) the national park, four hadis from inside the national park (INP) viz., Gaddehadi, Siddapura, Gonigadde and Murkal were chosen and three hadis viz., Chandanakere, Bommadu and Karekandi were selected from outside or periphery of the national park (ONP) for the study. The location of the study area, NNP of Virajpet taluk in Coorg district is given in Figure 2.

The study area NNP is situated in Virajpet taluk of Coorg district and covers a total area of 643.42 sq km. Of the total area, Core Zone and Tourism Zone consist of 192 sq km (29.86 per cent) and 110 sq. km. (17.09 per cent) respectively and Non-core area or buffer zone covers an area of 341 sq. km. (53.03 per cent).

There are 51 tribal settlements within the park comprising about 1974 households with a total population of 7389. The forest area is sparsely populated with a very low density of population. The average size of households is 3.74 and the density of population in INP is 11.49 per sq. km., which is less than the district's mean density of 118 per sq. km. The park provides shelter to diverse ethnic tribal groups viz., jenu kurubas, betta kurubas, paniyas, yeravas, soligas, and hakki pikkis, etc.

6.2. Occupational Status of Sample Households

The occupational status of the sample households of NNP indicates that 67 per cent and 52 per cent of households from INP and ONP respectively are plantation labourers working in coffee estates. It is evident from Table 1 that the proportion of cultivators is less (4 per cent) in INP compared to ONP (25.33).

TABLE 1

Occupational Status of the Sample Households in NNP

The Study Area	*Cultivators*	*Agricultural Labourers*		*Labourers in Forest Department*		*Others*	*Total*
		Coffee Plantation	*Other Crops*	*Salaried*	*Daily Wagers*		
INP	4 (4.00)	67 (67.00)	7 (7.00)	11 (11.00)	7 (7.00)	4 (4.00)	100 (100.00)
ONP	19 (25.33)	39 (52.00)	4 (5.33)	7 (9.33)	3 (4.00)	3 (4.00)	75 (100.00)
Grand Total	23 (13.14)	106 (60.57)	11 (6.28)	18 (10.28)	10 (5.71)	7 (4.00)	175 (100.00)

Note : Figures in parentheses represent percentages.

The proportion of households engaged as agricultural labourers other than plantation crops is 7 per cent and 5.33 per cent in INP and ONP accordingly. The majority of workers living inside the park are working as salaried class (11 per cent) and daily wagers (7 per cent), however, the proportion of labourers engaged in FD is less in case of ONP. Barely 7 per cent of households from both areas are engaged in other works. It is evident from the foregoing discussion that majority of the households from both the settlements are engaged in coffee plantation work.

6.3. Dependence of Local Communities on Forest

Resources

Local communities inhabited in NNP have depended on forest resources for eking out their livelihood for ages. An attempt is made in this section to quantify the extent of dependence on forest resources for a variety of direct and indirect benefits. NTFPs are defined as "biological materials other than timber that are extracted from natural forests for human use" (Gakou *et al.*, 1996). NTFPs play an important role in sustaining livelihood of forest-dwellers. As rightly pointed out by the FAO, "NTFPs contribute significantly to household food security and nutrition, generate additional employment and income, offer opportunities for processing enterprises, contribute to foreign exchange earnings and support biodiversity conservation and other environmental objectives" (Mallik, 2000). The collection of NTFPs and TFPs not only creates employment opportunities but also generates income to the tribal households. Self-consumption or own use of NTFPs is considered to be more significant for fetching non-cash income than for marketing them for cash income (Mallik, 2000). However, in recent years, tribals are harvesting a variety of forest products for commercial purpose in order to supplement their livelihood. The average money value derived from the collection of a range of NTFPs is estimated in the next section. Since many of the NTFPs do not have any market value, the prices of close substitutes are taken for estimating the money value of NTFPs.

6.4. Collection of NTFPs

The tribal households living in and around the NP collect an array of NTFPs and TFPs. According to Large-scale Adivasis Minor Forest Produce Society (LAMPS) based at Thithimathi, about 41 NTFPs are gathered in the park. However, prohibition of collection of NTFPs and absence of an appropriate organization to procure NTFPs have resulted in many of them being uncollected and utilized. The extent of

TABLE 2

The Extent of Annual Dependence of Households on Forests for Collection of NTFPs in NNP

Sl. No.	NTFPs	INP					ONP				
		% of families collecting the product (1)	Quantity per household (Kg.) (2)	Cash income (from sale) (3)	Non-cash income (from Self-consumption) (4)	Total value (Rs.) (3+4) (5)	% of families collecting the product (1)	Quantity per household (Kg.) (2)	Cash income (from sale) (3)	Non-cash income (from Self-consumption) (4)	Total value (Rs.) (3+4) (5)
1.	Food: Roots and Tubers	91	103.66	—	518.3	518.3	93.33	42.1	—	210.01	210.01
					(8.34)	(4.28)				(4.65)	(1.91)
2.	Honey	93	43.09	1677.27	261.77	1939.05	92	42.97	1618.66	314.99	1933.65
			(28.51)	(4.21)	(16.03)			(25.03)	(6.98)	(17.62)	
3.	Wild fruits	76	72.56	203.74	884.65	1088.4	88	52.71	245.1	545.54	790.65
			(3.46)	(14.21)	(9.00)			(3.79)	(12.09)	(7.20)	
4.	Medicinal Plants	62	13.77	—	96.39	96.39	30.66	9.45	—	66.15	66.15
				(1.55)	(0.79)				(1.46)	(0.60)	
5.	Fuel wood	100	43.19*	—	3671.15	3671.15	100	27.4	486.52	1842.47	2329
				(59.12)	(30.36)			(7.52)	(40.86)	(21.22)	
6.	Wild Spices	67	37.23	1294.41	194.78	1489.2	73.33	23.47	818.91	119.88	938.8
			(22.00)	(3.13)	(12.31)			(12.66)	(2.65)	(8.55)	
7.	Wild Nuts and Seeds	84	167.77	1405.9	104.03	1509.93	64	197.66	1668.82	110.11	1778.94
			(23.90)	(1.67)	(12.48)			(25.81)	(2.44)	(16.21)	

(Contd.)

TABLE 2 (*Contd.*)

		(1)	*(2)*	*(3)*	*(4)*	*(5)*	*(1)*	*(2)*	*(3)*	*(4)*	*(5)*
8.	Bees Wax and Gum	87	17.33 (22.10)	1299.75	— (10.75)	1299.75	72	21.69 (25.16)	1626.75	— (14.82)	1626.75
9.	Wild Meat and Fish	33	4.78	— (7.69)	478 (3.95)	478	37.33	5.53	— (12.26)	553 (5.03)	553
10.	Green fodder	0	—	—	—	— **	48	149.33 (16.55)	— (6.80)	746.65	746.65
	Grand Total	77 (100.00)	—	5881.07 (100.00)	6209.07 (100.00)	12090.14 (100.00)	69 (100.00)	—	6464.76 (100.00)	4508.8 (100.00)	10973.56 (100.00)

Note : *(Fuel wood in quintals) **(Bundle comprising 25 to 30 kgs of Green fodder).

dependence on a variety of NTFPs for self-consumption and for earning cash income by the local communities per annum from NNP is presented in Table 2 (Landscape table). The important NTFPs are grouped into 10 categories based on their nature and utility. They are: (1) Food, (2) Honey, (3) Wild fruits, (4) Medicinal plants, (5) Fuel wood, (6) Wild spices, (7) Wild nuts and seeds, (8) Bee wax and gum, (9) Wild meat and fish, and (10) Green fodder.

The households of INP and ONP derive income worth Rs. 12090.14 and Rs. 10973.56 per annum from the collection of a variety of NTFPs respectively. Of the total income, the share of cash income in the two settlements from the sale of NTFPs accounts for 48.64 per cent and 58.91 per cent respectively. The composition of non-cash income for self-consumption of forest products is estimated at 51.35 per cent in INP and 41.08 per cent in ONP accordingly.

Of the nine important categories of NTFPs collected from households of INP, four are gathered exclusively for self-consumption or domestic use and one product is solely for sale. The rest of the four products are collected for both self-consumption as well as for sale. Of the total income derived from the collection of NTFPs, fuel wood alone fetches more than 30.36 per cent of income, followed by honey (16.03 per cent); both products fetch about 46.39 per cent of income and rest of the NTFPs bring about 53.61 per cent of income to the households of INP. By and large all tribals including women are involved in collection of forest products. The households of ONP collect 10 categories of NTFPs, of which, four are gathered for domestic or self-consumption and five are collected for both sale and own use and the remaining one item is collected solely for sale. Of the total income generated from the collection of NTFPs, fuel wood, honey, wild nuts and seeds fetch 55.05 per cent of income and rest of NTFPs bring about 44.05 per cent of income. The dependence on NTFPs by the households of INP is largely for self-consumption (51.35 per cent) while the dependence of the ONP households is largely for earning cash income from sale (58.91%).

Thus, from the foregoing analysis, it is evident that, the households' dependency levels are varied depending upon the availability, requirement, marketability and more importantly free access to various NTFPs in the study area.

6.5. Collection of Timber Forest Products

Local communities collect TFPs with the permission of FD for meeting their basic requirements and not for sale. The households living in the study area of INP gather only three categories of TFPs while households of ONP collect four categories of TFPs for the purpose of domestic use. The dependence of households on TFPs of INP and ONP is about 87 per cent and 64 per cent respectively and they collect, on an average, 10.71 and 23.50 poles per annum. Timber used for domestic requirements alone fetches about 23.49 and 47.92 percentages of non-cash income to the households of INP and ONP respectively. The dependence of households on bamboo accounts for 91 per cent (INP) and 50.66 per cent (ONP) annually. Incomes earned from bamboo are estimated at 40.57 per cent and 29 per cent of the total income from TFPs. Collection of cane and fiber is accounted for 73 per cent in case of INP and 36 per cent in case of ONP. The income generated from cane and fiber is estimated at 35.94 per cent and 19 per cent respectively in INP and ONP. Only the households of ONP collect mulch and manure. About 45.33 per cent of households depend on mulch and green manure and they earn an income of 4.08 per cent to the total income from TFPs. The dependence on TFPs is high in case of ONP households (Rs. 2451.75) as compared to INP households (Rs. 2279.75). Local communities gather TFPs for various purposes like agricultural implements, house construction, wooden utensils, green manure, and fencing materials. Households were requested to give their first priority for collection of TFPs. Accordingly only their first choice is taken for analysis. Out of the total collection of TFPs, 63 per cent of INP households collect mainly for house construction and 23 per cent for fencing of houses and backyards to protect them from the menace of wild animals and 10 per cent of households collect timber for wooden utensils and only 2 per cent households for preparing agricultural implements. The demand for TFPs largely comes from households living in ONP. About 38.66 per cent of households collect TFPs mainly for making agricultural tools. The percentage of households collecting TFPs for the construction of houses, green manure and fencing material accounts for 24 per cent, 20 per cent and 12 per cent respectively.

6.6. Employment Generation from Forest Resources

Forest activities like collection of NTFPs and TFPs generate direct, indirect and self-employment opportunities for the forest dependent communities right through the year (Nayak, 2001). Collection of forest products is considered a major economic activity for local communities of NNP since they spend most of their time in extracting forest products as detailed in the preceding sections. The employment generated by forest resources has been estimated by evaluating the working days (man-days) spent by a household in collection of a range of forest products with respect to other economic activities (Table 3).

TABLE 3

Composition of Annual Employment of Households in Man-days

Sl.No.	Activities	INP			ONP		
		Mean Man-days	*S.D.*[1]	*C.V.*[2] *(%)*	*Mean Man-days*	*S.D.*	*C.V. (%)*
1.	Collection of NTFPs and TFPs	294.78 (52.78)	273.87	92.9	233.44 (40.19)	230.23	98.62
2.	Agriculture and allied activities	61.32 (10.97)	156.26	254.84	90.67 (15.61)	189.37	208.87
3.	Forest Department work	73.65 (13.18)	41.75	56.69	47.33 (8.14)	61.21	129.32
4.	Plantation work	120.51 (21.57)	117.14	97.2	187.87 (32.34)	261.47	139.16
5.	Others	8.23 (1.47)	22.95	278.89	21.45 (3.69)	42.94	200.16
	Total	558.49 (100.00)	100.63	82.22	580.76 (100.00)	99.38	63.28

Note : [[1]S.D.= Standard deviation. [2]C.V.= Co-efficient of variation (%)].
Figures in parentheses are percentages to the total.

It is found that annually about 294.78 (52.78 per cent) of mean man-days of employment are generated for each household of INP from collection of NTFPs and TFPs, out of the total annual employment of 558.49 man-days per household. The collection of NTFPs and TFPs generate 233.44 (40.19 per cent) of mean man-days of employment for the households of ONP, out of the total annual employment of 580.76 man-days per household. This indicates the importance of forest economy in providing employment opportunities to forest-dwellers. The generation of employment from forest-based activities is around 57.6 per cent of total man-days in other parts of the country as per the evidence of some studies (Mallik, 2000 and Prakash, 1999). Employment generated by agriculture and allied activities amounts for only 61.32 (10.97 per cent) mean man-days in INP and 90.17 (15.61 per cent) mean man-days in ONP. FD has provided seasonal and occasional employment opportunities to the households of INP and ONP. Employment generated from FD is estimated around 73.65 (13.18 per cent) and 47.33 (8.14 per cent) man-days per annum respectively. Another main source of employment is plantation work, which provides 120.51 (21.57 per cent) and 187.87 (32.34 per cent) man-days annually for households of INP and ONP respectively.

It is apparent from the foregoing analysis that employment generation from both NTFPs and TFPs is more stable (with variability of 92.90 in INP and 98.62 in ONP) than from other activities. Therefore, collection of NTFPs and TFPs is the major source of economic activity not only in generating employment opportunities but also income to the forest-dwellers of NNP.

6.7. Composition of Income from Forest Resources

Local communities of NNP are largely dependent on forest products for their livelihood and they derive annual income to the tune of Rs. 26,601.29 (INP) and 28,901.36 (ONP) per household from various sources (Table 4).

Out of the total annual income per household, income generated from collection of NTFPs and TFPs (for both own use and sale) in both the settlements of INP and ONP comprises of Rs. 14369.89 (54.02 per cent) and Rs. 13425.31 (46.45 per cent) correspondingly. While agriculture and allied activities contribute to the income to the tune of Rs. 947.78 (3.56 per cent) in INP, it is Rs. 3376.88 (11.68 per cent) in the case of ONP. The income derived by working in the FD constitutes Rs. 3682.50 (13.81 per cent) to the total income for households of INP and Rs. 2366.50 (8.18 per cent) for households of ONP. The second important contributor to income is plantation work and tribals earn

TABLE 4

Composition of Annual Income of Sample Households

Sl.No.	*Activities*	*INP*			*ONP*		
		Income	*S.D.* [1]	*C.V.* [2]*(%)*	*Income*	*S.D.*	*C.V. (%)*
1.	Collection of NTFPs and TFPs	14,369.89 (54.02)	2710.1	18.85	13,425.31 (46.45)	6335.17	47.18
2.	Agriculture and allied activities	947.78 (3.56)	1765.18	186.24	3,376.88 (11.68)	3335.38	98.77
3.	Forest Department	3,682.50 (13.84)	3799.14	103.16	2,366.50 (8.18)	1548.66	65.44
4.	Plantation	6,025.50 (22.65)	4475.41	74.27	7,272.22 (25.16)	4016.27	55.22
5.	Others	1,575.62 (5.92)	1745.98	110.81	2,460.45 (8.51)	2279.46	92.64
6.	All sources	26,601.29 (100.00)	1219.22	42.05	28,901.36 (100.00)	1845.88	52.69

Note : [[1] S.D.= Standard Deviation. [2] C.V.= Co-efficient of Variation (%)].
Figures in parentheses signify percentages to the total.

income from this activity mainly for meeting the household requirements. The income generated from plantation is estimated at Rs. 6025.50 (22.65 per cent) in INP and Rs. 7272.22 (25.16 per cent) in ONP. After the income derived from the collection of forest products, the income from plantation is considered the second highest source of income for households. Income received from other sources stands at Rs. 1575.62 (5.92 per cent) and Rs. 2460.45 (8.51 per cent) in both the settlements of INP and ONP respectively. The significant contribution to income from forest products is revealed by many studies also, for example, income from the collection of NTFPs in Kalahandi district in Orissa state stands at 52.2 per cent (Mallik, 2000) and 51.44 per cent in the study area of Uttara Kannada district in Western Ghats region in Karnataka (Prakash, 1999).

The above analysis clearly reveals that the collection of NTFPs and TFPs fetches the largest composition of income compared to other sources of income to the households of both INP and ONP. It can also be observed from the interpretation of co-efficient of variation analysis that income generated from the collection of forest products is more stable compared to plantation and agriculture and allied activities with a variability of 18.85 per cent and 47.18 per cent in both the settlements. This clearly shows that collection of forest products continues to play a pivotal role in the tribal economy by contributing a substantial and steady income.

6.8. Perceptions of Local Communities on the Importance of Forests

The elicitation of preferences and perceptions of forest dependent communities about the importance of forest resources is understood to serve many purposes, especially for evolving right policy decisions towards forest conservation and management through viable institutional arrangements. The study, therefore, makes an attempt in the following sections to elicit the views of local communities on a number of issues like importance of forests, deforestation, environmental role of forests, conservation and management of forest resources and so on. As a first step to explore the views of local communities, they were asked to place their value and preference on the importance of forest by assigning ranks in the order of importance attached to each of seven different issues highlighted in Table 5.

Of the 175 households from both INP and ONP, 146 households opined positively and assigned first rank to forests, as an important source of livelihood needs. The second most important issue of forests was the environmental role of forests, to which 83 households, out of

TABLE 5

Households' Perceptions on Importance of Forests

Importance of Forests	*Rank-1*	*Rank-2*	*Rank-3*	*Rank-4*	*Rank-5*	*Rank-6*	*Rank-7*	*Total*
(1)	*(2)*	*(3)*	*(4)*	*(5)*	*(6)*	*(7)*	*(8)*	*(9)*
Meet livelihood needs of local people	146(83.42)	19(10.85)	05(2.85)	03(1.71)	02(1.14)	00(00)	00(00)	175(100)
Environmental protection like soil conservation	09(8.18)	83(75.45)	07(6.36)	11(10.00)	00(00)	00(00)	00(00)	110(100)
Agricultural production and medicinal plants	10(9.52)	08(7.61)	77(73.33)	06(5.71)	02(1.90)	02(1.90)	00(00)	105(100)
Existence, cultural and religious values	06(6.45)	05(5.37)	09(9.67)	63(67.74)	07(7.52)	03(3.22)	00(00)	93(100)
Important for meeting the needs of future generation	12(15.00)	15(18.75)	10(12.50)	03(3.75)	40(50.00)	00(00)	00(00)	80(100)
Fostering wild animals and other habitats	02(2.66)	15(20.00)	03(4.00)	15(20.00)	05(6.66)	35(46.66)	00(00)	75(100)
Eco-tourism and generating income to the government	00(00)	00(00)	10(27.77)	03(8.33)	05(13.88)	05(13.88)	13(36.11)	36(100)

Note : Figures in parentheses represent percentages to the total.

110, assigned the second rank. Of the 105 households, 77 households assigned third rank to the issue of forests as being essential to agricultural production and medicinal plants. The fourth important issue, forest as the store of existence, cultural and religious values, was assigned fourth rank by 63 households out of 93 households, which responded. The fifth important issue of forests related to the needs of future generations. Out of 80 households, 40 assigned fifth rank to it. The sixth important issue cited was the role of forests in fostering wild animals and other habitats; 35 households out of 75 households assigned sixth rank to it. However, the issue of eco-tourism and generating income to the government did not impress households and hence, of the 36 households that responded, only 13 of them assigned the last rank.

From the foregoing analysis it is clear that the local communities are aware of the direct and indirect benefits of forests. However, they attach more importance to forests in terms of meeting their livelihood needs, followed by environmental protection. Therefore, the success of forest conservation through peoples' participation is linked to meeting the livelihood needs of households.

6.9. Impact of Deforestation on Livelihood Needs of Local Community

Forest resources have direct bearing on the life-support system of local communities. Local communities lose significant source of their livelihood (due to non-availability of an array of forest products) as a consequence of deforestation. In this section an attempt is made to capture local communities' perceptions in relation to the impact of deforestation on livelihood needs. All the households responded that deforestation had negative effects in many ways and they are categorized into seven issues. The respondents were asked to state their first choice or view on the consequence of deforestation and accordingly only their first choice is taken for analysis.

The responses are summarized in Table 6. Of the 100 sample households of INP, about 39 per cent perceived that the important outcome of deforestation is the direct loss of supply of NTFPs and TFPs and further they opined that it would have negative consequences on their well-being. Of the 75 households living on the periphery of the park, 22.66 per cent perceived that reduction in the availability of NTFPs and TFPs in the last few years was due to deforestation. About 23 per cent of households of INP and 17.33 per cent of households of ONP expressed their concern that deforestation might threaten the loss of environmental services and their livelihood needs. However, about 14 and 20 per cent of households expressed that livelihood might be

TABLE 6

Perceived Consequences of Deforestation on Livelihood Needs

(%)

Issues	*INP*	*ONP*	*Grand Total*
Households agree that there will be consequences	100	75	175
Reduction in supply of NTFPs and TFPs	39 (39)	17 (22.66)	56 (32.00)
Lose environmental services of forests	23 (23)	13 (17.33)	36 (20.57)
Livelihoods are threatened	14 (14)	15 (20.00)	29 (16.57)
Cultural and religious knowledge base is threatened	05 (05)	12 (16.00)	17 (09.71)
Decline of agricultural production and shortage of food	08 (08)	07 (09.33)	15 (08.52)
Blockade of income from forest resources	07 (07)	08 (10.66)	15 (08.52)
More time to be spent in collection of NTFPs	04 (04)	03 (04.00)	07 (04.00)
Total	100(100)	75 (100)	175 (100)

Note : Figures in parentheses represent percentages to the total.

threatened. Moreover, 5 per cent and 16 per cent of households believed that they would lose cultural and religious ethos and knowledge due to deforestation.

From the foregoing analysis it is evident that reduction in supply of both NTFPs and TFPs due to deforestation has affected their livelihood according to 32 per cent of households from both the settlements.

6.10. Perceptions about Peoples' Participation in Forest Conservation

6.10.1 Peoples' Willingness to Participate in Forest Conservation

It is evident from the results of earlier sections that local communities of NNP draw heavily on forest resources for their subsistence in their daily life. They also place high value on forest resources and their environmental role and they are also aware of detrimental effects of deforestation. Therefore, there is a vital need to test the supposition that there is an association between local communities' willingness to participate in forest conservation and access to forest resources for eking out their livelihood. The study makes an attempt in this section to test whether this relationship is justifiable or realistic one. An evaluation of this relationship however, helps to assess the welfare implications of conservation measures in NNP for local

TABLE 7

Households' Level of Participation in Forest Conservation

Willingness	*INP*				*ONP*			
	Access to Forest Resources				*Access to Forest Resources*			
	Access	*Not Access*	*Total*	*Chi-Value*	*Access*	*Not Access*	*Total*	*Chi-Value*
(1)	*(2)*	*(3)*	*(4)*	*(5)*	*(6)*	*(7)*	*(8)*	*(9)*
Willing	79(94.04)	05(5.96)	84(100)		55(90.16)	06(9.84)	61(100)	
Not willing	05(31.25)	11(68.75)	16(100)		04(28.57)	10(71.42)	14(100)	
Total	84(84.00)	16(16.00)	100(100)	39.435	59(78.66)	16(23.33)	75(100)	25.739

Note: Figures in parentheses represent percentages to the total.

communities because they have access to forest resources for years and to draw inferences for sustainable forest management. Therefore, households were requested to divulge their willingness or unwillingness to participate in forest conservation with and without access to forest resources (Table 7).

In respect of access to forest resources, of the 84 households, which responded, 79 households of INP were willing to participate in forest conservation and only 5 households were not willing to participate. Of the 16 households, 5 households were willing to participate although they had no access to forest resources and 11 households stated that they were not willing to participate unless they had access to forest resources. However, of the 75 sample households in case of ONP, 55 households were willing to participate with access to forest resources and only 4 households were not interested. However, 6 households were willing to participate in spite of not accessing forests products and 10 households were not willing to participate without access to forests.

In this study it is hypothesized that *"there is a positive linkage between peoples' willingness to participate in forest conservation and access to forest resources for meeting their livelihood needs"*. In order to test the validity of the above hypothesis, the chi-square test (χ^2) is used. The separate chi-square values for both the settlements indicate that calculated values of chi-square are greater than the table values (6.63) at 1 per cent level of significance at 1 degree of freedom. Hence, it can be inferred that peoples' willingness to participate in forest conservation depends on their access to forest resources for meeting their livelihood needs. Therefore, it may be summarized from the above analysis that peoples' willingness to participate in forest conservation measures is backed by economic compulsions or incentives, i.e. access to forest resources for meeting their livelihood needs.

6.10.2 People's Willingness to Participate by Extending Labour

The success of conservation programme depends largely on participation of local communities as observed by many studies. Therefore, an assessment of the extent of willingness to participate either by extending labour or paying money on forest conservation is required to evaluate the feasibility of conservation programme of the government. Therefore, the mode of participation of local communities in forest conservation of their own choice either by extending labour or willing to pay money is estimated.

The study has identified willingness or unwillingness to participate either by willingness to join or willingness to spend money towards forest conservation works as the dependent variable and demographic

TABLE 8

Households' Willingness to Participate (WTP) in Conservation of Forests by Extending Labour

Age Group	*INP*				*ONP*			
	WTP	*UWP*	*Total*	χ^2	*WTP*	*UWP*	*Total*	χ^2
(1)	*(2)*	*(3)*	*(4)*	*(5)*	*(6)*	*(7)*	*(8)*	*(9)*
Young age (18-30)	35(74.46)	12(25.54)	47(100)		27(75.00)	09(25.00)	36(100)	
Middle age (30-50)	25(75.75)	08(24.25)	33(100)		19(79.16)	05(20.84)	24(100)	
Old age (50 above)	13(65.00)	07(35.00)	20(100)		09(60.00)	06(40.00)	15(100)	
Total	73(73.00)	27(27.00)	100(100)	0.828	55(73.33)	20(26.66)	75(100)	1.833
Education Level								
Illiterate	44(75.86)	14(24.14)	58(100)		33(75.00)	11(25.00)	44(100)	
Up to 10th Standard	15(71.42)	06(28.57)	21(100)		14(73.68)	05(26.31)	19(100)	
PUC and above	13(61.90)	08(38.09)	21(100)		09(75.00)	03(25.00)	12(100)	
Total	72(72.00)	28(28.00)	100(100)	1.494	56(74.66)	19(25.33)	75(100)	0.013
Land Holding								
Landless Class	68(73.91)	24(26.08)	92(100)		37(74.00)	13(26.00)	50(100)	
Landholding Class	03(37.50)	05(62.50)	08(100)		12(48.00)	13(52.00)	25(100)	
Total	71(71.00)	29(29.00)	100(100)	3.136	49(65.33)	26(34.66)	75(100)	4.974

Note : χ^2- Chi-value), (Values in parentheses are percentages to the row total.

(age group, level of education) and economic (land size) variables as the explanatory. The local communities' willingness or unwillingness to participate in conserving forests by extending voluntary labour is examined in Table 8.

Local communities of INP were willing to extend labour, on an average, 13.53 hours per month by joining conservation work if communities were involved. The willingness of households to participate by extending labour in hours has been calculated by multiplying the duration of time (hours) with the prevailing agricultural wage rates of the study area (Rs. 75 for 8 hours). This is considered as an opportunity cost of extending labour in terms of hours for conserving forest resources. According to the estimation, local communities were prepared to make contribution in money terms (converted from labour time) about Rs. 126.84 per month. This worked out to Rs. 1522 per annum and it was estimated at 5.72 per cent of total annual income of a household. However, local communities of ONP were ready to extend 9.11 hours of labour per month and it was estimated in money terms of about Rs. 85.40 per month and Rs. 1025 per annum. The share of contribution towards conservation in money terms was 3.54 per cent of the total annual income of a household. The result shows that local communities living in INP were ready to contribute more in terms of labour hours towards conservation of forests compared to households of ONP. This difference is mainly due to the engagement of local communities of ONP in plantation work, agriculture and allied activities.

The study in this section has explored the association involving the age group of respondents and their willingness to participate in forest conservation by extending labour. A majority of respondents, about 73 per cent in INP and ONP, correspondingly under different age groups, were willing to participate in conservation of forests by extending labour; and on the contrary about 27 per cent of respondents were unwilling to extend their labour towards forest conservation. Among the different age groups, middle aged (30-50) respondents (75.75 and 79.16 per cent in INP and ONP) were more willing to participate in forest conservation by extending labour followed by young age and old age respondents. It is found from the above result that middle-aged family units showed impressive involvement in forest conservation by offering their labour.

The variable level of education of respondents and their willingness to participate in forest conservation by voluntary labour is examined. The participation of illiterate households in forest conservation compared to literate households stood at 75.86 per cent in

INP and 75 per cent in ONP. However, from both INP and ONP, 24.14 and 25 per cent of households were not paying attention to participate through extending labour in the study area.

The study explored the landholding and willingness to participate in forest conservation and it was found that among the households of INP and ONP 92 and 50 households did not possess land out of total sample households in both the settlements. Of the total sample households more than 73.91 and 74 per cent of households were willing to participate directly in forest conservation in both the settlements; however, 26.08 and 26 per cent of households were not interested. Among the land holding households 37.50 and 48 per cent of households in both the settlements were ready to participate in forest conservation and 62.50 and 52 per cent of households were not interested. It is evident from the above analysis that age and education are not associated with willingness to participate by extending labour in INP and ONP. However, land holding and willingness to extend labour is associated in case of ONP and *vice versa* in the case of INP. To sum up, it is observed that irrespective of education and age group, local communities were willing to participate in forest conservation. As rightly remarked by Thimma, a tribal youth, "forests are our life and conservation of forests is a way of life for us for ages and our lifestyles show the evidence of everything".

6.10.3 People's Willingness to Participate by Paying Money

An analysis is carried out in order to explore the willingness of the local communities of the NNP towards conservation of forest resources. Money is used as an indicator of preference in participation. The details in this regard are given in Table 9. Local communities of NNP were willing to pay Rs. 266.45 (on an average) per annum per household towards conservation of forests provided they were allowed to harvest forest products on a sustainable basis for meeting their bonafide needs. Among the different age groups, a maximum from young age households (40.42 per cent in INP and 41.66 per cent in ONP) were willing to pay money followed by middle age (33.33 per cent and 37.50 per cent) and old age group (20 per cent and13.33 per cent) households in both the settlements correspondingly.

Among different educated classes, households having pre-university education were willing to pay money in conserving forests in both the settlements as compared to illiterate households and households with having completed matriculation. However, across land holdings, about 50 per cent of households in both the settlements showed considerable interest in making payment for conservation

TABLE 9

Household's Willingness to Participate (WTP) in Conservation of Forests by Willing to Pay Money

Age Group	*INP*				*ONP*			
	WTP	*UWP*	*Total*	χ^2	*WTP*	*UWP*	*Total*	χ^2
(1)	*(2)*	*(3)*	*(4)*	*(5)*	*(6)*	*(7)*	*(8)*	*(9)*
Young age (18-30)	19(40.42)	28(59.57)	47(100)		15(41.66)	21(58.33)	36(100)	
Middle age (30-50)	11(33.33)	22(66.66)	33(100)		09(37.50)	15(62.50)	24(100)	
Old age (50 above)	04(20.00)	16(80.00)	20(100)		02(13.33)	13(86.66)	15(100)	
Total	34(34.00)	66(66.00)	100(100)	2.618	26(34.66)	49(65.33)	75(100)	3.878
Education Level								
Illiterate	10(17.24)	48(82.75)	58(100)		09(20.45)	35(79.54)	44(100)	
Up to 10th Standard	05(23.80)	16(76.19)	21(100)		03(15.78)	16(84.21)	19(100)	
PUC and above	09(42.85)	12(57.14)	21(100)		06(50.00)	06(50.00)	12(100)	
Total	24(24.00)	76(76.00)	100(100)	5.547	18(24.00)	57(76.00)	75(100)	5.453
Land Holding								
Landless Class	12(15.00)	68(85.00)	80(100)		11(35.48)	20(64.51)	31(100)	
Landholding Class	10(50.00)	10(50.00)	20(100)		22(50.00)	22(50.00)	44(100)	
Total	22(22.00)	78(78.00)	100(100)	11.42	33(44.00)	42(56.00)	75(100)	1.555

Note : (χ^2- Chi-value), (Values in parentheses are percentages to the row total).

programme. Thus, from the above, it is evident that age group, land holding size and educational levels do not influence the participation of sample households in forest conservation.

6.10.4 Households' Willingness to Participate in Conservation Activities

The elicitation of preferences of local communities is carried out in this section to examine the practicability of conservation programme by exploring type of forest conservation activity, in which the local communities were willing to participate. Table 10 shows that out of 175 households from INP and ONP, 25.71 per cent of them expressed their willingness to participate in preventing and extinguishing of forest fires that raze forests during summer season. About 16.57 per cent of households expressed their willingness to participate in afforestation and soil conservation programme. About 13.14 per cent of households were interested in guarding forests in preventing timber smuggling through night watch along with forest rangers and guards; and 12.57 per cent of households showed interest in preventing encroachment of forestland on the periphery of the park. It is evident from the above that local communities were prepared to assist FD in preventing forest fires, engage in afforestation programme and to guard forests in preventing timber smuggling if they were given such responsibilities.

TABLE 10

Households' Willingness to Engage in Forest Conservation Process

Forest Conservation Activities	*INP*	*ONP*	*Grand Total*
Prevention and extinguishing of forest fire	28 (28.00)	17 (22.66)	45 (25.71)
Afforestation and soil conservation	17 (17.00)	12 (16.00)	29 (16.57)
Night watch in preventing timber smuggling	12 (12.00)	11 (14.66)	23 (13.14)
Preventing encroachment of forestland	08 (08.00)	14 (18.66)	22 (12.57)
Stop grazing by domestic animals	07 (07.00)	10 (13.33)	17 (09.71)
Check poaching of animals	10 (10.00)	05 (06.66)	15 (08.57)
Assist forest officials in tracing wild animals	11 (11.00)	04 (05.33)	15 (08.57)
Regulate collection of NTFPs	07 (07.00)	02 (02.66)	09 (05.14)
Total	100 (100)	75 (100)	175 (100)

Note : Figures in parentheses represent percentages to the total.

6.10.5 Households' Forest Management Preferences

The forest management preferences of local communities in forest conservation through different institutional arrangements was examined by asking them to give their first priority. Six different institutional management options were given to make their choice for forest conservation (Table 11).

TABLE 11

Households' Forest Management Preferences

(%)

Forest Management Regimes	*INP*	*ONP*	*Grand Total*
Joint Protected Area Management (JPAM)	61 (61.00)	34 (45.33)	95 (54.28)
Village Panchayat Management (VFM)	21 (21.00)	17 (22.66)	38 (21.71)
Non-Governmental Agencies (NGOs)	06 (06.00)	15 (20.00)	21 (12.00)
Joint Forest Management (JFM)	08 (08.00)	03 (04.00)	11 (06.28)
Pure Government Management (GM)	03 (03.00)	02 (02.66)	05 (02.85)
Village Cooperative Society (VCS)	01 (01.00)	04 (05.33)	05 (02.85)
Total	100 (100)	75 (100)	175 (100)

Note : Figures in parentheses represent percentages to the total.

Out of 100 households surveyed in INP, 61 per cent of households favoured JPAM programme and they were ready to participate in the proposed programme it implemented in NNP. However, 21 per cent of households preferred Village Panchayat Management (VPM) of forests by declaring self-rule as per the constitutional provision of scheduled areas. Similarly, in ONP, out of 75 households interviewed, 45.33 per cent of households favoured JPAM. However, another 22.66 per cent respondents expressed their willingness to participate in forest management under VPM. About 20 per cent of households were willing to work with NGOs in forest conservation and management. Local communities are influenced by the ideology of a few NGOs, which have pledged support for the cause of adivasis and insisted for implementation of JPAM. It is evident from the above analysis that majority of local communities (54.28 per cent) were willing to work with the FD if JPAM was implemented in NNP. The next section deals with forest and wildlife policies and conservation of forests.

6.11. Forest and Wildlife Policies and Conservation of Forest

The state ownership of forest resources resulting in the genesis of forest and wildlife policies are perceived to be antagonistic and authoritarian. Implementation of these policies in forest conservation

TABLE 12

Households' Perceptions Pertaining to Nature of Forest Policy

Nature of Forest Policy	*Items*	*INP*				*ONP*			
		Conservation		*Total*	χ^2	*Conservation*		*Total*	χ^2
		Yes	*No*			*Yes*	*No*		
(1)	*(2)*	*(3)*	*(4)*	*(5)*	*(6)*	*(7)*	*(8)*	*(9)*	*(10)*
Forest Policy which supports local communities' livelihood needs and participation	Right	76	06	82		51	05	56	
Forest Policy which is indifferent to local communities' livelihood needs and participation	Wrong	05	13	18		05	14	19	
Total		81	19	100	40.40	56	19	75	31.44

Note : (Yes=Conservation), (No= Deforestation) and (χ^2= Chi-square value).

and management has had a major impact on various stakeholders and interest groups (Guha, 1994). The implications of such policies on the state of forest, environment and forest dwellers' livelihood have not been considered seriously for several decades. However, a paradigm shift from authoritarian policies to community-based programmes like JPAM and JFM has taken place with the implementation of recent Forest Policy (1988) and the Wildlife (Protection) Amendment Act, (2002) (Sen, 2004). Therefore, an attempt is made in this section to know the perceptions of local communities about the hypothesis whether *"wrong forest policies are responsible for the fast depletion of forest resources"*. To test the above said hypothesis, local communities were asked to state their views on the nature of forest policies and their implications for the state of forest resources (Table 12).

Majority of the households in INP (76 households) and ONP (51 households) agreed that forest policies that support local communities' livelihood needs and encourage their participation in forest conservation would contribute to forest conservation, however, only a few households opined that it would lead to deforestation. Local communities strongly opined that it would not be possible to conserve forests if forest polices were apathetic to livelihood needs and participation of local communities. In the study it is hypothesized that *"wrong forest policies are responsible for the fast depletion of forest resources.* It is assumed that wrong forest policies and fast depletion of forest resources are independent. The separate Chi-values for both the settlements indicate that calculated values of chi-square are greater than the table values at 1 degree of freedom and hence, null hypothesis is rejected (H_0) at 1 per cent level of significance. Thus, wrong forest policies and fast depletion of forest resources are dependent.

Therefore, it may be summarized from the above analysis that wrong forest polices are responsible for fast depletion of forest resources in the study area. Thus, in order to involve local communities in forest conservation, provisions have to be made in the forest conservation policies for harvesting of NTFPs on a sustainable basis for their livelihood.

7. A 'BLUEPRINT' FOR FOREST CONSERVATION WITH PEOPLES' PARTICPATION TO ACHIEVE SUSTAINABLE DEVELOPMENT IN THE STUDY AREA

Conservation of India's rich and mega biodiversity is absolutely essential for its immense contribution not only to India's long-term economic development but also to the entire world because, India's biodiversity is valued and recognized as one of the 12 Mega Biodiversity countries in the world. India's forests are also a homeland for millions of forest dwellers and other rural communities for eking out their livelihood and very subsistence and forests cannot be conserved without people's participation.

Historically forests in India have been conserved and managed by local communities by applying their age-old traditional environmental knowledge, considered as the best strategy evolved by local communities respecting the harmony of nature to lead a sustainable way of life. Local communities have traditionally managed forests in a sustainable way through self-restrained and regulated use of forest resources and also through practice and enforcement of social norms. Local communities of Coorg district, historically, are nature lovers and strong conservationists as they have been practicing worshiping of trees or SGs for centuries by leaving the entire stretch of habitat in its serenity. Their lifestyles are also incredibly fine tuned with nature. It is also evident from the present study that majority of local communities of the study areas have expressed their willingness to participate in forest conservation.

However, the successive authoritarian forest and wildlife polices have refused to accept the *"people's science"* in management of forests and alienated local communities from the management of forests. The appropriation of forests by the forest department has uprooted the traditional livelihood structures and socio-economic and cultural fabric of local communities. This has resulted in conflict between people versus forests and wildlife. The creation of national parks has had negative impact on local communities and their traditional forest based economy, especially where no provision has been made to accommodate their livelihood needs.

The Government Forest Act, 1865, was passed by the then British government was meant to empower the government to declare any land under tree as forests and punish those who violated the act. The Indian Forest Act (IFA), 1878, was designed to empower state to protect and control forests throughout India by overlooking community control. The IFA, 1927, was in no way different from earlier laws, in fact it imposed more stringent punishments to violators of the Act. The first Forest Policy (1952) after independence had showed apathy towards local communities for using and conserving forests at the cost of national interest while it permitted industrial exploitation. The Wildlife (Protection) Act, 1972, which aimed at protection of Protected Areas or National Parks, had no provisions for joint management of PA with local communities. However, the paradigm shift in policies took place with passing of the National Forest Policy, 1988 and Wildlife

(Protection) Amendment Act, 2002, emphasizing the need for conservation of forests and environment simultaneously by meeting subsistence requirements of local communities. However, JFM is restricted only to degraded forests and JPAM has not yet materialized on the ground. These lopsided polices have adversely affected environment and impoverished livelihoods of millions of local communities.

It is evident that the present forest and wildlife policies empower only the state to conserve and manage forests and there is little scope for people's participation. However, Dey, MoFE (Kothari *et al.*, 1996 and 1998), observes that without changing the present Wildlife (Protection) Act, 1972, adequate opportunities for people's bonafide requirements to be met from PA. He further points out that section 24(2)(C) allows local communities to enjoy their rights within sanctuaries at the discretion of the chief wildlife warden (CWLW) and section 29 allows human activities in specified zones of PA if they are not destructive to wildlife (Kothari *et al.*, 1996). Yet, the main bottleneck in allowing local communities to enjoy their bonafide rights is invoking of section 29, which says that any human intervention, which destroys PA, will not be encouraged. Further these policies have alienated local communities from their homeland and violated fundamental rights to their livelihood needs besides resulting in increasing conflicts between conservation priorities and livelihood priorities.

Therefore, if JPAM and JFM were to succeed, several changes in forest and wildlife polices are required to accommodate livelihood needs of stakeholders. Displacement of local communities from forestlands has impoverished their livelihood and destroyed their socio-cultural ethos. Therefore, displacement is impractical. The need of the hour is to accept the historically neglected TEK of local communities in forest conservation by reforming our attitudes or mindset in order to make reconciliation towards JPAM. Reconciliation can aim at substituting authoritarian, top-down and elitist forest policy with democratic, down-to-earth and community driven forest policy for ensuring both forest conservation and livelihood security of the local communities.

The existing experiences of JPAM have shown the way for implementing JPAM in NNP. Community based conservation (CBC) is successfully being implemented in BRT Wildlife Sanctuary in Karnataka. Bhimashankar Wildlife Sanctuary (BWS), in Western Ghat of Maharastra, is well managed with peoples' participation by forming Sanctuary Protection Committee (SPC). Kialadevi Wildlife Sanctuary in Rajasthan and Dalma Wildlife Sanctuary in Bihar are being successfully managed by the local communities by forming forest protection committees. The lessons of JPAM from abroad are also guiding principles for implementing community-based conservation of NNP. Annapurna Conservation Area in Nepal, Kakadu National Park in Australia, Khunjerab National Park in Pakistan are being successfully managed with people's participation.

It is evident from the above examples that formation of JPAM along with people's participation in NNP is the need of the hour. The feasible conservation strategies for NNP including zoning pattern of the park area, the nature and extent of control over park, management options, areas to be managed by local communities, institutional arrangements, local communities' rights over resources for meeting their bonafide needs and the role of forest department are proposed and presented in Figure 3.

The United Nations Educational, Scientific and Cultural

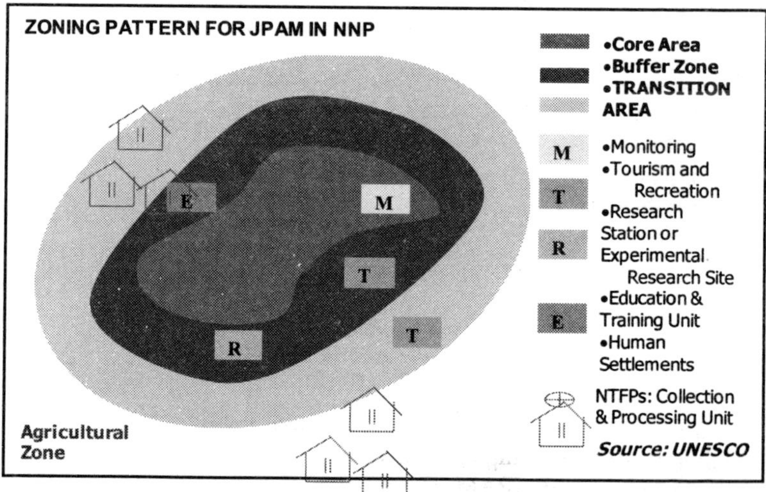

Figure 3 : The Proposed Joint Protected Area Management for NNP

Organization (UNESCO) launched the "Man and Biosphere" (MAB) programme in 1971 for conservation of Protected Areas (Batisse, 1997). It advocated conservation strategies for the involvement of local communities in conservation and management of national parks. The same model with necessary modifications to suit local conditions is recommended for conservation and management of NNP, as local communities are willing to participate in JPAM (along the outer layer of the park by retaining stake over forest products).

The total geogrphical area of NNP (643 sq. km.) should be legally categorized into three territorial components comprising core zone (192 sq km), buffer zone (110 sq. km.) and transition area (341 sq. km.). Core zone should be devoted to the sole purpose of biodiversity conservation for meeting long-term environmental objectives. The second territory, buffer zone surrounding the core zone, can be used for activities compatible with the conservation objectives, such as research, education, non-destructive resource use, recreation and eco-tourism. The third territory, a flexible outer transition area, constitutes the region where sustainable resource management practices are promoted and developed. JPAM should be operationalised in managing the park along with local communities' participation and cooperation and benefits derived from this area may accrue to stakeholders. Formation of joint management committee by the local communities will help in managing transition zone and agricultural zone and Forest Department will focus on the core zone. NTFPs collection and processing unit has to be set up out side the park to ensure reasonable prices and value addition for forest products. The unit can also create sustained employment opportunities to local communities. This down-to-earth conservation approach incorporates all the objectives including environmental conservation along with meeting essential livelihood needs. What is urgently required is policy framework for creating legal or institutional mechanism for operationalising JPAM through cooperative agreements between various stakeholders. Therefore, policy and legal constraint that come in the way of JPAM will have to be removed for enlightened conservation approach.

8. POLICY IMPLICATIONS

The study has exposed the high reliance of local communities on NNP for a wide variety of forest products. This demonstrates that they have a stake in ensuring the productive development of forests for meeting their own usufructs needs. Local communities have also expressed their willingness to participate in forest conservation. Therefore, it is not correct on the part of the government to expect the local communities to participate in the onerous task of forest conservation without giving them a stake over forest resources. Thus, launching of JPAM in NNP can move in the right direction to achieve twin objectives: biodiversity conservation and meeting stakeholders' bonafide needs in a sustainable way. This participatory and decentralized approach can pave the way for responsibilizing and empowering local communities in realizing forests for the people and by the people.

Strengthening of usufructuary rights of local communities by establishing a permanent institutional mechanism for harvesting of

NTFPs on a sustainable basis and processing of NTFPs is the best viable and sustainable option to improve the living standard of the local communities along with forest conservation. The overexploitation of forest resources may be discouraged by creating alternative employment opportunities in afforestation programmes, eco-tourism, fire prevention, forest-guarding, etc. Food-for-work programme and targeted public distribution system (TPDS) need to be further strengthened in NNP to ease the heavy dependence on forests.

Stringent regulations over forest resources use have resulted in increased conflicts between forest authorities and forest dwellers and they have provoked further illegal exploitation of forest resources. Therefore, present forest and wildlife policies have to be amended to encourage people's participation in forest conservation as local communities are against these policies. Local communities of NNP have opposed vehemently their displacement. Forced displacement threatens their socio-cultural and traditional livelihood structure. Forced eviction violates human rights to existence and right to livelihood. Thus, livelihood needs of local communities have to be met sustainably in their aboriginal place by involving them in forest conservation.

Eco-tourism should be encouraged in non-core areas of national park with people's participation. Income generated from eco-tourism should be shared between forest department and local communities.

However, in order to reduce over dependence on forest resources owing to growing population in the long-run as per the local communities' demand, they can be displaced to the periphery of the park in Virajpet taluk; wastelands may be distributed on the periphery of the park to displaced communities for eking out their livelihood. Frequent forest fires are destroying flora and fauna irreversibly in the national park in recent years. Hence, cooperative fire prevention and control programme should be launched with people's involvement. The World Bank sponsored Eco-development programme may be continued for some more years without insisting local communities to vacate their homelands. The government should give thrust to educate local communities about the role of forests in providing tangible and intangible benefits and their total economic value. Nature trips and workshops can be conducted to create awareness on forest conservation.

References

Agarwal, Anil (Ed.) 1992. The Price of Forests, Proceedings of a Seminar on the Economics of the Sustainable Use of Forest Resources, Centre for Science and Environment, New Delhi.

Batisse, Michel (1997). Biosphere Reserves: A Challenge for Biodiversity Conservation and Regional Development, *Environment*, Vol. 39, No. 5, pp. 7-33.

Bhattarai, Madhusudan and Michael Hammig (2001). Institutions and Environmental Kuznets Curve for Deforestation: A Cross-country Analysis for Latin America, Africa and Asia, World Development Vol. 29, No. 6, pp. 995-1010.

Brown, Gardner (1997). Management of Wildlife and Habitat in Developing Countries in in Partha Dasgupta, and Karl-Goran Maler. (Ed.) 1997. *The Environment and Emerging Development Issues*, Vol. 1 and 2 Clarendon Press, Oxford, UK.

Chopra, Kanchan (1998). Economic Aspects of Biodiversity Conservation: Micro and Macro Strategies for Intervention, Economic and Political Weekly, Vol. 33, No. 52, Dec. 26-Jan. 1. 1998. pp. 3336-40.

Chopra, Kanchan, and G.K. Kadekodi (1999). Operationalising Sustainable Development: Economic-Ecological Modelling for Developing Countries, Sage Publications, New Delhi.

Central Statistical Organization, Ministry of Statistics and Programme Implementation, Government of India (1999). Compendium of Environmental Statistics, New Delhi.

Centre for Monitoring Indian Economy Pvt. Ltd. (CMIE) (2002). Economics Intelligence Service, Agriculture, December 2002, Mumbai.

Dang, Himraj (1991). Human Conflict in Conservation, Protected Areas: The Indian Experience, Vikas Publishing House Pvt. Ltd., New Delhi.

Dasgupta, Partha, and Karl-Goran, Maler (1997). The Resource-Basis of Production and Consumption: An Economic Analysis in Partha Dasgupta, and Karl-Goran Maler. (ed.), 1997. *The Environment and Emerging Development Issues*, Vols. 1&2 Clarendon Press, Oxford.

Fernandes, Walter, *et al.* (1998). Forests, Environment and Tribal Economy: Deforestation, Impoverishment and Marginalisation in Orissa, Indian Social Institute, New Delhi.

Gakou, Mamadou and Force, Jo Ellen (1996). Learning With Farmers For Policy Changes in Natural Resource Management, Forests, Trees and People, Newsletter No. 31, 1996.

Jeffery, Roger, and Nandini Sundar (Ed.), 1999. A New Moral Economy For India's Forests? Discourses of Community and Participation, Sage Publications, New Delhi.

Lynch, Owen J. (1992). "Securing Community Based Tenurial Rights in the Tropical Forests of Asia", Issues in Development, November, World Resource Institute, Washington.

Karlsson, B.G. (1999). Eco-development in Practice: Buxa Tiger Reserve and Forest People, *Economic and Political Weekly*, Vol. 34, No. 30, July 24-30.

Kothari, Ashish, Neena Singh and Saloni Suri (1996). People and Protected Areas: Towards Participatory Conservation in India, Sage Publications, New Delhi.

Kothari, Ashish *et al.* (ed.) (1998). Communities and Conservation: Natural Resource Management in South and Central Asia, Sage Publications, New Delhi.

Mahanty, Sanghamitra (2000). Negotiating Agendas in Biodiversity Conservation: the India Eco-Development Project, Karnataka, in Luca

Tacconi (ed.) Biodiversity and Ecological Economics: Participation, Values and Resource Management, Earthscan Publication Ltd. London.

Mallik, R.M. (2000). Sustainable Management of Non-timber Forest Products in Orissa: Some Issues and Options, *Indian Journal of Agricultural Economics*, Vol. 55, No. 3, July-Sept. 2000. pp. 384-97.

Markandya, Anil, and Nick Dale (Ed.) (2001). Measuring Environmental Degradation, Edward Elgar, UK.

Nadkarni, M.V. *et al.* (1989). Political Economy of Forest Use and Management, Sage Publications, New Delhi.

Nadkarni, M.V. (1996). Forests, people and Economics, *Indian Journal for Agricultural Economics*, Vol. 51, Nos. 1 and 2, Jan.-June.

Nadkarni, M.V. (2001). Poverty, Environment and Development in India, (Chapter 2), in Adrian Hayes and M.V. Nadkarni, *Poverty, Environment and Development in India*, UNESXO Principal Regional Office for Asia and the Pacific, Thailand, pp. 25-89.

Nathan, Dev and Govind Kelkar (2001). Case for Local Forest Management: Environmental Services, Internalization of Costs and Markets, *Economic and Political Weekly*, Vol. 36, No. 30, July-28-August 3.

Nayak, Bibhudatta (2001). Economic-Ecologic Values of an Indian Forest: A Case Study, *Indian Journal of Agricultural Economics*, Vol. 56, No. 3, July-Sept. 2001, pp. 325-34.

Pearce, David, and Jeremy J. Warford (1993). World Without End: Economics, Environment, and Sustainable Development. Published for the World Bank, Oxford University Press, Washington DC USA.

Perrings, Charles (2000). The Economics of Biodiversity Conservation in Sub-Saharan Africa: Mending the Ark, Edward Elgar, UK.

Pearce, David, and Dominic Moran (1994). Economic Value of Biodiversity, Earthscan Publishers, London.

Pearce, David (1998). Economics and Environment: Essays on Ecological Economics and Sustainable Development, Edward Elgar, UK and USA.

Pearce, David, Anil Markandya, and Edward B. Barbier (2000). Blueprint for a Green Economy, Earthscan Publications Ltd., London.

Prakash, Surya (1999). An Economic Analysis of Non-Timber Forest Products (NTFP) in the Tribal Economy in the Western Ghats Region of Karnataka, *My Forest*, Vol. 35 (3), pp. 173-83.

Raymond, Noronha (1997). Common-Property Resource-Management in Traditional Societies, in Partha Dasgupta, & Karl-Goran Maler (ed.) 1997. *The Environment and Emerging Development Issues*, Vols. 1&2 Clarendon Press, Oxford, UK.

Reddy, V. Ratna, *et al.* (2001). Forest Degradation in India: Extent and Determinants, *Indian Journal of Agricultural Economics*, Vol. 56, No. 4, Oct.-Dec., pp. 631-51.

Repetto, Ropert (1997). Macroeconomic Policies and Deforestation, in Partha Dasgupta, and Karl-Goran Maler. (ed.) 1997. *The Environment and Emerging Development Issues*, Vols. 1&2 Clarendon Press, Oxford

Roba, Adamp Wario (2000). Costs and Benefits of Protected Areas: Marsabit Forest Reserve, Northern Kenya, in Charles Perrings, The Economics of Biodiversity Conservation in Sub-Saharan Africa, Edward Elgar, UK.

Saxena, N.C. (2000). Research Issues in Forestry in India, *Indian Journal of Agricultural Economics*, Vol. 55, No. 3, July-Sept. 2000. pp. 359-83.

Serageldin, Ismail (1993). Making Development Sustainable, Finance and Development, *I.M.F. and World Bank*, Vol. 30, No. 4, December 1993. pp. 6-10.

Singh, Katar (1999). Sustainable Development: Some Reflections, *Indian Journal of Agricultural Economics*, Vol. 54, No. 1, January-March 1999, pp. 06-41.

Sharma, Narendra, P. (ed.) (1992). Managing the World's Forests: Looking for Balance Between Conservation and Development, Kendall/Hunt Publishing Company, Dubuque, Lowa.

Stolton, Sue and Nigel Dudley, (ed.) (1999). Partnership for Protection: New Strategies for Planning and Management for Protected Areas, IUCN and Earthscan Publications Ltd., London.

Sukdev, Pavan (2000). "Money Does Grow on Trees", *The Economic Times*, Nov. 28, Bangalore.

Wood, Alexander, *et al.* (2000). The Root Causes of Biodiversity Loss, Earthscan Publications Ltd., London, UK.

World Bank (1992). World Development Report: Development and the Environment, Oxford University Press, Published for the World Bank, Oxford New York.

World Bank (1996). India Eco-Development Project, Project Document, Global Environmental Facility, South Asia, Report No. 14914-IN.

World Bank (2003). World Development Report: Sustainable Development in a Dynamic World—Transforming Institutions, Growth, and Quality of Life, Oxford University Press, Published for the World Bank Oxford, New York.

World Commission on Environment and Development. (WCED) (1997). Our Common Future, Oxford University Press, New Delhi.

World Resource Institute (1998). World Resources: A Guide to the Global Environment, Environmental Change and Human Health, 1998-99, Oxford University Press, New York.

World Resource Institute (2002). World Resources—People and Ecosystems, 2000-01, *A Guide to the Global Environment*, New York, OUP.

Chapter 15

The Complexities of Local Commons and Forest Dependency in Arunachal Pradesh

A. MITRA AND A. IBOTOMBI SINGH

ABSTRACT

The survival of the communities living on forest fringes or within forest in the Eastern Himalayas, particularly Arunachal Pradesh depend their daily needs on the community forest. However, institutional arrangements evolve along with changes in the structure of political economy. During transitions from common or community property rights regime to private property regime, the institutional structures of community-based institution for natural resource management are expected to change as well. The implications of these changes will not be uniform across the different rural classes, primarily because as theoretical and empirical evidences suggest, privatization of commons is often accompanied by increasing inequality in the distribution of resources. Therefore, the present study is an attempt to analyse the forest dependency in rural Arunachal Pradesh and its differences in between poor and non-poor households. It also makes an enquiry that the institution-based CPR management results in extraction of more or less forest products and this form of forest management would improve the welfare of poor households.

INTRODUCTION

It is being widely recognized now that participatory, decentralized and local level management of natural resources through long-enduring community-based institutional arrangements may have some advantages over state-controlled and market-based alternatives. This is because, the survival of the communities living on forest fringes or within forest depends their daily needs on the community forest. However, institutional arrangements evolve along with changes in the structure of political economy. During transitions from common or community property rights regime to private property regime the institutional structures of community-based institution for natural resource management are expected to change as well. The implications of these changes will not be uniform across the different rural classes, primarily because as theoretical and empirical evidences suggest, privatization of commons is often accompanied by increasing inequality in the distribution of resources. The key questions that need to be investigated, in such a context are: Are there any differences in consumption value from community forest between poor and non-poor households? Does the institution-based CPR management results in extraction of more or less forest products? Does this form of forest management improve the welfare of poor households?

In order to analyse the management of common property forest resources in Eastern Himalayas, two districts in the State of Arunachal Pradesh that is Lohit and Anjaw districts were selected on the basis of the area of forest coverage as well as to represent two important forms of property rights structure i.e., community and private property ownership since the State has witnessed large-scale changes in the property rights structure as well as natural resource base. Therefore, community forest offers a unique opportunity to study these enquiries in detail. The rest of the paper is organized into four sections. Section - I deals with the methodology of the study. A brief profile of the study area is outlined in Section II. The Role of Village Council and its management are presented in section III. The Consumption from Community Forest is discussed in section IV. The last section summarizes the conclusion that emerged from the study and some policy recommendations that could form the genesis for further research work.

SECTION I

METHODOLOGY

The present study is mainly empirical in nature. The data which are used and incorporated in different chapters of the study were collected from primary as well as secondary sources. The State level information has been collected from Arunachal Pradesh Government offices like, Directorate of Economics and Statistics, Directorate of Agriculture, Arunachal Pradesh Remote Sensing and Application

Centre, Directorate of Rural Development, Directorate of Environment and Forests and Reports and Publications such as National Sample Survey Organisation, Central Statistical Organisation, Directorate of Census, North-Eastern Council, Human Development Report of Arunachal Pradesh, etc. These data have been used for describing the State level picture of CPRs of Arunachal Pradesh. Moreover, the data from the same survey have been incorporated to deal the general background of the State as well as Lohit and Anjaw districts which are taken up for detailed investigation. In order to complete the district level background such data have been supplemented by information collected from district office of Economics and Statistics. In order to have a clear picture of the study area, a location map of Lohit and Anjaw districts of Arunachal Pradesh is shown in Map 1.

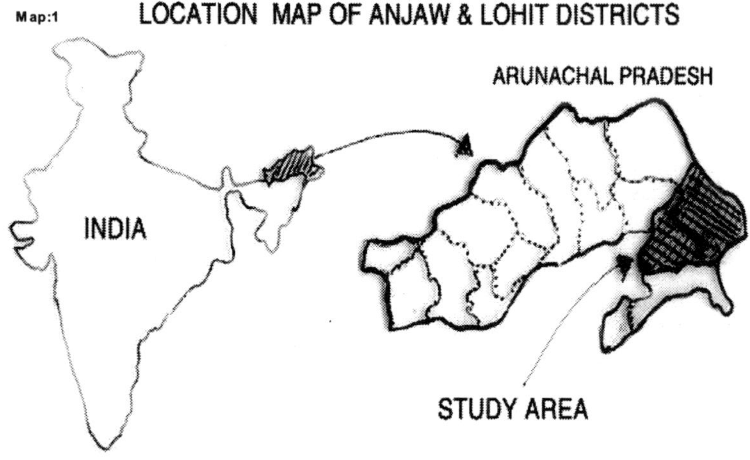

Map 1

In the field study the preliminary unit of observation is the household of a village. A multi-stage sampling design is used for selection of households of the selected villages. The different stages under the technique are as follows:

Stage I	:	Selection of districts
Stage II	:	Selection of sub-divisions
Stage III	:	Selection of villages
Stage IV	:	Selection of households

At Stage I, two districts namely, Anjaw district and Lohit district were selected by purposive sampling on the basis of forest coverage as well as to represent the two important forms of property rights structure i.e., community and private ownership of forests. At Stage II, all the sub-divisions of both the districts i.e., Hayuliang (which is the only sub-division in Anjaw district), Tezu and Namsai (which are the two sub-divisions in the present Lohit district) were considered to represent the entire area. At Stage III, seven villages (three villages from Hayuliang sub-division, two villages from Tezu sub-division and four villages from Namsai sub-division) were selected on the basis of purposive sampling. The selected villages from Hayuliang sub-division were Brailiang, Kaniliang and Kasanglat. Danglat and Loiliang were the villages selected from Tezu sub-division and from Namsai sub-division, Chireng, Karhe, Soolungtoo and Tillang Kyong were selected. The details are shown in the Figure 1.

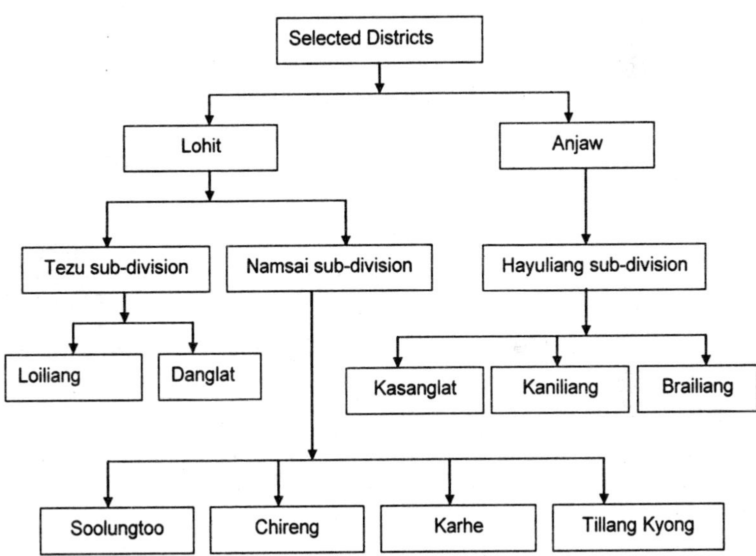

Figure 1 : Design of Sample Survey

The main reasons for selecting the above mentioned villages were based on (i) altitude and (ii) distance of villages from the urban centres i.e., Tezu and Namsai so as find the degree of dependency on community forest resources. The selection of these villages was taken into consideration on the assumption that the villages nearer to the urban centres may have less dependency on forests because of market access and some exit options in terms of outside earning opportunities in non-farm sector. At the same time, there are also variations of altitude among the surveyed villages. For example, the altitude of Brailiang,

Kaniliang and Kasanglat villages are within the range of around 1000-1100 meter above the sea level (the altitude of Goiliang is found at 975 meter above the sea level) and the distance from the nearest urban centre i.e., Tezu is within the range of around 120-128 kilometer. Danglat and Loiliang villages are within the range of around 200-220 meter above the sea level and the distance from Tezu is on an average of 10 kilometer. Soolungtoo village is found at around 140 meter above the sea level and the distance from the nearest urban centre i.e., Namsai is around 10 kilometer. Finally, Chireng, Karhe, and Tillang Kyong villages are found at the range of 680-800 meter above the sea level and the distance from the nearest urban centre i.e., Namsai is within the range of around 17-40 kilometer.

TABLE 1

Sample Households in the Surveyed Villages of Lohit and Anjaw Districts

Name of Village/ District	*Altitude (approximate in meters)*	*Total Number of Households (As per the record of Gaon Burah or Village Chief)**	*Number of Households Surveyed*		
			Poor	*Non-poor*	*Total*
Kasanglat (Anjaw)	1100	28	17	11	28
Kaniliang (Anjaw)	1050	12	8	4	12
Brailiang (Anjaw)	1000	15	9	6	15
Loiliang (Lohit)	220	327	52	23	75
Danglat (Lohit)	200	165	32	18	50
Soolungtoo (Lohit)	140	105	22	18	40
Chireng (Lohit)	680	12	9	3	12
Karhe (Lohit)	780	56	32	24	56
Tillang Kyong (Lohit)	800	24	15	9	24

Note : *Since the village level, household data for 2001 is not available till date.

In the next stage, altogether 312 sample households were selected by a stratified random sampling technique of which the total number of surveyed households from Anjaw district stood at 55 sampled households and from Lohit district, it was 257 sampled households. This was done on the basis of proportion of population of both the districts. These sampled households were further divided into the relatively poor and non-poor households. This was done by compiling a census of village households with Participatory Rural Appraisal (PRA) techniques. The participants in the PRA exercise were asked to categorise all households into two different household groups namely poor and non-poor based on the criteria as to what the villagers consider as important for assessing an individual's socio-economic position in the village. Fox (1983), Richards *et. al.*, (1999), and Adhikari (2001) used the similar criteria for categorising households into different groups. This categorisation should be understood in relative terms since all the households in the study area are subsistence *jhumias* and farmers with few households having earning opportunities outside agriculture and CPRs. The details are shown in Table 1.

The sample size consisted of 41.94 per cent of the total households. In small villages all the households were surveyed. In relatively bigger villages like Loiliang, Danglat and Soolungtoo villages of Lohit district, an attempt was made to survey 20 to 35 per cent of the households.

SECTION II

A BRIEF PROFILE OF THE STUDY AREA

There were a number of administrative changes before coming into existence of the present Lohit and Anjaw districts. The present Lohit district has inter-district borders with Anjaw district on the North East, Lower Dibang Valley district on the North-West and Changlang district on the South and inter-state boarder with Assam on the West. And Anjaw district has international boarders with China on the East and the North and Myanmar on the South-East. It has inter-district boarders with Lohit district on the West and Upper Dibang Valley on the North West.

The tribes like the *Taraons* or *Tawarahs* or *Digaru Mishmis* which is a sub-tribe of the Mishmis, the Khamptis and the Singphos inhibit in Lohit district while the Mishmis sub-tribe *Kamans* or *Miju Mishmis* and the Zekhrings (*Meyor*) tribe are the major group of people who dwell in Anjaw district. The Mishmis follow animistic way of life and worship different deities, and many more living and non-living objects. They have very strong cultural practices of offering efficacious things. The Khamties and the Singphoes are religious and pious tribe, and they are known for their hospitality. The Zekhrings live in the hilly terrain and banks of the Lohit River in Walong and Kibithoo area of Anjaw district. They follow the Mahayana sect of Buddhism blended with traditional animist beliefs.

The present Lohit district covers a geographical area of 5,212 square kilometer (6.22 per cent of the total geographical area of the State) spread over to 206 villages and two Census towns. It has a population of 1,25,086 people (which is 11.39 per cent of the State population) as per the 2001 Census. On the other hand, Anjaw district covers an area of 6,190 square kilometer (7.39 per cent of the State area) with 281 villages but no Census towns. The population of the district in 2001 Census is 18,441 people (which is around 1.68 per cent of the State population). The population density of Lohit and Anjaw districts are recorded at twenty four and three people per square kilometer respectively while the literacy rate of Lohit district is 62.42 per cent and Anjaw district 19.2 per cent as against the State average of 54.74 per cent. Porter tracks and mule paths were the main road communication that existed in the area of Lohit district during the pre-independence and early 1950s. With the initiative of the State government, the length of roads (both surfaced and unsurfaced) constructed so far in Lohit district (including Anjaw district) is around 806 kilometer only during 2002-03 and the road density of both the districts is 9.63 per cent per 100 square kilometer. The number of villages having road connectivity is 40.70 per cent as on 1997.

The villages selected for the study are Kasanglat, Kaniliang, Brailiang, Loiliang, Danglat, Solungtoo, Chireng, Karhe and Tillang Kyong. Out of these villages, the first three villages are inhabited by the *Kamans* or *Miju Mishmis* and are located in Anjaw district under Goiliang Circle of the administrative unit of Hayuliang sub-division and the remaining six villages are located in Lohit district. Out of the six villages of Lohit district, the first two villages are inhabited mainly by both the *Taraons* or *Tawarahs* or *Digaru Mishmis* and the *Kamans* and are located under Tezu Circle of the administrative unit of Tezu sub-division. Chireng, Karhe and Tillang Kyong are inhabited by the *Kamans* or *Miju Mishmis* and are located under Wakro Circle of the administrative unit of Namsai sub-division while the last village, Solungtoo is inhabited by the Khampti tribe and is under Namsai Circle of the administrative unit of Namsai sub-division. Regarding the altitude, all the three villages under Hayuliang sub-division are found at the range of around 1000-1500 metre. The villages under Tezu sub-division are located at 200-250 metre while the villages under Namsai sub-division like, Chireng, Karhe, and Tillang Kyong villages are located at around 700 metre and Solungtoo at around 140 metre above the sea level. Kasanglat village is situated in the steep mountainous slope and eastern bank of Dav River while Kaniliang and Brailiang villages are situated in the upper flat slopes on the southern bank of Brai River, which are tributaries of the Lohit River. On the other hand, Loiliang is situated in plain lands near the foothills. Danglat and Sotungtoo villages are situated in plain lands while Chireng village is situated on the foothills and Karhe village on the lower slope of the Patkai Mountain. However, Tillang Kyong is situated in a picturesque landscape vale surrounded by the lower ridges of the Patkai Mountain from all directions. The villagers, in Anjaw district, mainly practice jhum cultivation while the villages, in Lohit district practice both permanent as well as jhum cultivation. Agriculture is the main source of livelihood of the people of surveyed villages. Taking all the sample households together, we find that the average size of holdings is around 4.54 hectare. However, there are variations among the villages surveyed. The details are shown in Table 2.

TABLE 2

Socio-economic Characteristics of the Surveyed Villages

Name of Village	*Average Size of Households*	*Average Years of Schooling*	*Average Size of Operational Holdings (in ha.)*	*Average Value of Livestock Reared (in '000 rupee)*
Kasanglat	4.93	2.89	5.16	10.21
Kaniliang	4.17	0.95	4.31	7.49
Brailiang	5.53	1.26	3.28	8.90
Loiliang	4.08	4.32	4.04	6.39
Danglat	3.6	2.33	3.96	10.66
Soolungtoo	5.98	2.64	5.85	13.46
Chireng	6.08	2.00	5.06	7.40
Karhe	7.63	1.23	4.69	12.62
Tillang Kyong	8.38	1.16	4.73	12.76

Source : Survey Data, 2005-06.

If we undertake an analysis of other social infrastructural facilities like, connectivity, health, education, etc. available to these villages, it is found that there is a high variation among these indicators. The details of such infrastructural facilities of the surveyed villages are given in Table 3.

TABLE 3

Infrastructural Facilities in the Surveyed Villages (Distance in kilometer)

Name of Village	*Jeepable Road*	*Bus Stop*	*Post-Office*	*Market*	*Primary Health Centre*	*Pre/ Primary School*	*Middle School*	*Secondary School*	*Higher Secondary School*
(1)	*(2)*	*(3)*	*(4)*	*(5)*	*(6)*	*(7)*	*(8)*	*(9)*	*(10)*
Kasanglat	4	20	20	20	6	6	20	20	20
Kaniliang	8	25	25	25	5	5	25	25	25
Brailiang	10	30	30	30	7	7	30	30	30
Loiliang	0	0	12	12	0	0	0	0	12
Danglat	0	0	8	8	0	0	0	8	8
Soolungtoo	0	0	10	10	0	0	0	0	10
Chireng	2	2	2	2	2	2	2	2	59
Karhe	5	9	9	9	9	0	9	9	75
Tillang Kyong	26	39	39	39	39	0	39	39	99

Source : Survey Data, 2005-06.

As Table 3 indicates that around 33 per cent of the surveyed villages have jeepable road facility. Out of these villages, Loiliang, Soolungtoo and Danglat villages in Lohit district have better health, education and connectivity facilities than any other village. As these villages are in plain areas and nearer to the district headquarter, these villages have better accessibility. However, the average distance from villages to the jeepable road in the study area is found to be 6.11 kilometer. The average distance from the surveyed villages to the nearest bus stop is around 14 kilometer. All the surveyed villages do not have higher secondary school, but Loiliang and Soolungtoo villages are the surveyed villages which have a secondary school. None of the surveyed villages in Anjaw district has either a primary or middle school. For sending the children to primary school, the villagers have to come down at Goiliang and for middle school it is at Hayuliang, the administrative sub-division of Anjaw district. Among the surveyed villages in Lohit district, Loiliang, Soolungtoo and Danglat villages have both primary and middle schools in their villages. However, Soolungtoo, Karhe and Tillang Kyong villages have only primary school.

The health services in rural areas are provided through community health sub-centres, primary health centres, hospitals etc. However, among all the nine surveyed villages, only two villages have the primary health centres. In some surveyed villages like Kasanglat, Kaniliang, Brailiang, Soolungtoo, Karhe and Tillang Kyong, the users have to walk more than 5 kilometres for availing themselves of the medical services

TABLE 4

Descriptive Statistics of Infrastructural Facilities in the Surveyed Villages

Distance from	*Max.*	*Min.*	*Mean (in km.)*	*Standard Deviation (in km.)*	*Coefficient of Variation (in per cent)*
Jeepable Road	26	0	6.11	8.28	73.78
Bus Stop	39	0	13.89	14.98	92.72
Post office	39	2	17.22	12.09	142.44
Primary Health Centre	39	0	7.56	12.26	61.63
Primary School	7	0	2.22	2.95	75.37
Middle School	39	0	13.89	14.98	92.72
Secondary School	39	0	14.78	14.27	103.54
Higher Secondary School	99	8	37.56	32.49	115.58

Source : Survey Data, 2005-06.

from primary health centres. A descriptive statistics of infrastructural facilities of villages surveyed are given in Table 4.

Table 4 shows that the distance from post office of the surveyed villages is found to have a high degree of variation followed by the distance from higher secondary school, secondary school, middle school, bus stop, primary school, jeepable road and primary health centre respectively.

SECTION III

THE ROLE OF VILLAGE COUNCIL AND ITS MANAGEMENT OF COMMUNITY FOREST

Contrary to the bureaucratic belief that local people living in forest fringes are enemies of forest, a growing awareness among the scholars and practitioners that the action of local people greatly determine the success and failure of schemes regarding natural resource management is heartening. The role of local people at the local level is crucial for the sustainable development of forest resources on the one hand and on the other, their dependence on products harvested from forest is understandably necessity at large for their survival. Yet, lack of clear-cut rights on forest products make them vulnerable to discretionary treatment at the hands of government officials. The Village Council—a council of village elders having different status plays an important role to protect the age-old customary laws of different tribes of Arunachal Pradesh and governs their society. It is known by different names in different tribes such as Kebang in Adi tribes, Bulyang in Apatani tribes, Kebeya in Tawarah (Digaru Mishmi) tribes, Pharari in Kaman (Miju Mishmi) tribes, etc. In Lohit and Anjaw districts of Arunachal Pradesh, the Village Councils are known as either Kebeya or Pharari and these institutions had no written form of norms and rules but they were transmitted verbally from one generation to another generation. The Council is headed by the *Village Chief* or *Gaon Burah* and it plays an important role for decision-making, conflict resolution and collective action on CPR management in the surveyed villages. However, its role was found to be declining particularly in the context of privatization of resources in the surveyed villages. In fact, the Council is basically rule structures regulating the behaviour of tribal societies. It represents values, customs and culture of the local people.

Depending on the variation in social infrastructure, the ownership rights to land and to forest differ from one district to another. As far as land ownership is concerned, community owned land is found in Anjaw district while both private and community ownership to land prevail in Lohit district. However, the right to use and the right to transfer of land,

TABLE 5
Institutional Arrangements in the Surveyed Villages

Resource ownership	*Name of Village*								
	Kasanglat	*Kanilian*	*Brailiang*	*Loiliang*	*Danglat*	*Solungtoo*	*Chireng*	*Karbe*	*Tillang Kyong*
(1)	*(2)*	*(3)*	*(4)*	*(5)*	*(6)*	*(7)*	*(8)*	*(9)*	*(10)*
1. Forest Land									
(a) Owned Collectively	√	√	√	√	X	X	√	√	X
(b) Owned Collectively with private ownership of village	X	X	X	X	√	√	X	X	√
(c) User rights to all members of village	√	√	√	√	X	X	√	√	X
(d) Open Access	√	√	√	X	X	X	X	X	X
(e) Use rights to villagers but others pay entry fee	X	X	X	X	X	√	X	X	X
(f) User rights to other villages of the same community	√	√	√	X	X	X	X	X	X
2. Cultivated Land									
(a) Owned Collectively	√	√	√	X	X	X	√	√	X

(Contd.)

TABLE 5 (*Contd.*)

(1)	*(2)*	*(3)*	*(4)*	*(5)*	*(6)*	*(7)*	*(8)*	*(9)*	*(10)*
(b) Owned Collectively with private ownership of village	X	X	X	√	√	√	√	√	√
(c) User rights to all members of village	√	√	√	√	X	√	X	X	X
(d) Open Access	X	X	X	X	X	X	X	X	X
(e) Use rights to villagers but others pay entry fee	X	X	X	√	√	√	X	√	√
(f) User rights to other villages of the Same Community	X	X	X	X	X	√	X	X	X
3. Village Commons other than Forest									
(a) Owned Collectively	√	√	√	√	√	√	√	√	√
(b) Owned Collectively with private ownership of village	X	X	X	X	X	X	X	X	X
(c) User rights to all members of village	√	√	√	√	√	√	√	√	√
(d) Open Access	√	√	√	X	X	X	√	√	X
(e) Use rights to villagers but others pay entry fee	√	√	√	√	√	√	√	√	√
(f) User rights to other villages of the same Community	√	√	√	X	X	X	√	√	√

Note : The tick (✓) mark indicates the presence of attributes scale.
Source : Survey Data, 2005-06.

in principle, are solely based on the occupancy as well as the right to inheritance among these tribes. Presently, rules are slowly changing due to the modernisation process as well as the shift from non-monetised economy to monetised economy. In some exceptional case of land transfer, an interesting case was observed during the field study trip that a piece of land was under the occupancy of a Nepalese woman who happened to be settled in the 1950's at Tezu town. When she wanted to return to her native village, she sold the land in cash to *a Kamans* through mutual understandings and negotiations but could not sell it to non-locals. Similar cases are also observed of selling land by non-locals to other tribes of Arunachal Pradesh particularly in urban areas. However, such cases are not yet reported to any legal agencies. The right to sale or transfer of land is generally allowed within the community itself, but inter-tribe selling or buying is yet to be found in these surveyed districts.

Forests are an indispensable part in the socio-economic and religious activities of these communities. It is, in general, owned collectively by them and they heavily depend on it for their survival. However, the emergence of private ownership to forests is of recent phenomenon due to the introduction of monetised market economy and it is found in the villages of Danglat, Soolungtoo and Tillang Kyong of Lohit district. But, in all the villages of Anjaw district and Loiliang, Chireng and Karhe villages of Lohit district, forest ownership is bestowed by tradition to the community or village council which is called by the *Taraons* or *Tawarahs* as *Kebeya* or *Pharari* by the *Kamans*. In other words, the ownership right to village common other than forests rests mainly with community in Lohit district and both the villages common as well as the forest rests with community in Anjaw district.

Regarding the quality of forests, it also differs from one administrative sub-division to other sub-division in these districts. The quality of forests in the surveyed villages is shown in Table 6.

Table 6 indicates that the four villages viz., Kasanglat, Kaniliang, Brailiang and Tillang Kyong have dense forest while the remaining villages have degraded forests within 2 kilometers from the villages. On the other hand, within 5 kilometers from the village, Loiliang is the only village which has a moderately dense forests and Danglat and Soolungtoo villages have degraded forests. However, Kasanglat, Kaniliang, Brailiang Chireng, Karhe and Tillang Kyong villages have dense forests within this limit. But, within 10 kilometers from the villages, all the villages except Soolungtoo and Danglat have dense forests. Trees are mainly used for fuelwood, fencing as well as construction of houses in all the surveyed villages of Anjaw district and

TABLE 6

Quality of Forests in the Surveyed Villages

Name of Village	*Within 2 kms. from the Village*			*Within 5 kms. from the Village*			*Within 10 kms. from the Village*		
	Dense	*Moderate*	*Degraded*	*Dense*	*Moderate*	*Degraded*	*Dense*	*Moderate*	*Degraded*
(1)	*(2)*	*(3)*	*(4)*	*(5)*	*(6)*	*(7)*	*(8)*	*(9)*	*(10)*
Kasanglat	√			√			√		
Kaniliang	√			√			√		
Brailiang	√			√			√		
Loiliang			√		√		√		
Danglat			√			√		√	
Soolungtoo			√			√		√	
Chireng			√	√			√		
Karhe			√	√			√		
Tillang Kyong	√			√			√		

Note : The tick (√) mark indicates presence of the attributed scale.
Source : Survey Data, 2005-06.

Loiliang, Danglat and Soolungtoo in Lohit district. However, in Chireng, Karhe and Tillang Kyong villages under Wakro Circle of Namsai sub-division, Lohit district, bamboos are mainly used for these purposes as it is abundantly available in the vicinity of the villages. A general perception amongst the villagers about the forest cover was that 10 to 20 years back, there was dense forest surrounding the villages. They are of the opinion that illegal felling of trees for commercial purpose as well as high dependency on forest products for household consumption such as firewood, house construction materials, fencing, etc. are the main factors for the degradation of forests. In all the three villages of Anjaw district, they reported that they still collect firewood and other household materials within 2 kilometers. In the villages of Lohit district, villagers are now collecting these items within 5 to 10 kilometer distance from the surveyed villages except Chireng, Karhe and Tillang Kyong.

SECTION IV

CONSUMPTION FROM COMMUNITY FOREST

Community forest plays an integral part in rural household economics. The myth, belief and tradition are centred around the forest and its environment and their dependency on it becomes crucial due to many socio-economic factors like predominance of *jhum* cultivation, lack of road connectivity and transport facility, non-availability to market access, traditional belief, etc. Under such circumstances, obviously the rural households are compelled to depend on different forest products. On an average, 55.73 per cent of the consumption expenditure of the households in the study area were derived from the community forest irrespective of poor and non-poor households. The relative dependence i.e., percentage of consumption expenditure of households from common property forest resources in different surveyed villages shows the heterogeneity in their consumption pattern. Table 7 reveals the annual consumption of forest products in the surveyed villages of Anjaw and Lohit districts.

It was found that the firewood consumption emerged as the most significant contribution to total consumption expenditure of rural households probably because of the traditional dependence on firewood due to climatic conditions as well as the non-availability of alternative fuels. Further, the socio-economic determinants have direct effect on the nature and level of the benefits derived from commons. However, non-poor households used community forest mainly for extraction of timber and other commercial forest products while the poor households used it mainly for collection of firewood, bamboo, and other minor forest products for self-consumption (Figure 2).

TABLE 7

Annual Consumption of Forest Products in the Surveyed Villages

(*in per cent*)

Name of Village	*Economic Criteria*	*Firewood*	*Timber*	*Bamboo & Cane*	*Leafy Veg.*	*Others*
(1)	*(2)*	*(3)*	*(4)*	*(5)*	*(6)*	*(7)*
Kasanglat	Poor	84.16	8.75	4.13	2.62	0.34
	Non-poor	72.18	22.77	1.61	0.95	2.49
Kaniliang	Poor	85.10	4.11	6.66	4.13	0
	Non-poor	51.96	43.08	0.80	1.84	0.31
Brailiang	Poor	93.18	3.39	2.38	1.04	0
	Non-poor	75.77	17.40	0.99	1.44	4.40
Anjaw District	Poor	87.11	6.25	3.99	2.46	0.18
	Non-poor	70.38	24.33	1.30	1.23	2.76
Loiliang	Poor	90.80	4.53	2.27	2.41	0
	Non-poor	73.02	22.16	1.44	1.44	1.94
Danglat	Poor	80.93	13.59	4.24	1.23	0
	Non-poor	70.92	23.93	2.83	1.4	0.92
Soolungtoo	Poor	83.18	11.13	3.38	2.31	0
	Non-poor	67.41	29.33	1.94	1.32	0
Chireng	Poor	87.35	4.73	5.07	1.82	1.02
	Non-poor	43.23	50.65	0.30	1.89	3.92
Karhe	Poor	83.43	13.44	2.83	0.30	0
	Non-poor	67.31	28.19	2.48	0.12	1.90
Tillang Kyong	Poor	87.79	6.45	4.64	0.55	0.57
	Non-poor	80.87	7.32	3.83	0.50	4.28
Lohit District	Poor	85.71	9.29	3.37	1.50	0.13
	Non-poor	68.79	26.24	2.28	0.93	1.76

Source : Survey Data, 2005-06.

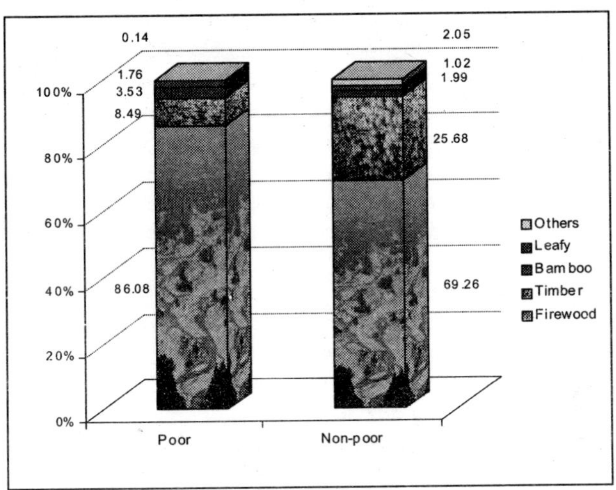

Figure 2 : Consumption of Different Forest Products in Anjaw and Lohit Districts (in per cent)

The overall consumption of firewood from community forest appeared as 78.17 per cent of total consumption value from commons, of which the poor derived 86.08 per cent and the non-poor 69.26 per cent respectively. The consumption of firewood was relatively high in Anjaw district than that in Lohit district particularly among the poor and non-poor households. The firewood consumption of poor households in Anjaw was 87.11 per cent of the total consumption value from commons while in Lohit district, it was 85.71 per cent. In case of non-poor in Anjaw district had 70.38 per cent and in Lohit district it was found as 68.79 per cent. The highest dependency of firewood among the poor households was observed in Brailiang village of Anjaw district and the lowest in Danglat village of Lohit district among the surveyed villages. Among the non-poor households, Tillang Kyong village had the highest percentage of firewood consumption to total consumption expenditure of households while Chireng village of Lohit district being the lowest. In general, rural households in Anjaw district had higher dependency of firewood than their counterparts in Lohit district.

With regard to the degree of variability in firewood consumption between poor and non-poor, it is found that the coefficient of variation (CV) of overall firewood consumption for the poor and the non-poor was estimated as 34.39 per cent and 43.56 per cent respectively in the study area. In Anjaw district, we found the CV of firewood for poor households as 26.41 per cent and for non-poor households, it was 28.21

per cent. Similarly, it was found in Lohit district that the CV of firewood consumption for poor households as 20.56 per cent and for non-poor households, it was 29.32 per cent. Thus, as far as firewood consumption is concerned, although the variation is low but there is greater variance in non-poor households both in Anjaw and Lohit districts. The relatively high variability in firewood consumption of non-poor is probably due to availability and affordability of alternative fuel, accessibility of market, etc. among some of the non-poor households.

Timber consumption (16.57 per cent) emerged as the second highest contributor to consumption from community forest. The poor households derived timber consumption of 8.49 per cent from community forest while the non-poor resulting 25.67 per cent in the study area. Among the poor households, Danglat village of Lohit district had the highest timber consumption while Brailiang village of Anjaw district had the lowest consumption in the surveyed villages. In case of non-poor households, Chireng village had the highest share of timber consumption and Tillang Kyong village of Lohit district was the lowest. The extent of variability in timber consumption from community forest in the study area, it was found that the CV for overall timber consumption of poor and non-poor households was 123.84 per cent and 106.95 per cent respectively. In Anjaw district, the CV for poor households in timber consumption was 83.22 per cent and it was 71.88 per cent for non-poor households. Similarly, in Lohit district, it was found that the CV for poor and non-poor households in timber consumption was as high as 132.86 per cent and 116.76 per cent respectively. Thus, the poor households in both Lohit and Anjaw districts had higher variation in timber consumption than the non-poor households though the level of variation was higher in Lohit district.

Regarding the consumption of minor forest products (5.26 per cent), the overall CV for the poor and non-poor households in Anjaw district was 111.75 per cent and 79.24 per cent respectively while in Lohit district it was 139.31 per cent for poor and 121.09 per cent for non-poor households. Similarly, the consumption of leafy vegetables, eatable roots and leaves, the non-poor households had high variability than the poor households both in Anjaw and Lohit districts. Moreover, in the other products like honey, medicinal herbs and seeds and wild animals hunt, the poor households had higher variability than the non-poor households did. Thus, in general, the dependency of poor households on commons was relatively higher than the non-poor households on firewood, bamboo, leafy vegetables, eatable roots and leaves while the dependence of non-poor is relatively higher than the poor households on timber and other groups of forest products like honey, medicinal herbs, seeds and wild animals hunt which can be marketed.

The relative dependency i.e., percentage of consumption expenditure of households from common property forest resources in the surveyed villages of Anjaw and Lohit districts, shows another dimension of this study. The results are presented in Table 8.

TABLE 8

Percentage of Consumption Expenditure of Rural Households from 'CPRF' in the Surveyed Villages

District	*Name of Village*	*Percentage*
Anjaw District	Kasanglat	68.10
	Kaniliang	68.80
	Brailiang	62.55
Total		66.04
Lohit District	Loiliang	44.23
	Danglat	54.53
	Soolungtoo	41.15
	Chireng	75.03
	Karhe	68.16
	Tillang Kyong	64.63
Total		52.44

Source : Survey Data, 2005-06.

From Table 8, we find that there are some variations of dependence on community forest (in terms of consumption expenditures) among various surveyed villages of Lohit and Anjaw districts of Arunachal Pradesh. The dependency on community forest was relatively high in the surveyed villages of Anjaw district than those in Lohit district. This may be due to two factors, namely, (i) concentration of high dense forest area in Anjaw district than but in Lohit district (where the percentage of forest area is relatively low), and (ii) relatively high altitude of villages of Anjaw district than those of Lohit district. Thus, we find the evidences that dependency was declining where stock of forest resources was relatively less and degraded whereas dependency increased where such stock was relatively high. Let us now turn our analysis on the forest dependency among different categories as per poor and non-poor.

TABLE 9

Percentage of Consumption Expenditure of Rural Households from 'CPRF' in the Surveyed Villages on the basis of Different Income Categories

District	*Income Group*	*Percentage*
Anjaw	Poor	69.02
	Non-poor	64.03
Lohit	Poor	62.40
	Non-poor	44.13
Total	Poor	64.03
	Non-poor	48.62

Source : Survey Data, 2006-07.

Table 9 shows that around 64.03 per cent of the total consumption expenditure of relatively poor households were derived from community forests. On the other hand, 48.62 per cent of the total expenditure was derived from community forests for relatively non-poor categories of rural households. Thus, like the previous findings, we find that the forest dependency, no doubt, decreased with the increase in income but the dependence on community forest for the non-poor household was also quite significant. In other words, the poor derived more benefits from the community forest mainly on non-timber forest products but the dependency on community forest for the non-poor households was quite significant in timber as well as other commercial forest products. This may due to the easier access to intermediate forest products which benefits non-poor households.

Thus, the socio-economic determinants have direct effect on the nature and level of the benefits derived from commons. However, non-poor households used community forest mainly for extraction of timber and other commercial forest products while the poor households used it mainly for collection of firewood, bamboo, and other minor forest products for self-consumption. For example, the consumption of relatively poor households consisted more of firewood (86.06 per cent), followed by timber (8.49 per cent), and minor forest products (5.43 per cent) which mainly comprised of bamboo and leafy vegetables, eatable roots and leaves, etc. On the other hand, the consumption of relatively non-poor households comprised more of firewood (69.26 per cent) followed by timber (25.68 per cent), and minor forest products (5.06 per cent). Therefore, the consumption of firewood and timber emerged as significant contributors to relatively non-poor households. In fact, the study shows that as the income of the household increases, the percentage consumption of timber also increase but the consumption of other forest products relatively reduces. That means the relatively poor households were depending on community forest for firewood, bamboo and other minor forest products like leaves, medicinal herbs and seeds, vegetables and roots but the relatively non-poor households exploited the community forest more for timber consumption and other forest products which can be marketed.

SECTION V

CONCLUSION AND POLICY IMPLICATIONS

The study found that the poor households consumed much less value from community forest than the non-poor households. On an average the poor households derived their consumption expenditure of around rupees 22,000 thousand per annum from community forest while the non-poor households derived around rupees 33,000 per annum respectively. Thus, in terms of absolute contribution to the total consumption expenditure of the rural households, community forest contributed more to the non-poor households as compared to the poor households. At the same time, the contribution of the community forest to the household economies was significant in the study area. On an average, 55.73 per cent of the consumption expenditure of the households in the surveyed villages was derived from the community forest. The relatively poor households derived 64.03 per cent of the total consumption expenditure from community forest. On the other hand, 48.62 per cent of the total consumption expenditure was derived from the same source for relatively non-poor households.

Thus, we find that the forest dependency decreased with the increase in total consumption expenditure of the households. It was also found that the poor households had higher consumption in firewood and the non-timber forest products like bamboo, leafy vegetables, eatable roots, and leaves than the non-poor households did while the non-poor households had high consumption in timber and the non-timber products like honey, medicinal herbs and seeds and wild animals than the poor households. The share of firewood consumption to consumption value from community forest of the poor households was 86.08 per cent and for timber consumption, it was 8.49 per cent. But, minor forest products accounted for 5.43 per cent of the consumption value from community forest of which bamboo had a share of 3.53 per cent, leafy vegetables, eatable roots and leaves had a share of 1.76 per cent 0.14 per cent respectively. In case of the non-poor households, the share of firewood to consumption value from community forest was

69.26 per cent followed by timber (25.68 per cent), bamboo (1.88 per cent), leafy vegetable (1.02 per cent), and honey, medicinal herbs/seeds and wild animals (2.05 per cent). Thus, the non-poor households exploited the community forest more of timber than the poor households. The study also shows an interesting fact that the extraction of market-oriented products from community forest did not necessarily decrease with the increase in income. Instead we find the evidence that the exploitation of market-oriented products is increasing with the increase in income in the surveyed villages of Lohit and Anjaw districts of Arunachal Pradesh where the stock of natural resources are high. The study shows that the prevailing CPR management in the surveyed villages led to the emergence of 'Neo-rich' class in the study area. This has created not only the inequality among the rural households but also created environmental degradation.

Thus, the foregoing analysis of prevailing CPR institutions has clearly revealed that the traditional mechanism which maintained balance between man and environment has weakened. In fact, the tribes of the surveyed villages had their own system of forest conservation and management system because they were virtually connected with the preservation of community forest which reflected the age-old pattern of coexistence between man and nature. For example, the majority of the tribes in the study area had strong belief about the presence of supernatural power inside the forest and hence, they often took to perform special rituals associated to it. However, these beliefs are shrinking as the people are coming in contact with modern belief system.

In fact, the traditional property rights formation which was largely collective is being replaced by private property rights over land and forest. In the traditional phase, apart from the State control over cultivable land and forest, three different types of ownership of land were present in the study area i.e., collective ownership by village communities, clan ownership and private and individual ownership. These ownership rights are not properly recognized by the State but informal ownership through habitual users' rights exhibit a great deal of diversity and adaptability. The collective ownership is not just being replaced by private ownership; the meanings and function of the traditional property rights structure have also undergone a significant transformation. In our surveyed villages, we noticed the emergence of private property rights over land and forest and hence it is the most fundamental challenge before us to design a new or dynamic CPR management system so that the forest resources can be used on equity and sustainable basis.

Many collective action theorists have advocated that if a group of individuals is placed in a situation in which they could all mutually benefit by complying the rule of restricted CPRs use, they will not cooperate without external enforcer. Therefore, these scholars have strongly recommended for a change in CPR structure wherever the resources are the nature of commons.

On the other hand, the scholars associated with 'new institutionalism' have strongly criticized the argument presented by the proponents of change in the property rights of CPRs. They have advocated that user community can use and manage the CPRs in a sustainable way without any external coercion, defying the above mentioned classifications of approaches to solve the CPR problem, it has been realized that there is no universal solution to the problem as the success or the failure of a management system which depends on a great variety of situation specific factors. An institutional option best fits in one situation may not be applicable in other settings. Hence, an attempt is made to explore the best-fit institution to manage the CPRs in the study area by taking into account the location specific factors.

The forest area of the surveyed villages can be classified into three, namely, (i) State Forest (ii) Unclassified State Forest, and (iii) Community or Clan Forest. The villagers claimed that the Unclassified State Forest is also Community Forest. In our study area, the Government's initiative to protect and develop CPRs by extending the State Forest Programme like Joint Forest Management was found.

However, it is found that illegal extraction of CPRs (firewood, timber, bamboo, etc.) is a common phenomenon in the surveyed villages. Thus, the conversion of CPRs into State property had failed to fulfil its objective to regulate the commons. At the same time, the empirical evidences from the surveyed villages showed that the growing tendency of privatization of CPRs had led to an increasing inequality in the distribution of natural resources.

Hence, as an alternative to both market and State controlled institutional arrangements, local level decentralized management of CPRs should be emphasized. In this connection, it is noted that the tribes, in the study area, had their own system of forest management where the age-old traditional institution headed *Gaon Burah* or *Village Chief* played an important role in management of CPRs in terms of resource allocation mechanism, sanction and enforcement rules. However, much of its importance has lost due to modernization and introduction of monetary economy. Further, the importance of the age-old traditional institution is at stake due to the introduction of village Panchayat system and increasing interference of the State functionaries.

Our theories and experiences based on the case studies show that in many situations, though not always co-ordination and leadership problems play a dominant role. When the poor households overexploit local resources, they are aware of the ecological impact of their actions. It is often because they face acute survival constraints which lead them to discount streams of future benefits heavily. They generally need externally provided economic incentives to conserve the resources. An External catalytic role by the State via local level institutions can play a significant role. Hence, in the context of Arunachal Pradesh, the sustainable CPR management is only possible through decentralized management with active participation of local stakeholders. The CPRs can never be managed unless the rural households are made to feel that they have a say in their matter.

The study reveals that CPRs play a very important role in the economy of the surveyed villages although the consumption value from community forest is higher for the non-poor households in absolute terms yet in relative terms, the poor households' dependency on community forest is very important and crucial for their survival. Hence, there is urgent need to form sustainable management of CPRs, particularly the forests in order to avoid 'the tragedy of commons'.

Secondly, there is an urgent need to strengthen the age-old traditional system of village council headed by *Gaon Burah* or *Village Chief* in addition to the village Panchayat with specific purpose of management of these resources.

Finally, the success of restoration and sustainable use of CPRs depends on the overall socio-economic development of the rural masses.

Hence, it calls for higher agricultural productivity through modernization of agriculture along with the expansion of horticulture so as to reduce the degree of forest dependency to some extent. It should be noted that any policy intervention to ensure livelihood security of rural communities in Arunachal Pradesh has to be based on a properly framed institutional strategy particularly in relation to the CPR management.

References

Adhikari, B. (2001): 'Socio-Economic Heterogeneity and Income Distribution: Evidence from Common Property Resource Management', *Journal of Forestry on Livelihoods*, Vol. 1, pp. 22-24.

Baland, J. and J. Platteau (1996): *Halting Degradation of Natural Resources: Is there a Role for Rural Communities* (Rome: FAO).

Bromley, D.W. (1989): 'Property Relations and Economic Development: The Other Land Reforms', *World Development*, Vol. 17, No. 6, pp. 867-77.

Chopra, K., G.K. Kadekodi and M.N. Murty (1989): 'Peoples Participation and Common Property Resources', *Economic and Political Weekly*, pp. 189-95.

Choudhury, S.D. (1978): *Arunachal Pradesh District Gazetteers: Lohit District*, Directorate of Information and Public Relations, Government of Arunachal Pradesh, (Shillong).

Chowdhury, J.N. (1982): *Arunachal Through the Ages: From Frontier Tracks to Union Territory*, (Shillong).

Ciriacy-Wandtrup, S.V. and R.C. Bishop (1975): 'Common Property as a Concept in Natural Resources Policy', Natural *Resources Journal*, Vol. XV, pp. 713-27.

Coase, R.H. (1960): 'The Problem of Social Cost', *Journal of Law and Economics*, Vol. III, pp. 1-44.

Das, Gurudas (1995): *Arunachal Tribes in Transition* (New Delhi: Vikash Publishing House)

Dayton-Johnson, J. (2000): 'Determinants of Collective Action on the Local Commons: A Model with Evidence from Mexico', *Journal of Development Economics*, Vol. 62, pp. 181-208.

Demestz, H. (1967): 'Towards a Theory of Property Rights', *American Economic Review*, Vol. 62, No. 2, pp. 347-59.

Di Falco, S. and C. Perrings (2003): 'Crop Generic Diversity, Productivity and Stability of Agro-Ecosystems: A Theoretical and Empirical Investigation', *Scottish Journal of Political Economy*, Vol. 5, No. 2, May.

Edwards, V.M. and N.A. Steins (1998), 'Developing an Analytical Framework for Multiple Use Commons', *Journal of Theoretical Politics*, Vol. 10, No. 3, pp. 347-83.

Fox, J.M. (1983): *Managing Public Lands in a Subsistence Economy: The Perspective from a Nepali Village*, Unpublished Ph.D. Dissertation, University of Wisconsin, Madisen.

Gordon, H.S. (1954): 'The Economics of a Common Property Resource: The Fishery', *Journal of Political Economy*, Vol. 62, No. 2, pp. 124-42.

Government of India (2001): *State Forest Report*, Forest Survey of India, Ministry of Environment and Forest, New Delhi.

——, (2001): *Statistical Abstract, India*, Directorate of Economics and Statistics, Ministry of Agriculture and Cooperation, New Delhi.

——, (2004): *Forest and Wildlife Statistics*, Forest Survey of India, Ministry of Environment and Forest, New Delhi.

Government of Arunachal Pradesh (1987): *Statistical Abstract*, Directorate of Economics and Statistics, Shillong.

——, (1987): *Statistical Handbook of Lohit District*, Economics and Statistics Department, Tezu.

——, (1992): *Statistical Handbook of Lohit District*, Economics and Statistics Department, Tezu.

——, (1994): *District Census Handbook, Lohit*, Census of India, 1991, Series 3, Arunachal Pradesh, Directorate of Census Operations, Itanagar.

——, (1995): *Statistical Abstract of Arunachal Pradesh*, Directorate of Economics and Statistics Department, Itanagar.

——, (1998): *State Domestic Product of Arunachal Pradesh*, State Income Unit, Directorate of Economics and Statistics Department, Itanagar.

Government of Arunachal Pradesh (2000): *Statistical Abstract of Arunachal Pradesh*, Directorate of Economics and Statistics Department, Itanagar.
———, (2001): *Statistical Abstract of Arunachal Pradesh*, Directorate of Economics and Statistics Department, Itanagar.
———, (2002): *Statistical Abstract of Arunachal Pradesh*, Directorate of Economics and Statistics Department, Itanagar.
———, (2002): *Statistical Handbook of Lohit District*, Economics and Statistics Department, Tezu.
———, (2003 & 2004): *Statistical Handbook of Lohit and Anjaw District*, District Statistical Office, Tezu.
———, (2005): *Arunachal Pradesh Human Development Report*, Department of Planning, Itanagar.
———, (2005): *Statistical Handbook of Lohit District*, Economics and Statistics Department, Tezu.
———, (2007): *Draft Tenth Five Year Plan*, Department of Planning, Itanagar, Retrieved on 2nd Jan, 2007 from http://www.arunachalpradesh.nic.in.
Hardin, G. (1968): 'The Tragedy of Commons', *Science*, Vol. 162, pp. 1243-48.
Iyengar, S. (1989): 'Common Property Land Resources in Gujarat: Some Findings about Their Size, Status and Use', *Economic and Political Weekly*, Vol. XXIV, No. 25, pp. A67-A77.
Jodha, N.S. (1986a): *The Decline of Common Property Resources in Rajasthan, India* (London: Overseas Development Institute).
———, (1986b): 'Common Property Resources and Rural Poor in Dry Regions of India', *Economic and Political Weekly*, Vol. XXI, No. 27, pp. 1169-82.
———, (1990): 'Rural Common Property Resources: Contribution and Crisis', *Economic and Political Weekly*, Vol. XXV, No. 26, pp. A65- A78.
Kadekodi, G.K. (2001): 'Environment and Development', in Rabindra, N. Bhattacharya (ed.), *Environmental Economics: An Indian Perspective*, (New Delhi: Oxford University Press).
———, (2004): *Common Property Resource Management*, (New Delhi: Oxford University Press).
Mishra, S.N. (1983): 'Arunachal's Tribal Economic Formations and Their Dissolution', *Economic and Political Weekly*, Vol. XVIII, No. 43, pp. 1837-1846.
Mitra, A. (1998): 'Environment and Sustainable Development in Hilly Regions of North-East India—A Study of Arunachal Pradesh', *International Journal of Social Economics*, Vol. 25, No. 24, pp. 196-206.
———, (2006): 'Transformation of Arunachal's Economy: Opportunities and Challenges for a Less Developed State', *Indian Development Review*, Vol. 4, No. 2, p. 433.
Mitra, A. and K. Chattopadhyay (2003): *Environment and Nature-based Tourism: An Endeavour at Sustainability*, (New Delhi: Kanishka Publishers).
Mitra, A. and D.K. Mishra (2005): *Report on Environment, Property Rights and Rural Livelihood Strategies: A Study on Arunachal Pradesh*, (New Delhi: Indian Council for Social Science Research).
NCAER (1967): *Techno-Economic Survey of NEFA*, New Delhi.
NEC (1982-83): *Annual Plan*, Shillong.
———, (2002): *Basic Statistics of North-Eastern Region*, Shillong.

Ostrom, E. (1986): 'An Agenda for the Study of Institutions', *Public Choice*, Vol. 48, pp. 3-25.

———, (1990): *Governing the Commons: The Evolutions of Institutions for Collective Action*, (Cambridge: Cambridge University Press).

———, (2000): 'Reformulating the Commons', *Swiss Political Science Review*, Vol. 6, No. 1, pp. 29-52.

Pasha, S.A. (1992): 'CPRs and Rural Poor: A Micro Level Analysis', *Economic and Political Weekly*, Vol. XXVIII, No. 46, pp. 2499-2503.

Richards, M., K. Kanel, M. Maharjan and J. Davies (1999): *Towards Participatory Economic Analysis by Forest User Groups in Nepal*, (London: Overseas Development Institute).

Steins, N.A. and V.M. Edwards (1999), 'Collective Action in Common Pool Management: The Contribution of a Social Constructivist Perspective to Equity Theory', *Social and Natural Resources*, Vol. 12, pp. 539-52.

Steins, N.A., V.M. Edwards and Niels Rolling (2000), 'Redesigned Principles for CPR Theory', *The Common Property Resource Digest*, No. 53, pp. 1-4.

Tang, S.Y. (1991): 'Institutional Arrangements and the Management of Common-Pool Resources', *Public Administration Review*, Vol. 51, No. 1, pp. 42-51.

UNDP (2001): *Human Development Report*, New York.

Wade, R. (1987): 'The Management of Common Property Resources: Collective Action as an Alternative to Privatisation or State Regulation', *Cambridge Journal of Economics*, Vol. 11, pp. 95-106.

Weitzman, M. (1974): 'Free Access *v.* Private Ownership as Alternative Systems for Managing Common Property', *Journal of Economic Theory*, Vol. 8, No. 2, pp. 225-34.

Yanggen, D. and T. Reardon (2001): 'Kudzu-improved fallows in the Peruvian Amazon', in A. Angelsen and D. Kaimowilze (ed.), *Agricultural Technologies and Tropical Deforestation*, CABI Publishing in Association with Centre for International Forestry Research, Bogor, pp. 213-29.

Chapter 16

Non-Timber Forest Products and Socio-Economic Characteristics

An Empirical Study in Drought Prone Area of West Bengal

JYOTISH PRAKASH BASU

ABSTRACT

In India, forests play a vital role in the sustenance of about 170 million people and 250 million cattle living around 31 million hectares of forests in 1,70,000 villages. India has implemented a very large Joint Forest Management (JFM) program covering nearly 85,000 villages and more than 17 million hectares. Under JFM, the village community gets a greater access to a number of Non-Timber Forest Products (NTFPs) and a share in timber revenue in return for increased responsibility for its protection from fire, grazing and illicit harvesting. This approach has resulted in active participation of the people in forest conservation and also resulted in higher incomes for local people. Given this backdrop, the objective of the paper is to examine the factors affecting the collection of non-timber forest products. This paper is an empirical study based on data collected through field survey. This study covers two villages of Kalaberia and Kalyanpur located in the drought prone District of Bankura, South West Bengal, consisting of 65 households in 2008. The analysis has theoretical foundation in the household economy to explore how socio-economic characteristics influence the collection of non-timber forest products. Econometric analyses suggest that household labor allocation decision for non-timber forest products collection is dictated by various socio-economic variables. In this paper

multivariate regression model has been applied to estimate the factors affecting the collection of non-timber forest products. Staying hours in the forest as dependent variable while age, gender, caste, education, household members, land holding, distance between home and forest have been taken independent variables. The results of the study revealed that these socio-economic factors are affecting the livelihood of the forest dependent communities. This paper has important policy implication for forest conservation, sustainable forest management practices, and rural development, environmental sustainability.
Keywords: Non-timber forest products, Livelihood, labor allocation decision, econometric model, institution and socio-economic factors.

INTRODUCTION

Under Joint Forest Management (JFM), the village community gets a greater access to a number of Non-Timber Forest Products (NTFPs) and a share in timber revenue in return for increased responsibility for its protection from fire, grazing and illicit harvesting. The United Nations Food and Agriculture Organization (FAO, 2002) approximates that 80% of the developing world relies on non-timber forest products (NTFPs) for nutritional and health needs. A recent study conducted by the World Bank revealed that in some 54 cases evaluated, forest income averaged 22% of total income (Vedeld *et al.*, 2004). Research by Cavendish (2000), Mahapatra *et al.* (2005) and Sverrer and Olsen (2005) suggest that while total NTFP income tends to increase as other sources of income increase, the share of total income derived from NTFPs tends to be highest for the poorest households. The contributions that non-timber forest products (NTFPs) can make to rural livelihoods, and the fact that their use is less ecologically destructive than timber harvesting, have encouraged the belief that more intensive management of forests for such products could contribute to both development and conservation objectives, and have led to initiatives to expand commercial use of NTFPs (Arnold and Perez, 2001). The collection of some forest produce throughout the year, such as mahua, tamarind, kendu, mango, jackfruit, sal leaves and seeds, vegetables and roots, gave the household an additional source of both food and income. Dwindling availability of forest-produce—food, fuel, medicinal herbs, etc. has deprived them of a supplementary source of both income and food (Rao and Rana, 1997). Since NTFPs are essential for poor households' well-being, it follows that any program or policy to encourage community involvement in forest management will have to consider the role of women.

Some studies have contended that JFM program may hurt rather than help poor women and villagers who depend on fuelwood. Sundar

(2000) and Sarin and others (1988) argue that JFM helps well-off villagers who can secure alternative sources of fuelwood but burdens poor villagers. Agarwal (2001) focuses on 'participatory exclusion' of women within JFM regimes. She finds that women bear significant costs associated with loss of access to forests, additional time spent to collect fuelwood and fodder, use of alternative inferior fuels, scarce supply of fuelwood even after forests have regenerated, and inequitable distribution of community benefits.

Given this backdrop, the objective of the paper is to examine how socio-economic factors influence the collection of non-timber forest products.

The paper is organized as follows. Section 1 discusses the importance of non-Timber Forest Products in the household economy in India and West Bengal in particular. A simple static household decision-making model is discussed in Section II. Section III discusses data and methodology. The results of the econometric model are presented in Section IV. Section V presents concluding remarks.

SECTION I

IMPORTANCE OF NTFPS AND HOUSEHOLD ECONOMY

Forests and forest resources, say Non-Timber Forest Products (NTFPs), play an important role in the viability and survival of tribal households in India, because of the importance of forests in their social, cultural and economic survival (Tewari, 1989). Non-timber forest products are important to JFM efforts for a number of reasons. First, NTFPs are integral to the lifestyle of forest-dependent communities. They fulfil basic requirements, provide gainful employment during lean periods and supplement incomes from agriculture and wage labour. Medicinal plants have an important role in rural health (Prasad and Bhatnagar, 1991). In parts of West Bengal, communities derive as much as 17 percent of their annual household income from NTFP collection and sale (Malhotra *et al.*, 1991). According to J.Y. Campbell (1988, cited in Tewari and Campbell, 1995), small-scale forest-based enterprises, many of which rely on NTFPs, provide up to 50 percent of the income for about 25 percent of India's rural labour force.

Second, NTFPs have a decided advantage over timber in terms of the time needed to achieve significant volumes of commercially valuable production. Timber production is a long-term endeavor, and in many areas timber harvesting may not be ecologically desirable. Moreover, many NTFPs become available even in the earliest stages of rehabilitation of degraded forest areas. Third, at the national level over 50 percent of forest revenue and about 70 percent of forest export revenue comes from NTFPs, mostly from unprocessed and raw forms (Tewari and Campbell, 1997; Prasad, Shukla and Bhatnagar, 1996).

Recent studies conducted in the tribal regions of Bihar, West Bengal and Karnataka offer further empirical evidence for the extent of dependence of tribal households on NWFP collection. For example, in two southern districts of Bihar, 41 percent of the families collect mahua flowers; 31 percent collect kendu leaves used in making indigenous cigarettes (or bidi); 23 percent of the families collect mushrooms and mahua seeds; 55 percent of the families collect tamarind; and 31 percent of the families depend on the collection of wild brooms (Rao and Singh, 1996). Even higher rates, ranging between 56-73 percent, have been recorded in Midnapore district of West Bengal (Rao and Singh, 1996, Hegde *et al.* 1996), in a study of Soliga households; found that the income contribution from the collection of NTFPs is disproportionately greater than the time spent in collecting the products. Their study indicated that households living on the periphery of the forest spend 39.25 percent of their time in collection and realize 47.63 percent of their income from NTFPs; for tribal living closer to the forest, the figures are 54.46 percent and 60.44 percent respectively. The critical role that NTFPs play in the livelihood strategies of tribal households is highlighted by the very favorable income returns to the time spent in collection, and the stability of income from NTFPs.

Dependence on forests and common property resources increases as a household becomes economically marginalized. Ramamani (1988), in a study of tribal economy in Srikakulam District in the Eastern Ghats of Andhra Pradesh, disaggregated tribal dependence on forests. The more marginal a tribal household, the greater is the proportion of its income from forests. Data indicated that sub-marginal and marginal tribal households accrue 35 to 36 percent of their income from forest produce. As poverty increases, women become more prominent in ensuring the survival of households by assuming greater responsibility to provide resources from forests and common lands. The importance of NTFPs for the very poor tribal households has been well documented by other studies as well (Hedge *et al.*, 1996; Godoy *et al.*, 1995). In Andhra Pradesh, the poor obtain 84 percent of their fuel supplies from common property resources, and are employed for 139 days to collect products from common property resources (Jodha, 1992). The potential NTFPs of the forests in South-West Bengal are sal leaves, dry leaves, mushrooms, various medicinal plants, sal, kendu leaves, etc. Among these green sal leaves are most important and many poor villagers depend upon this for their daily subsistence. It is found from the study conducted by Mishra and others (2005) that the average income from

NTFPs in the region of Midnapore district of West Bengal is Rs. 550 per household per year. There is large diversity of yearly income from NTFPs. In Ranibandh-Kamalpore region of Bankura district of West Bengal income from NTFPs is above Rs. 8 lakhs for the whole year and Vedua in the same district it is as low as Rs. 3500 (1 US $ = Rs. 45.00). Throughout India, collection of kendu leaf generates part time employment for 7.5 million people—a majority of them tribal women (Arnold, 1995). Women, according to studies in Uttar Pradesh, derive a greater proportion of their income from forests and common lands; poor women derive 45 percent of their income from forests and common lands as opposed to 13 percent for men (FAO and SIDA, 1991). Fernandes and Menon (1987), based on a study done in Orissa, reported that women generally walk 3 to 4 hours into the forests and work 15 hours per day, whereas men work 11 hours a day. The studies all indicated that women spend more time and labor in forest-related activities, and depend on forests not only to meet subsistence needs but also for income. Interestingly, the considerable role played by women in ensuring food security, and in the provision of cash income from the sale of NTFPs gives women a higher status in tribal societies (Sarkar, 1994). The women have to spend major part of their time and have to walk long distances daily to collect fuel wood, fodder and other Non-Timber Forest Products (NTFPs) from forests. Of nearly 170 million people living in and around forests in India, more than half of them are tribal and depend on non-timber forest products (NTFPs). The non-timber forest products such as mahua, kendu, mango, jackfruit, sal leaves and seeds, vegetables and roots, food, fuel wood, fodder, mushroom, honey, medicinal herbs, etc. are very important contributors to the well-being or livelihood of villagers. More time spent in collection of forest products, often leads to less time for other activities, such as agriculture, which results in lower levels of income or subsistence.

SECTION II

A SIMPLE STATIC HOUSEHOLD DECISION-MAKING MODEL

In this section we formulate a model which captures household behaviour with regard to labour allocation, with differential returns to time spent in collection and time spent in other wage employment. The household derives utility from the consumption of non-collected goods as well as from the direct consumption of goods it collects for this purpose from the Common Property Resource. The assumptions are given below.

- Let w be the returns to a given employment available for time T_o (where T_o is thus effectively an employed labour constraint).
- Assume that the rest of the time $(T-T_o)$ is spent on leisure, collections for consumption and collections for sale. With pm as the returns per unit time for collection, the model is set up as follows:

  The utility function is given by:

  $U = f(X, L, C)\ ;\ U_x, U_L, U_c > 0;\ U_{xx}, U_{LL}, U_{cc} < 0$ (1)

  and $C = F(a_c, T_c)$ (2)

  Where

  X: Consumption of non-collected goods with a price of unity

  L: leisure time

  C: Consumption of goods collected from the commons etc. $= T_{cc}/T_c$ where $T_{cc} + T_{cs} = T_c$

Distinguishing between time spent in collection for consumption (T_{cc}) and time spent on collection for sale (T_{cs}) implies: $T_c = T_{cc} + T_{cs}$ where, T_c is total time spent on collection activities. Let a_c be the proportion of time spent in collection for self-consumption out of the total time spent in collection activities (T_c). Collection is proportional to time spent; hence consumption C can be assumed to depend on a_c and T_c, the total time spent on collection. Therefore, a_c is treated as the decision variable with the time constraint being such that:

$T = T_c + T_o + L$ (3)

The full income budget constraint is defined in terms of a time constraint as:

$I + p^m(T-T_o) + wT_o = p^mL + X + a_cp^mT_c$ (4)

In this constraint I represents non-labour income (for instance remittances, interest income on assets).This constraint reflects the fact that given a labour employment constraint of T_o, the household allocates the rest of its time $(T - T_o)$ between leisure (L) and collections from commons (T_c), whether for sale or consumption. Thus, the returns to T_o are evaluated at w, while the returns to time spent on collections and leisure are evaluated at p^m.

The Lagrange function is set-up as follows to derive the optimality conditions.

$£=U(X, L, C) + ë\{I + p^m(T-T_o) + wT_o-p^mL-X- a_c\, p^m\, T_c\}$ (5)

The first order conditions on the decision variables imply:

$£x = U_x - ë = 0$ (a)

$£_L = U_L - ëp^m = 0$ (b)

$£a_c = U_c C a_c - ë p^m T_c = 0$ (c)

$£T_c = U_c C_{Tc} - a_c ep^m = 0$ (d)

$£ë = \{I + p^m(T - T_o) + wT_o - pmL - X - a_c p^m T_c$ (e)

from (a) and (b) : $Ux/U_L = 1/p^m$

$e = U_x = U_L/p^m$

from (c) and (d): $Uc\, C\, a_c - ë\, p^m\, T_c = U_c C_{Tc} - a_c\, ëp^m$ Since,

$Uc\, C\, a_c / ë\, p^m = T_c$, This leads to the following:

$a_c ep^m = U_c C_{Tc} - UcCa_c + ep^m UcCa_c/ep^m$ (6)

$a_c e_p m = U_{cCTc}$ (7)

The left hand side gives the gain in terms of the proportion of time spent in collection for consumption, evaluated at the market rate of return. This is equated to the (right hand side) gain in utility (in terms of X) from a unit increase in X through increased access to marketed commodities arising from the returns (sale value) out of collection time (T_c) that the household is able to sell.

This system would therefore solve for an optimal a_c where,

$a_c = a_c(p^m, w, T_0, I)$ (8)

It follows that $(1-a_c)$, the time spent on collection for sale, say call it

$U^* = U^*(p^m, w, T_0, I)$ (9)

The analytical model helps to locate household characteristics such as access to labour and product markets that determine the opportunity cost of their time and hence the time spent on collection for sale. Analytically, households are distinguished between on the basis of the choice set available to them, which in turn is determined by access to product and labour markets.

SECTION III

METHODOLOGY AND DATA

West Bengal, one of the states of India is the study area. The total recorded forest area in West Bengal is 1.19 million hectares, which constitutes 13.38% of the geographic area. The study was conducted in the district of Bankura in West Bengal in 2008. This district is one of the drought-prone districts of West Bengal. Agro-climatically, this district mainly consists of red and laterite soil zone, parts of it extending into coastal saline zone (parts of Midnapur). In this district, ten years (1995-2004) average maximum temperature is 44.3 degree Celsius and minimum temperature is 7.9 degree Celsius. The amount of actual rainfall is below the normal rainfall. This district is poor in social sectors (International Institute for population Sciences, India, 2006).

The field study was conducted in 2008 in two different villages say Kalaberia and Kalyanpur in the district of Bankura, in the Sonamukhi block, West Bengal. These villages are in the drought prone areas. 65 households have been selected randomly (35 households from Kalaberia and 30 from Kalyanpur). The field work combined interviews and discussions with the local people and interviews with local experts and school teachers and other knowledgeable elders in the village. A total of 65 structured household interviews were conducted, representing about 20% of households. Socio-economic data like age, caste, education, land holding, occupation and the time spent for the collection of non-timber forest products and the specification of NTFPs like leaves and medicinal herbs, food for livestock, fruits, fuelwood and honey. The socio-economic data of the sample households are shown in the Appendix A. The rainfall and temperature diagram in the drought prone district of Bankura is also shown in the Appendix B. The district is poor in social sector is in the Appendix C.

SECTION IV

ANALYSIS OF ECONOMETRIC MODEL AND RESULTS

Definition of Independent Variables

Land holdings (LH): Operational land holdings (in acre)

Household size (HHS): No. of people in work force

Caste: measured in dummy (if SC, ST = 1, otherwise 0)

Education (EDU): Education of respondents measured as number of school years.

Gender (GEN): Sex of respondent (Male = 1 and female = 2)

Distance between forest and household home (DIST): measured in km.

Dependent variable: Time spent for the collection of non-timber forest products (the average time spent per day by the member is 3.5 hours).

The regression equation is given by

$$T = \alpha + \beta 1\, LH + \beta 2\, HHS + \beta 3\, CASTE + \beta 4\, EDU + \beta 5\, GEN + \beta 6\, DIST + \Psi \qquad (1)$$

Results and Analysis

Let us analyze the results of the above study in the following way:

TABLE 1

Econometric Results

Independent variables	*Expected sign*	*Coefficient estimate*	*S.E.*	*t- ratios*	*P-values*
constant		1.63	0.45	3.65*	.003
Land holding	+	0.18	0.09	2.07*	0.039
Household size	+	0.21	0.14	1.56	0.120
Caste	-	-0.39	0.17	-2.27*	0.024
Education	-	-0.43	0.11	-3.87*	0.001
Gender	+	0.64	0.19	3.34*	0.0008
Distance between forest and household home	+	0.08	0.05	1.71	0.088

Significant at 5% level.

First, caste is negatively associated to the collection of non-timber forest products. This indicates that lower caste households extract less firewood from the commons. Caste of an individual influences cultural attitude towards food, bathing and rituals which might derive demand for fuel woods.

Second, the fuel wood collection shows an interesting gender pattern as female headed households extract less from the common forest than that of male headed households. One important policy observation from the analysis is that women and children are not only labour force engaged in fuel wood collection. This observation is also similar to that of Amacher *et al.* (1993), who observed that women are not the sole collectors of fuel wood.

Third, education shows a negative relationship with fuel wood collection from the commons as increasing education level makes fuel wood collection increasingly unprofitable. This means that education level of the family is negatively related to forest dependency.

Fourth, households with larger landholdings are more inclined to the CF for their increasing demand for fuel wood.

Fifth, the coefficient of distance to the forest is positive. Though it is not significant, it appears that distance could not explain the effort for fuel wood collection. This may be due to the fact households residing near to the forest could not harvest unauthorized forest products since there is strict rules and penalties for the rule violators.

Lastly, though not significant, household's size is positively associated with fuel wood collection. A family with a larger labor force can mobilize household's labor in collecting more forest extraction activities than household's with a smaller labor force.

SECTION V

CONCLUSION AND POLICY SUGGESTION

The above analysis has theoretical foundation in the household economy to explore how socio-economic characteristics influence the collection of non-timber forest products. Econometric analyses suggest that household labor allocation decision for non-timber forest products collection is dictated by various socio-economic variables. In this paper multivariate regression model has been applied to estimate the factors affecting the collection of non-timber forest products. Staying hours in the forest as dependent variable while age, gender, caste, education, household size, land holding, distance between home and forest have been taken independent variables. The results of the study revealed that these socio-economic factors are affecting the livelihood of the forest dependent communities. This paper has important policy implications for forest conservation, sustainable forest management practices, and rural development, environmental sustainability.

References

Arnold, J.E.M. (1995) : Socio-economic benefits and issues in non-wood forest product use. In *Report of the International Expert Consultation of Non-Wood Forest Products,* Food and Agriculture Organization of the United Nations. Rome. pp. 89-123.

Byron, N. and Arnold, M. (1999) : What futures for the people of the tropical forests? World Development, Vol. 27, No. 5.

Cavendish, W. (2000) : Empirical regularities in the poverty-environment relationship of rural households: Evidence from Zimbabwe, *World Development* 28: 1979-2003.

FAO and SIDA(1991): *Restoring the Balance : Women and forest resources,* Food and Agriculture Organization of the United Nations, Rome.

FAO. (1995) : *Report of the International Expert Consultation on Non- Wood Forest Products,* Food and Agriculture Organization of the United Nations. Rome.

Heltberg, R. (2001): " Determinants and Impacts of Local Institutions for Common Resource Management", *Environment and Development Economics* 6: pp. 183-208.

Hobley, M. (1996) : *Participatory Forestry: The process of Change in India and Nepal,* London: Overseas Development Institute.

Iqbal, Mohammad (1991) : *Non-timber forest products: a study of their income generation potential for rural women in North-West Frontier Province (Pakistan),* Planning and Development, Department and International Labour Organization. Peshawar.

Jodha, N.S. (1992) : *Common property resources : A missing dimension of development strategies,* World Bank Discussion Papers 169, Washington, D.C.

Kumar, S. (2002): "Does participation in common pool resource management help poor? A Social cost-benefit analysis of Joint Forest Management in Jharkhand, India", *World Development*, 30(5), *Land Economics,* 76: 213-32.

Lise, W. (2000) : "Factors Influencing People's Participation in forest management in India", *Ecological Economics* 34: pp. 379-92

Poffenberger, Mark (1990b) : *Joint management of forest lands: experiences from South Asia.* Ramamani, V.S. (1988) : *Tribal economy: Problems and prospects.* Chugh Publications. Allahabad, India.

Rajan, Rashmi Pachauri.(1995) : Sal leaf plate processing and marketing in West Bengal. In Jefferson Fox (ed.) *Society and non-timber forest products in Tropical Asia,* Occasional Paper No. 19, Environment Series, East-West Center, Honolulu, Hawaii,. pp. 27-36.

Rao, A. Ratna, and B.P. Singh. (1996) : Non-wood forest products contribution in tribal economy. *Indian Forester,* 122(4): 337-41.

Sarin, Madhu.(1995b) : Regenerating India's forests: reconciling gender equity with joint forest management, *IDA Bulletin,* 26(1): 83-91.

Sundar, N. (2000): "Unpacking the 'Joint' in Joint Forest Management", *Development and Change,* Vol. 31, pp. 255-79.

Svarrer, K. and Olsen, C.S. (2005) : The economic value of non-timber forest products: A case study from Malaysia. *Journal of Sustainable Forestry,* 20: 17-41.

Tewari, D.N. (1989) : *Dependence of tribals on forests,* Gujarat Vidyapith. Ahmedabad .

Vedeld, P., Angelsen, A., Sjaastad, E. and Berg, G.K. (2004) : Counting on the Environment. Forest Incomes and the Rural Poor, World Bank Environment Department, Environmental Economic Series, Paper #98. June, Washington D.C., *World Development,* 27: 789-805.

Appendix A

Socio-economic Indicators of the Two Forest Dependent Villages in Bankura District

Indicators	*Kalaberia Village (% of hhs)*	*Kalyanpur Village (% of hhs)*
Age		
Less than equal to 20	4	21
21-30	17	16
31-40	43	31
41-50	28	16
51-60	8	16
Total	100	100
Caste		
General	79	5
SC	19	11
ST	2	84
Total	100	100
Gender		
Male	62	79
Female	38	21
Total	100	100
Land holdings (in acre)		
Landless	17	37
0-1	45	47
1.01-2.5	28	11
2.51-5.00	9	5
More than 5	2	0
Total	100	100
Educational status		
Illiterate	19	47
Literate	21	21
Primary education	15	5

	Kalaberia Village (% of hhs)	*Kalyanpur Village (% of hhs)*
Secondary education	30	27
Higher secondary	11	0
Graduate	4	0
Total	100	100
Occupation		
Business	4	0
Housewife	26	5
Farmer	32	16
Service	8	5
Day labour	30	74
Total	100	100
Collection of different types of NTFPs		
Fuel wood	100	100
Food	28	84
Building Materials	32	100
Grazing for livestock	34	53
Medicine	11	32
Sal leaves	34	84
Sal seeds	11	53
Kendu	17	68
Mahua	30	100
Mushroom	57	100
Tasar	6	11
Others	4	0

Appendix B

Trends in rainfall and temperature in the District of Bankura, one of the Drought Prone District of West Bengal

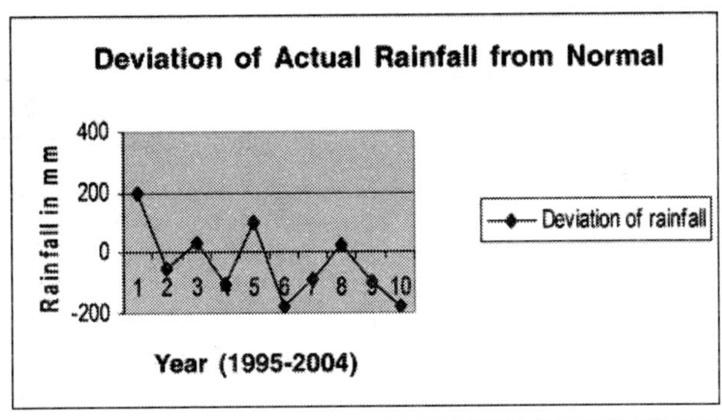

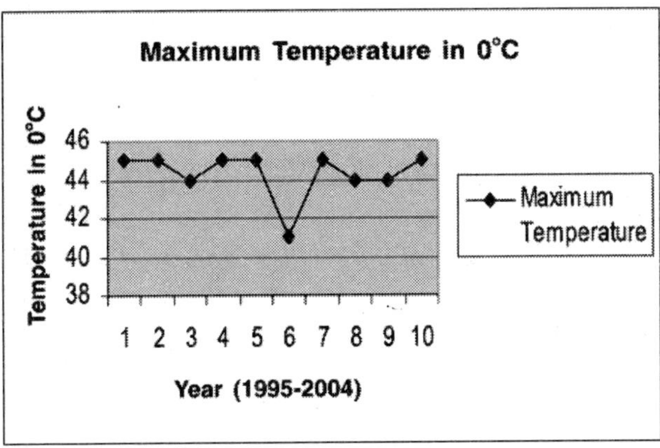

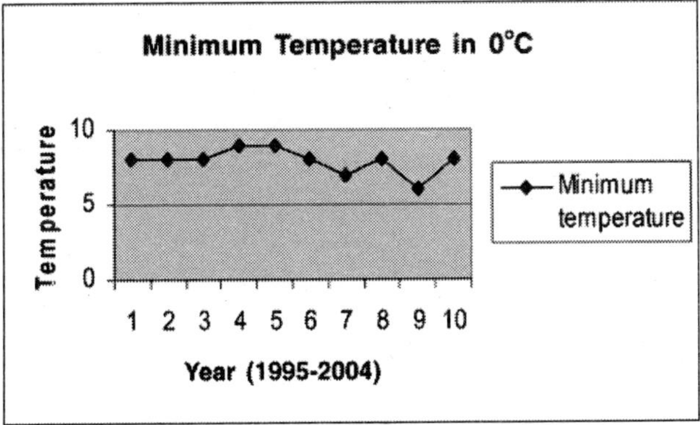

Appendix C

Socio-economic Indicators of the District of Bankura in all India Perspective

Indicators	%	*Rank in India*	*Index value*
% of households using safe drinking water	71.1	373	0.69878
% of households with toilet facility	11.9	538	0.07741
% of Electrified households	27.7	451	0.25508
Under-five mortality rate per 1000	79.0	122	0.83776
% of female literacy rate	49.8	341	0.40364

Source : International Institute for Population Sciences, India, 2006.

Chapter 17

Participatory Forest Management (PFM)

A Powerful Tool in Contemporary Forest Management in India

NABAGHANA OJHA

ABSTRACT

Joint Forest Management (JFM) popularly called as Participatory Forest Management (PFM) was introduced in India in the form of guidelines issued by the Ministry of Environment and Forests (MoEF), on June 1, 1990 in view of the demand for sustainable forest management. What led to the formulation of the guidelines was a series of success stories of participatory forest management in West Bengal (Arabari area), Haryana (Sukhomajri Project) and Gujarat (south Gujarat initiatives by AKRSP and others). The guidelines envisaged the involvement of NGOs, village communities and forest department in management, protection and regeneration of (degraded) forests. It is more than two decades that JFM has been operational in the country. Many studies have been made to look into the ups and downs of JFM. Series of issues were flagged at different points of time. In the meantime powerful legislations like PESA in 1996, Scheduled Tribes and Other Traditional Forest Dwellers (Recognition of Forest Rights) Act, (Forest Rights Act or just FRA hereafter) in 2008 came into force to provides the communities a political space in forest governance. In the other hand, Government of India has been implementing series of schemes and programmes to increase the green cover of the country through JFM committees. There are huge amount of money being pumped into the forestry sector keeping the climatic issues in mind.

The article is based on a study done in central Indians states to examine certain pertinent issues behind the functioning of JFM. Based on the findings and keeping the importance of PFM in mind some broad recommendations have been made that would help to reshape JFM in the country.

1. INTRODUCTION

Hundreds of millions of people in India rely on forest products; both timber and non-timber, for their own subsistence and also for raising a cash income for survival. It is estimated that out of approximately 260 million people living below poverty line, more than 100 million are partially or wholly dependent on forest resources for their livelihood. Nature's bounty, particularly the forests that ranked the natural resource on top priority met most of their needs. Local management systems were prevalent in those areas where indigenous people exhibited larger needs on forests. In course of time, this precious natural resource came under the direct control of the state. As a result, there have been frequent conflicts among forest-dwellers and the state. Over the past few years, planners and forest administrators began developing new strategies to reduce the conflict between state agencies and rural groups. By the late 80s, many programmes of government had begun to recognize the fact that the livelihood of the tribal people are fully or partially dependent on forests and attempts were made to facilitate the emergence of collaborative forest management. The National Forest Policy (NFP), 1988, with its emphasis on close association of tribals with protection, preservation and development of forests and recognition of their customary rights on forests, became a major instrument for the change in the forest management model. It opened the path to involve the communities in the forest management process. The National Forest Policy, 1988 and the 1st June 1990 Joint Forest Management (JFM) resolution began reshaping the policy environment, acknowledging the need to give greater rights and authority to communities. This arrangement sought the involvement of village communities and voluntary agencies in the regeneration of degraded forestland. The programme sought to establish management 'partnerships' between local forest-dependent communities and the state for the sustainable management and benefit sharing of forests. Under JFM, the forest department (mostly a local forest official) and the village community enter into an agreement to constitute a committee to jointly protect and manage the forest. The key understanding in the arrangement is that in return for protecting forestland and helping in the regeneration and management of the said forest, people will be entitled to usufruct rights over fuel wood, timber, NTFPs as well as a significant share (that varies from state to state) of poles/timbers (in cash or in kind) on final harvesting. But the community and the forest department are expected to share not just the benefits, but the responsibilities too. In other words, JFM strategies bring local forest users and the forest department on a common platform for the management of forest.

The institutional framework and the procedure of operation of JFM vary from state to state as each state has issued its own JFM resolution. The JFM programme, over the past decade, has spread all over the country. The area under JFM as of September 2003 is 17,331,955.12 ha. and the number of groups involved 84,632.

More than 20 years have already been passed since the 1st guidelines of Ministry of Environment and Forests were issued. Though the implementation of the programme leaves a lot to be desired, it has certainly had some positive results. There is credible evidence that forest cover and density in JFM areas has increased to some extent. But a review of literatures on participatory forest management presents a very grim situation.

In a study on factors influencing people's participation in forest management in Bihar, Hariyana and Uttar Pradesh, (Lise 2000) found that the first consideration for people's participation in forest management is social indicators, with economic indicators coming next. The study also notes that high dependence on forests and good forest quality enhances people's participation. It argues that when people's dependence on forests is high, their interest in forests is likely to be greater, inducing people to participate in forest management and protection activities.

Another study in Orissa (Sarap, 2004) states many of the local VSS groups are extremely weak institutions and are beset with a number of problems. The study estimated that as many as half of the total VSS formed are dormant and inactive. Problems have arisen from a number of factors. In order to fulfil the formation targets of VSS, many indigenous CFM groups have been persuaded by the forest department to register themselves as VSS groups. So, in fact, it is simply a change of label in many cases and many of the converted VSS groups do not like the management arrangements followed.

In the central Indian states, there are thousands of self-initiated protection groups that are protecting several hundred thousands of hectares of state-owned forests. They are primarily in the states of Orissa, undivided Bihar and Madhya Pradesh. With the expansion of JFM, there is evidence of conflicts in the interface between state forest departments and many self-instantiated forest protection groups (Sarin,

1997). The latter very often refuse to participate in the JFM programmes as final felling of mature forest (a central agreement of the JFM programme) and sharing with the forest department is not acceptable to them.

There is a provision in the JFM resolution that in order to involve the local community in the decision-making process, a detailed micro plan needs to be developed in a participatory approach. The plan is expected to be prepared keeping the needs of all sections of the community in mind. Unfortunately, however, the experience of developing detailed micro plans has been pretty discouraging. It is seen that the plans are done in a very superficial way. Under the JFM programme, it is not the community but the FD, which plays the lead role in deciding management activities and formulating a plan to achieve them. It is commonly felt that micro plans tend to reflect FD agendas rather than community needs. They are drafted in a traditional silvicultural format (Viksat, 2001).

From various scholarly studies and official reports it is apparently clear that devolution of management responsibilities was a much felt one and need of the hour. This 'devolution of management responsibility' has remained more as an abstract concept for which it has not gained the tangible results as has been desired, specifically in the context of JFM. Impressions emerge that the idea of devolution of management responsibilities of forests to the forest dwelling communities is yet to take shape in the JFM programme. Reasons are many. The roles and responsibilities of different stakeholders (community, civil society organisations and forest department) in JFM have not been clearly established. This paper is an effort to assess the extent of participation of poor forest dependent communities in joint forest management. It also tries to examine the nature of involvement of the forest department with the community and process of implementation of the programme. At another level, the paper also seeks to identify the difficulties in the participatory process of forest management that affect the forest dependent population, especially in the central Indian states.

The field study, on which this paper is based, was undertaken in five states, i.e. Orissa, Chhattisgarh, Bihar, Jharkhand and West Bengal during 2004-05. These five are the most forested states of India with greater concentration of tribal population.

In most of the areas of these states the communities have been protecting and managing the forest for quite some time with their self, initiated rules and regulations. These communities had initiated forest protection in response to growing forest product scarcities and threats of

TABLE 1

Forest Fringe Villages and their Population

Sl. No.	*State*	*Total number of villages*	*Villages having forest*		
			Number	*Forest Area (ha.)*	*Population*
(1)	*(2)*	*(3)*	*(4)*	*(5)*	*(6)*
1.	Bihar including Jharkhand	67,513	17,044	2,502,137	11,205,120
2.	Madhya Pradesh including Chhattisgarh	71,526	29,294	6,715,840	19,953,453
3.	Orissa	46,989	29,302	1,779,953	15,934,768
4.	West Bengal	37,910	8,571	614,682	8,399,279
	Total	253,026	96,414	16,730,159	76,811,573

Source : Forest Survey of India.

TABLE 2

Forest Coverage (area in sq. km)

State of forest	*Orissa*	*Chhattisgarh*	*Jharkhand*	*Bihar*	*West Bengal*
Geographic Area	155,707	135,191	79,714	94,163	88,752
Very Dense Forest	288	1,540	2,544	76	2,303
Moderately Dense Forest	27,882	37,440	9,137	2,951	3,742
Open Forest	20,196	17,018	11,035	2,531	6,298
Total	48,366	55,998	22,716	5,558	12,343
Percent	31.06%	41.42%	28.50%	5.90%	13.91%

Source : State of Forest Report, FSI, Dehradun.

exploitation by outside groups. After the formation of JFM many self-initiated forest protection committees were converted to JFM committees. After 2000 more number of JFM committees were constituted in the state followed by the central resolution of 2000. The progress of JFM in the states under study is mentioned in Table 3.

TABLE 3

Progress of JFM in the Study States

State	*Number of JFM Groups*	*Area under JFM (ha.)*
Bihar	493	267,240.94
Chhattisgarh	6,881	2,846,762.16
Jharkhand	3,358	847,967.93
Orissa	15,985	821,504.00
West Bengal	3,892	604,334.00
Total	30,609	5,387,809.03

Source : Bahuguna *et al.*, 2004.

2. JFM POLICY IN THE STATES

Ministry of Environment and Forests issued a set of guidelines on 1st June, 1990 concerning the involvement of village communities and voluntary agencies in the protection, development and management of forests through a programme called joint forest management (JFM). Over a period of three years, the operation of the guidelines through individual state resolutions has thrown up several conflicts and issues. At the time of the issuance of the guidelines, the West Bengal and Orissa governments were already carrying out JFM activities with regulations developed on their own. In February 2000, MoEF issued revised guidelines specifying women's representation in JFM groups (minimum 33% in executive committees and 50% in the general body). It also specified that JFM micro-plans must conform to the silvicultural prescriptions in the working plans. While suggesting that all village organisations participating in JFM be registered as societies to provide them an independent legal identity, the new guidelines also recommended that these societies be called 'JFM committees' across the entire country (MoEF, 2000), irrespective of their diverse histories, legal status, institutional structures, the diversity of ecosystems within which they fall and the livelihood systems and cultural diversity they harbour. In 2002, another set of guidelines was issued by the Ministry of Forests and Environment.

> **Orissa Resolution at a Glance**
>
> **August 1988**—State govt. came out with a resolution to involve the local community in forest protection and conservation.
>
> **1990**—Another resolution provided for inclusion of protected forests in the ambit of community protection and conservation.
>
> **1993**—Comprehensive resolution passed by state govt. that mentioned the term JFM for the first time.
>
> **1994**—1993 resolution modified for enhancing the allotted area to the committees
>
> **1996**—Another resolution passed and the very important step to notify the JFM areas as village forest taken.
>
> At present, the JFM programme is implemented through 1993 resolution.
>
> **2008**—Another resolution passed and the purview of JFM expanded to protected areas with the involvement of EDC. This resolution is now in place.

Orissa has pioneered the concept of Joint Forest Management (JFM) in the country. Even before the adoption of National Forest Policy, 1988, Govt. of Orissa came out with a resolution in August 1988 soliciting participation of local communities in regeneration and management of Reserve Forests. The programme was extended to Protected Forests in 1990. In July 1993, Govt. of Orissa came out with a comprehensive resolution on JFM providing for an institutional framework at all levels. It also prescribed a clear-cut benefit sharing mechanism for the first time. In August 1994, the July 1993 resolution was modified to enhance the area to be allotted to each VSS. In 1996, the state government came out with another resolution, but the old framework is still operational in the state.

In 1990, the Bihar government passed a resolution authorizing the organization of village forest management and protection societies. Since 1993, the forest department has taken a number of active measures to implement the government order. The guidelines for community forest protection are very specific regarding organizational structure and function, with limited flexibility to incorporate the diverse range of local organizations—many of them formed informally—to streamline forest use. In 2000, the state government came out with another resolution providing for two types of JFM committees in Bihar. Eco-Development Committees are to be constituted in villages within a radius of 5 km from the forest nearer to sanctuaries and national parks. Similarly, Village Forest Management and Protection Committee or Forest Committees are to be constituted in villages within a radius of 5 km from reserve forest.

Bihar and Jharkhand Resolution at a Glance

1990—Bihar govt. came out with a resolution to authorize village forest management and protection societies.

1993—Another resolution passed to take more active steps on JFM process

2000—A fresh resolution categorized two types of committees for reserve forest and protected forest

2001—Jharkhand govt. came out a new resolution providing for two types of committees under JFM.

After it became a separate state, Jharkhand formulated a new resolution on JFM in 2001. Even a cursory look makes it clear that the Jharkhand resolution on JFM is based on the premise that people are to blame for the destruction of the forests. The objective, therefore, is to gradually diminish the people's dependence on forests. The Resolution to introduce JFM was clearly made unilaterally by the Forest Department. There was no scope in it for consultation with the people, no effort at consensus building and no mechanism for dissemination of information. The resources required for the villagers' needs are to be extracted from the protected forests according to the micro/annual-working plan. It seems that the Jharkhand resolution is simply an extension of the Bihar JFM resolution.

West Bengal, the state that has played a pioneering role in seeking and obtaining people's participation in conservation and development of forests, has earned many laurels ranging from Indira Priyadarshini Briksha Mitra Award for individuals and clusters of FPCs to the international Paul Getty award for the forest protection committees of

WB Resolution at a Glance

1989—State govt. came out with a resolution to involve the local community in forest protection and conservation.

1996—This resolution involved EDCs to protect and mange protected areas.

the state for their contribution to resuscitation of large tracts of degraded forests in the south-western parts of the state. The forest protection movement in West Bengal started in the early 70's at Arabari in Meidapur district in WB, where 618 families of 11 villages were motivated to rejuvenate 1,186 hectares of degraded *sal* forests by roping in their participation through employment generation activities and sharing of NTFP from such forests. This was followed by the Government's decision in 1987 to give a 25% share of usufructs of the intermediate and final yield to the forest fringe communities. The decision was the much-needed shot in the arm that led to the formation of FPCs in this tract. Soon, the state Government passed another resolution on adoption of JFM for managing forest resources in different agro-climatic zones of the state. The first West Bengal state resolution providing official legitimacy to community management groups, however, was not passed until June 1989; about fifteen years after experiments in joint forest management had started. Through the resolution issued in June 1996, EDCs were constituted seeking cooperation of the fringe people through their participation in protection and development of wildlife protected areas (sanctuary and protected areas). All government notifications on the JFM resolutions, including the composition of FPCs and EDCs, the duties and functions of members of FPCs and EDCs, usufructury benefits etc., were published in the local language and circulated amongst the targeted communities.

Madhya Pradesh too has played a pioneering role in implementing the JFM programme. The Government of M.P. issued the first resolution in this regard in 1991. According to the resolution, JFM in MP was to be implemented through two different committees. Communities residing in the vicinity of forests sensitive to damage (close forest areas where the canopy density is greater than 0.4) were to be organized into Forest Protection Committees (FPCs). In contrast, Village Forest Committees (VFCs) were to be constituted in degraded forest areas with open forests with a canopy density of less than 0.4.

The order was amended on 4th January 1995. The new order clarified the differences between the two types of committees and provided a revised set of benefit-sharing arrangements to be implemented in the state. A revised resolution passed in 2000 made provision for three kinds of committees, i.e. Forest Protection Committees (FPC) for protection of well-stocked forests, Village Forest Committees (VFCs) for regenerating degraded forest areas and Eco-development Committees (EDCs) in and around Protected Areas (PAs) with a view to ensuring bio-diversity conservation in National Parks and Sanctuaries. The Committees were to be constituted within a radius of 5 km from the periphery of forests.

MP and Chhattisgarh Resolution at a Glance

1991—MP becomes the first state to come out with a resolution to implement JFM programme.

1995—Fresh resolution passed to revise the benefit sharing mechanism in the state.

2000—An amended resolution provides for 3 types of committees—for degraded, high and protected forests of the state.

2001—Chhattisgarh comes out with new resolution to implement JFM.

2002—Two types of committees, VFCs and FPCS, are functioning in the state.

After the formation of Chhattisgarh, the state passed a new resolution in 2001, which was amended in 2002. According to the resolution, every family that is a member of the JFM committees will be entitled to receive Nistar forest produces subject to their availability. In the new resolution, Eco-Development Committees (EDCs) were merged with Forest Protection Committees (FPCs). All forest committees, as per the resolution, were eligible to get 100% of the forest produces obtained from time to time from mechanical thinning and cleaning of rehabilitated areas and cleaning of bamboo clumps in degraded forests as per the prescriptions of micro-plan/working plan, on payment of expenditure incurred on harvesting. Forest products equivalent to 15% of the value of the timber/bamboo harvested or its cash equivalent would be given to the concerned FPC after deducting the expenditure incurred on harvesting from the total value of forest products. Forest produces equivalent to 30% of the value of the total amount of timber/bamboo obtained after final felling in areas of plantation/rehabilitation on degraded forests was earmarked for the Village Forest Committee (VFC) after deducting the expenditure incurred on harvesting.

3. FOREST PROTECTION AND MANAGEMENT

3.1. Evolution of Forest Protection

There are evidences that thousands of villages in India have been involved in protection and management of forests decades before. In Orissa alone there are about 10000 villages protecting and managing forests for their livelihood security and environmental protection. The village communities, organised as primary groups like youth clubs and village forest protection committees, started protecting nearby forest areas for different reasons with local school teachers, foresters and influential leaders emerging as key motivators in the process. In the tribal districts of south Bihar, communities have also become increasingly engaged in forest protection since the early 1970s. Many villages had initiated forest protection groups in the 1970s in response to growing forest product scarcities and threats of exploitation by outside groups. In West Bengal and MP it is also the informal groups that have been successfully involved in forest protection and management long before the implementation of JFM.

3.2. Forest Protection and Management System

In the states under the study, three major processes have been followed by the committees for protection of forest. The first one is through a watchman, the second through a group of committee members working on a rotational basis and the third one through social fencing. Where a watchman has been appointed for protection of forest, he/she is being paid from the committee fund. There is little provision for payment by the forest department. In most of the study committees members of the executive committee take most of the routine decisions on forest protection. The committee is responsible for organising regular patrols of the area by members on a rotational basis. Village problems and forest protection matters are discussed in meetings of committee. In each month the committees meet at least once. The minutes of the meeting are maintained by the secretary in most cases. In return for protection, the villagers have rights to collect fallen wood, fodder grasses and other non-timber forest products.

Felling of trees in the regenerated area is strictly prohibited. The committee has conventional powers to detain anyone caught breaking rules and may impose fines also. Grazing is also prohibited in plantation and regenerated area. In case of illegal tree cutting, the fine is imposed based on the decisions of the executive body. In some cases the forester is consulted. The money collected as fines from illegal grazing and tree logging is deposited in the account of committee. This shows that the villagers have high stakes in protecting the forests. The detailed rules and regulations developed by the committees are mentioned in Table 4.

TABLE 4

Forest Management Rules

States	*Rules and regulations*
Orissa	▪ The committee has to decide whether the protection arrangement should be through watchman or through committee members. ▪ Payment to the watchman is made from the community fund and whenever necessary, people contribute membership fees. ▪ Initially, grazing is not allowed. ▪ Cleaning operations are undertaken whenever there is a need for it. The people, however, have to quote their requirements in advance. ▪ Meetings of the forest protection committee are to be conducted once a month. ▪ The decision to punish an offender lies with the key leaders of the forest protection committee. ▪ Any household, which collects fuel wood from the forest, has to take the permission of the president and also has to deposit a token amount in certain cases. ▪ The VSS (under exceptional circumstances) may allow members to access the forest for timber.
West Bengal	▪ Protection of forest through a group of committee members (3-4 members) on a weekly basis. In North Bengal, protection is through a watchman paid from the community fund. ▪ Prior permission of committee is required for collecting fuel wood, timber, poles, etc. from the forest for personal use. ▪ Cleaning and thinning activities are to be done through forest department. ▪ For construction of houses, VSSs can give a maximum of 10 poles of Sal to each family. For the rest, they need to pay or purchase from other forests. ▪ Every member of the committee can collect NTFP available in the forest. ▪ The forestland can't be used as grazing purpose.
Chhattisgarh	▪ The working committee members jointly decide the way of forest protection. ▪ In some places where watchman is recruited, payment is made from the fund made available by the forest department. ▪ Fines could be imposed in consultation with the local forest official. ▪ Meetings have to be organized every month with provision for holding emergency meeting at the time of urgency.

	▪ Conflict management is through a mutual basis in most cases.
Jharkhand	▪ The protection of forest is done by the members of the committee on a rotational basis. ▪ The committee can charge fines up to Rs. 500 on anybody involved in illicit felling of trees in the forest, depending upon the nature of the offence. ▪ Prior permission of committee is required for collecting fuel wood, timber, poles etc from the forest for peersonal use. ▪ For construction of houses, VSSs can give a maximum of 10 poles of Sal to each family. ▪ Every household can collect NTFP available in the forest as per its requirement. ▪ Forestland can't be used for grazing. ▪ Species like Char, Mahua, Amla, Karanj and Neem will be protected and not disturbed till they bear fruit. ▪ The benefits will be distributed equally among all families. ▪ Each member will participate in the protection of forest as and when required by the VSS.
Bihar	▪ In most of the committees, a watchman is recruited for protecting the forest. He is paid from the fund collected by the committee members. ▪ Nobody is allowed collection of living trees. ▪ Fines could be imposed on anybody involved in illicit felling of trees in the forest, depending upon the nature of offence. ▪ The members of the committee have to take prior permission for collecting fuel wood, timber, poles, etc. from the forest for their own use. ▪ Grazing is provided in the plantation area.

3.3. Benefit Sharing

In many parts of the states under the study, communities historically possessed a range of customary rights over forest resources, which were recognized under earlier forest acts. In the past decades, national and state governments have curtailed these rights. Forest-user communities across the country, dependent on forests for fuelwood, fodder, small timber and NTFPs, have accessed forest products under different rights regimes. During the pre-JFM period, communities in some states accessed forest products under rights and concessions provided under settlements. In others, communities illicitly extracted forest produce, with or without the knowledge of the FD field staff. However, under the JFM programme, residents of forest-fringe villages have been provided access to forest produce to meet their basic needs of fodder, fuel wood and NTFP. In lieu of this, people are protecting and managing the forests with the FD. Regular voluntary patrolling by villagers is a common practice while the number of members in the patrol party is seen to vary between 2 to 10 in the five states included in the study. Both sides have gained in the bargain. While the FD has benefited by way of a considerable reduction in its workload in forest protection, the participating communities are now entitled to their legitimate share of fuel wood in the form of dry and fallen twigs and leaves from the forests. Fuel wood generated from various silvicultural operations also supplement fuel wood supply in the states. In the pre-JFM period, people either had no grazing rights or had limited grazing rights. However, instances of uncontrolled and excessive grazing in forests were common, accelerating forest destruction. With JFM, there is now a ban on uncontrolled grazing and only rotational grazing is allowed. These practices have helped the regeneration and survival of vegetation in forests and in increasing the supply of fodder grasses. All NTFPs, barring a few nationalized products, are now available to the people free of royalty in all these states. People have a right to collect even nationalized products like Kendu leaves, Sal seeds, etc. but

Box 1

Mr. Batakrishna Padhi, the then range officer of Badampahad range in Mayurbhanj district, played an important role in motivating the people to constitute VSSs in the village and assured all possible help, legal or otherwise, to them. In 1997-98, two persons from the neighbouring village, Upperbeda entered into the forest protected by Dublabeda VSS and illegally felled some trees. On their way back, they were caught by the VSS members, who informed the range officer about the incident immediately. Unfortunately for them, the officer was out on mobile duty and could not reach in time. In the meanwhile, around 500 villagers of Upperbeda rushed to the spot and got the culprits released. As Dublabeda is a small village in comparison to Upperbeda, they were scared of being overwhelmed by the 500-strong mob. As soon as he got information about the incident, the RO rushed to the spot and immediately captured 10 persons who were taking the lead and locked them in a room. He then issued an ultimatum to the mob that if they did not come forward for a compromise, the captured persons would be handed over to the police. He also threatened to use his revolver if they did not relent. The threat did the trick. Some influential persons of Upperbeda village tried to calm down the mob and came forward for a compromise. The villagers cite several such instances where the concerned RO reached the spot and took control of the situation immediately whenever needed.

have to deposit their collection with the agency responsible for procurement and earn a prescribed wage. Some states have introduced their own incentives. In South-west Bengal, for instance, people get 25% of the net profit from cashew. VFCs are also entitled to a share in the timber harvest in varying proportions.

3.4. Conflict Resolution

At several places, JFM has resulted in increased inter and intra-community conflicts. In the study sites, it was found that intra-community conflicts mainly originate from inequitable distribution of costs and benefits among different sub-groups (class, caste, gender, etc.) within the community. In Harekrishnapur VSS of Dhenkanal forest division, Orissa, there are two major hamlets and the executive committee has representation from both. Perhaps due to personality clashes between executive body members from both the hamlets and the apparent inefficiency of one executive body member from the smaller hamlet tensions have cropped up in the village. But a group of households from the smaller hamlets has made an appeal with the forest department to be allowed to from another VSS. Till the time of writing the case study the issue was yet to be resolved. Conflicts over resources are rarely pure resource conflicts. They are indicative of the broader power struggles.

Inter-community conflicts arise due to the involvement of outsiders, basically the adjacent villages in cutting of trees. When noticed by the committee protecting the forest, an attempt is usually made to resolve the case through very minimal fines. In most cases, there is utter confusion, requiring the intervention of the FD. Once these committees were brought under the fold of JFM, they became highly dependent on the forest department for resolution of any conflict.

There was a specific commitment from the forest department during the implementation of JFM that it would lend a helping hand to the communities in resolving their conflicts. But in practice, the number of conflicts has continued to mount. The forest department has more cases pending with it than it has resolved. The main functionaries of the committees are being harassed several times at different levels. Usually, a smuggler caught by the community while felling trees or otherwise, is later handed over to forest department officials if there is any violence. But the smugglers invariably manage to get released by paying a fine or using their political connections. The community effort thus goes in vain. The people of Dublabeda village under Rairangpur forest division are convinced that all JFM committees would soon become defunct. And they have sound reasons to back up their theory too. The JFM policy, they point out, does not empower the committee to take any penal measures against the mafia. There is also no clear statement in the resolution on how the forest department will penalize them either. There have been instances where the DFO and ACF have not backed up action taken by lower level forest officials. No wonder then that forest officials at the ground level avoid taking risks and booking criminals. The villagers were of the opinion that a pro-active forester can inspire the committee to protect forests with vigour.

4. ACHIEVEMENTS AND ISSUES

4.1. Major Achievements after JFM

Despite several limitations, JFM has managed to create a great deal of awareness on participatory forest management among people belonging to different walks of life such as people living in villages located close to forests, NGOs, CBOs, researchers, etc. Field level forest officials, who initially opposed JFM on the ground that the people don't have the ability to protect forest, have now embraced the programme. In many places, especially in the forested area, people now are more conscious about the need for forest protection. Irrespective of the support of the forest department, villages/communities have come forward to protect and manage forests on their own. The attitude of the people, especially forest-dwellers, towards forest officials has also changed significantly in many places. The same foresters, who once apprehended people for felling trees, now come to the people with a completely different mindset and talk about participatory forest management.

- A sea change in the relationship between local communities and forest department field officials as well as an attitudinal change towards forests have been reported from different study areas.
- In several areas, introduction of JFM has resulted in reduction in area under encroachment and fall in the rate of fresh encroachment.
- Implementation of JFM programme has resulted in increase in the income of participating communities at several places. Several externally assisted projects have laid emphasis on employment generation and creation of productive community assets as part of the entry point of development activities.
- Some states like Chhattisgarh and West Bengal have set-up exemplary JFM committees that have been involved in forest protection and management. Most of the new committees

formed recently visit those places to learn forest management activities. Ranisagar in Chhattisgarh is one such committee. In some committees, the forest area initially protected under JFM has increased. The interest of the people in protection of forest has encouraged forest department to allot more area to these committees. As a result, their income has been increased to some extent and the pressure on forest has also reduced considerably.

- In some places, the forest protection initiatives of the committees have inspired the adjoining villages to protect forests. Later on, they have requested the forest department to form JFM in their village and allot the forest area officially. An example of this is the JFM committee in Korra village, 2 km away from Dokal in the Dhamtari forest division of Chhattisgarh. After formation of FPC in Dokal, the villagers of Korra started protecting the nearby forest area. Soon, other surrounding villages too started protecting forests since they knew that the villagers of Dokal would never allow them to enter the forest protected by them. The promise of usufruct benefits too played a part in persuading them to take up protection. They consulted the line department to form committees in their villages. The forest department finally constituted JFM committee in the village in 2000. This is the only case found in Keregaon area where neighbouring villages have been positively influenced after the formation of committee in one village. The committee members of Dokal village have never faced any major case of theft or felling of trees in their forest after the formation of committee.
- The forest has got a new lease of life after the protection initiatives of the JFM committees. Due to protection of the forest, the ecosystem of the area is well maintained and basic household requirements are easily met. Forest vegetation, which had all but vanished, has now taken a new shape. A number of species have been regenerated.
- A significant number of mandays have been generated in forest development activities. With the help of external assistance schemes, more development activities in the forestry sector have been undertaken with the involvement of local communities. With FDA fund, the committees have got good assets like cooking utensils, tent materials, etc. In some committees like Dokal in Chhattisgarh, the fund

generated through forestry activities has been used by the committee for procurement and trade of NTFPs and agricultural produces also.

- In Chhattisgarh, the fund generated by the committees is a great asset for them. At the time of urgency, money is borrowed by the villagers from this fund with very minimal interest. With limited resources, people earlier had no option but to depend on the forest for fuel wood collection, fodder, poles and timbers. But now, their involvement in protection has minimized the pressure on forest.
- In Bihar, the committee members predict that after some years of protection of forests, the production of NTFPs—especially Mahua—would increase manifold and will play a vital role in the economy of the concerned villages. They consider Mohua as the priority species for conservation.

4.2. Issues in Participatory Forest Management

4.2.1. General Issues

Some issues in JFM are state-specific, while others are universal in nature. The most important issue in JFM today is the absence of any legal standing of JFM. JFM is neither incorporated in the mainframe of forest law nor has a separate law been enacted to give a legal standing to it. The program is being implemented under a Government resolution that remains subject to change.

Many VSSs are unaware of their status, as they don't have the copies of the Memorandum of Understanding and the registration certificates. The records of the VSSs are with the Forester as he is the member secretary of VSS. According to the resolution, if a VSS performs its duty properly for 5 years, then it is eligible for getting the usufruct benefits/rights. The local forest officials are often not in a position to give a clear picture on the registration of the VSSs. Their usual refrain is that they have forwarded the applications for registration and MoU to higher ups in the department and are waiting for the response. Even where VSSs have been formed, meetings are seldom organised regularly by the forest officials. This being the situation in respect of a mandatory requirement, the less said about training programmes for VSS leaders or members, the better.

Micro-plans have not been prepared in the vast majority of the VSSs. Wherever it is prepared, it has not been implemented. Currently, micro-plans are being prepared especially for the VSSs, which are members of Forest Development Agencies, by Foresters without adequate support from higher officials. No thorough training programmes have been organised for these foresters on how to prepare micro-plans with community participation. These are not really comprehensive forest protection, management and conservation plans. Most of these plans have been prepared to meet the requirements of plantation, soil and moisture conservation work in the forest. It is basically a description of activities to be taken up through VSS. VSSs are apparently seen as beneficiaries of employment generation programmes and also as contributors of voluntary labour for forest protection and management. The technical and management responsibility still lies with the forest officials. No efforts have been made so far to transfer the technology to the VSSs so that they can take up operations like silviculture, disease control, etc. on their own.

There are also a couple of issues relating to equity and the democratic functioning of VSS. The executive body is hardly elected through a participatory process. The executive body has a tenure of two years. But the same body continues for years together without any election in a majority of the VSSs. The real decision-making power is with the Forester, with or without a handful of village elites. The participation of women and weaker sections in the decision-making process has been extremely limited in many VSSs in the study area.

4.2.2. Specific Issues

Sustainability of JFM Institutions

The sustainability of JFM institutions is a major challenge in JFM. It is observed that in most of the cases, forest departments tried to form JFM institutions, as it was their mandate rather than out of any conviction in its efficacy. Various commitments were made at the time to lure the villagers into JFM. But many of these promises have remained unfulfilled. In some cases in Orissa, the committees were not even aware about the policy of JFM. They were not sure whether it was a resolution or an act. In a cluster level meeting held in Paschimeswar, Dhenkanal forest division, Orissa during December 2005 the members of different committees asked the forest official present about the legal standing of JFM, especially whether the programme is run through a resolution or an act, policy, etc. Absence of legal standing is a serious setback on the sustainability of the programme. The pumping in of funds to JFM committees through external assistance schemes like FDA has seriously affected the cohesion of the committees. These developmental activities, which in any case will not last long, have created fissures where none existed.

The selection of committees, which would avail of these development funds, is highly speculative and often leaves a lot of scope for heartburn in those who lose out. Even assuming that the funds are utilized for the purpose for which they are meant, inflow of easy outside money creates a negative impact on the sustainability of the institutions. The case of a committee under Dhenkanal forest division would illustrate how. Discussion with the office-bearers of the committee revealed that they are highly dependent on forest officials. All records, files, passbooks, etc. are being kept and maintained by the beat officer. The decisions are taken more or less by the forest officials. And this particular committee is considered one of the active committees in the whole of Dhenkanal forest division! The members were convinced that the committee would collapse the moment the funds stop pouring in from the forest department. So much for sustainability! The focus of the forest department has been all wrong. It is yet to learn that pumping in money from outside has never guaranteed the sustainability of institutions. Instead of working towards institutional capacity building, it has chosen to invest money that has created more dissension than cohesion.

Non-cooperation or poor participation of forest officials in the process, especially where no

Box 2

Ranisagar village in Chhatisgarh is situated 38km away from the district headquarters, Raigarh and 6 km from the Kharsia Range. One of the early benefits of JFM in the village was wage employment. The Forest Department facilitates villagers' access to a wide range of developmental schemes such as construction of all weather roads, wells and check dams. The village has benefited substantially from village infrastructure development. The Range officer of the Kharsia Range admits that the forest has been protected only because the department provides them wages and secures their livelihoods. "If we won't support them financially, they will never protect the forest", he says.

Neighbouring non-project villages receive nothing except the money they collect through fines or at best a monthly salary for one watchperson. This has led to a lot of heatburn and suspicion. Where pre-existing voluntary protection was replaced with payment, the department's money reduced psychological controls. As long as the villagers contribute themselves, the psychological restraints were much stronger. *The committee may well become non-functional the day the government stops giving it money.*

development funds are available, is a major stumbling block in its sustainability. It was found from the field study that the forest officers don't attend the meetings of the committee even after several requests, both verbal and written. In a cluster level meeting, 11 VSSs in Dhenkanal forest division decided that they would meet the DFO and ask him the reason for the continuous absence of the forest officer who is a part of the process. They also decided that the concerned forest officer would not be allowed to attend the meetings of the committee, if he continued to remain absent. This goes to show the seriousness with which the department takes the sustainability of JFM institutions.

Management under JFM Programme

Under the JFM programme, it is not the communities, which play the lead role in deciding management objectives and formulating a plan to achieve them. Micro plans tend to reflect FD agendas rather than community needs. The agendas of the forest department like working plans are drafted as traditional sylvicultrural formats. There is a big gap between the micro plan and the working plan. These plans are drawn up in a very cursory, superficial way. The committee members have no idea of the provisions in the micro plan, its year of formulation, etc. Even the forester, who is the secretary of the committee, is frequently not aware about the details of the micro plan though he is deeply involved in developing the micro plan. Another fact that emerged during the study is the non- involvement of communities in forest management activities. For forest development activities like cleaning, thinning, etc., they are totally dependent on the forest department. The capacities of the committee members are not being built-up for taking up such management activities at the committee level.

Conflict between JFM and CFM

Besides the JFM programme promoted by the FD, another movement to conserve forests has gained ground at the grassroots level. There are several villages which have been protecting forest patches adjoining their territories, on their own, without any outside assistance. They are commonly referred to as Self-Initiated Forest Protection Groups or Community Forest Management (CFM) groups.

Over the last 50 years, thousands of such informal groups have emerged locally, primarily as a response to scarcity of forest resources. With the expansion of JFM, there are increasing evidence of conflicts in the interface between state forest departments and self-instantiated forest protection groups. The latter very often refuse to participate in the JFM programmes as final felling of mature forest (a central agreement of the JFM programme) and sharing with the forest department is not acceptable to them. Even when the villagers sign the JFM agreement, their traditional rights are submerged under the rules of protection and sharing laid down by the forest department. At the same time, most of the self initiated strategies lack a legal status, as all forests are state owned.

Generally, these community groups received little support from the state forest department. In fact, it is found that the villagers are distrustful of the forest department and in many cases had banished departmental field staffs from their areas with threat of physical harm.

In undivided MP, there are several instances of conflict among foresters and user groups. The present JFM framework does not recognize the widespread community forest management initiatives in the state. Some of them are more than two decades old and hence precede JFM. These are products of history and have features of JFM. But there are myriad institutional and management diversities that do not quite fit into JFM. They have successfully rejuvenated forest resources and zealously protected their freedom and resisted undue interference from FD. There is a lot to be learnt from these initiatives. For long, they have been agitating for legal recognition. Though the Government of India guidelines of February 2000 has recognized these initiatives, it is still up to the state government to rejuvenate these age-old informal organisations.

Impact of External Assistance on JFM

External financial assistance to the JFM committees for forestry operations or plantations happens to be both a motivating and a de-motivating factor. When there is external financial assistance to the committee for any purpose, the community usually becomes very active about utilizing the resources as per the prescribed way despite the mandatory allegations of misappropriation of funds by the front liners of the committee. When the fund is finished, the community goes idle and keeps waiting for further funds. In the absence of long-term funding, the planning of the community gets derailed. In certain cases, communities appoint paid guards to keep vigil over exploitation of forest by outsiders, including smugglers, using part of the external grant. With the cessation of such grants, communities usually withdraw guards since they cannot pay him anymore. Although the community as a whole guards the forests—usually on family term basis—the appointment of a paid guard is often perceived as more contributory than anything else. Sustainability of the institutions should be given more focus and alternatives should also be put in place after the end of grants.

Lack of Transparency

In most of the study sites, the committee members were found to be very reluctant to share general information about the committee in the initial round of discussion. They sought the permission of the beat guard, who is the secretary of the committee, before discussing developing activities done through the committee. This was particularly evident in case of committees, which get external funds. There is an urgent need to increase transparency at the field level, especially in areas where large amounts of funds are being provided for JFM through special projects.

4.3. Role of Different Players

Role of Women in Forest Management

Within the broad framework of community involvement in the management of forest resources, gender issues have received widespread attention. Acceptance of the need for active involvement of women is reflected in the provisions made in the JFM resolutions of some states and their subsequent modifications. However, policy provisions by themselves are not enough to ensure women's participation in community institutions, including JFM committees. Their participation in the committees is perfunctory. It is the male members who take all the decisions as has always been the case. Participation of women in the decision making process is still a matter of concern in the male dominated society. In some cases, the participation of women is by default. Women of adjacent villages, if caught by the watchmen of the committee while felling trees, often allege that they are ill-treated by the

TABLE 5

Provisions for Women's Participation in JFM Orders

State	*Eligibility for general body membership*	*Women's representative in management committee*
Orissa	I male and 1 female per household	Minimum 3 women out of 11 to 13 members
Chhattisgarh	I male and 1 female per household	Minimum 2 women. 1 EC member per 10 families
Jharkhand	1 representative per household	Minimum 3 to maximum 5 women out of 15-18 members
Bihar	1 representative per household	Minimum 3 to maximum 5 women out of 15-18 members
West Bengal	Joint membership of husband and wife	Nothing specified

male members of the committee. To avoid such situations, most committees involve women in patrolling duty. This was found to be the case in all the state sunder study. But this, unfortunately, was not how women's participation was desired in the first place.

In recent years, there have been a number of instances of village women in many states taking over local management of natural resources that affect their daily lives most acutely. There are many reasons for this.

There are several instances where women have been actively involved in protection and management of forest resources in Orissa. Such examples could be found in Bolangir district. In Jharkhand's Santhal Parganas, about 25 women's groups have taken up forest protection on their own. But when they sought formal recognition under JFM, they were told by the forest department that JFM rules did not permit women-only groups. The question is: will the policy regime ever recognize those initiatives. Another matter that needs to be studied in depth is the very limited role they play in JFM despite their keen interest in forest management.

There are several reasons for the low participation of women in the decision making process under JFM initiatives. Traditionally, village assemblies and councils have excluded women. Most of the contemporary forest management initiatives draw upon this tradition and don't call women to the meetings. Secondly, the presence of women in meetings is restricted by social and cultural barriers. The norms of acceptable behaviour, notions about appropriate spaces and assumptions about the capabilities and roles vary widely for men and women. Women's effectiveness is also restricted by their limited experience in public speaking and illiteracy.

From the study, it is also found that the reasons for low attendance in group meetings include lack of interest, inability to spare time due to other work, lack of information, etc. In most of the villages, JFM meetings are convened by the president (in most cases a male) alone or in consultation with a group of village elites (who also happen to be executive committee members). He decides the date, venue and time of the meetings making sure that they suit the convenience of the forester. Since the timing and the venue of JFM meetings are all-male decisions, most committees covered in the study invariably call group meetings only in the afternoon or evening–when men are either free or have returned from work. But women are usually busy at that time in domestic activities. Consequently, they are effectively excluded from JFM meetings. The meetings are generally held in a place, which is convenient for the male members.

Role of NGOs

The collaboration between the NGOs and forest department leaves a lot to be desired. Instances of successful collaboration have been very limited. The role of NGOs in participatory forest management is very negligible. There seems to be a power struggle between forest officers and NGOs, with the forest department not willing to relent control over the vast areas, which hold the potential of community empowerment and poverty alleviation. The forest department doubts the abilities of the NGOs to be involved in forest regeneration. Consequently, the involvement of NGOs in the JFM programme is at best token and tentative or restricted to NGOs being accorded the soft jobs of community mobilisation, training, etc. NGOs in general have failed to gain the confidence of the forest department as they lack uniformity in approach to solve the important issues of access and control over resources. While some NGOs are talking of the crucial role JFM plays in conservation and management of degraded land, there exist quite a few other NGOs who oppose the fact. They are fighting against the issues like mining in forest area, benefit sharing, etc.

While the forest department focused its attention on internal changes and building new relationships with forest users, it has paid less attention to non-government organizations (NGOs). In most of the study areas, the role of NGOs in the process of JFM is very limited. The states where NGOs are working on forestry, their focus is limited to community based forest management. In West Bengal, though IBRAD and RKM are working on JFM, their role is confined to capacity building of foresters and JFM committees. In Bihar, there is virtually no NGOs working on JFM in the areas covered in the study. It seems that the forest department has a very limited understanding of the role of NGOs in rural development. For the future success of the programme, it is be crucial for the forest department to continue supporting NGOs in the many activities under JFM.

The way Forward

JFM is too diverse to allow generalised conclusions about whether it is successful or replicable. The policies and practices relating to JFM should change and the frontline forest officials should be given adequate flexibility and scope to implement the programme. The target chasing approach should be dispensed with. The capacity of the forest officials needs to be enhanced and basic minimum resources made available to them to carry out the activities. Any information relating to participatory forest management should flow down to the level of the Forest Guard. The lower level forest officials should be in a position to decide the nature and extent of the JFM programme in their area of operation. They should undertake forest protection and management activities involving communities based on the local needs. Adequate time should be provided to undertake a process of social mobilization, which would ensure that there is a clearer understanding of the duties and responsibilities. Recent guidelines of the state governments should be published in local dialects and disseminated to the committees by civil society organisations or forest officials so that the communities, including the marginal class of people, know their duties and responsibilities and those of the forest department. Lower level forest officials consistent guidance of their seniors in establishing JFM. According to the forest officials working with the communities, it is difficult to motivate villagers for protection of forest without giving them incentives. The programme has to be planned in such a way that the VSS gets some short-term benefits from forest protection. They need not wait for 20-30 years to get their share from the final harvest. Some specific suggestions, which need to be discussed in a larger context, are discussed below.

1. Lack of tenurial security is the major concern in the whole framework of joint forest management. A big drawback at the governance level is the frequent amendments in policy and misplaced emphasis on target achievement of programmes, ignoring the strengthening of the processes. This creates an information gap at many levels and at the same time raises apprehensions of several kinds. It is, therefore, worthwhile to think of a complete policy on JFM over a longer duration, not just in the form of resolutions with only a limited scope for frequent amendments so as to restore mutual trust among stakeholders and community processes. The JFM resolution should be incorporated in the state forest acts or in the main frame of Indian Forest Act. There needs to be widespread debate on the possibility of incorporating JFM in the context of PESA. The best way to enhance the interest of the community in protection is to ensure that JFM committees should feel secure. Similarly, the self-initiated forest activities of the communities should be legally recognized though the issue is being addressed in the recent central guideline. The type of legal recognition, impacts of legal recognition, results of legal recognition need to be discussed in a broader context. Again in line with Forest Rights Act the entire JFM process need to be

revamped. It should be designed in such a way that the local communities will have a space for governance.

2. Institutional sustainability, which is a major issue in the present context, needs to be debated more. Strategies should be devised so that the external funding sources do not have a negative impact on other committees. The selection of committees for availing the benefits needs to be done with sincerity and on merit. The roles and responsibilities and functions of the committee should be clearly circulated among the office-bearers. While raising awareness in the community about their roles and responsibilities remains the primary task, a far more important issue is building the capacity of the community and its institutions. Lack of proper planning and community involvement in the process of planning, mismanagement of village resources and above all the conflict of faith among stakeholders are the other major bottlenecks in the JFM programme. Inadequate participation of forest dependent communities such as the landless, artisans and women, is a grey area. To overcome it, adequate measures must be taken to improve their participation and means of livelihood.
3. Alternative arrangements should be thought of for those committees, which are dependent on forest for fuel wood, poles, etc. *Solar chullas*, plantation of trees for fuel wood are some of the viable options that emerged during the study. Civil society organizations, in consultation with the local people, should develop a detailed plan of action for alternative livelihood options for the community. The possibilities of improving the economy of the community need to be explored in the forestry sector. They should pressurize the forest department to invest in the sector and intervene. The whole idea is to minimize pressure on the forest.
4. Conflict resolution is another grey area in JFM. The policy document should empower the local communities to take any penal measures against the offenders. The other alternative might be prompt steps by the forest department to penalize the offenders involved in illicit forestry activities.
5. In case of Chhattisgarh and Jharkhand, where there is a strong base of community initiatives in protection of forest, efforts may be made to document the forest management practices. A clear and precise database needs to be developed

to document the practices, area under protection, resource mobilisation, conflict management, benefit sharing etc. Civil society organisations, which are working on forestry in different states, may work closely with the community and the forest department to define the roles of the community in the forest management process. They should advocate with the forest department for devolving management responsibilities to the communities. They should also help the forest department in developing forest management tools for involvement of communities in forest management.

6. Regular support from the forest department will definitely create more interest among the forest protection groups. More than money, the support required is legal and moral powers from the department to protect forest. In case of highly degraded forests, there is a need for financial support to raise plantations and treat the land and forest. The JFM should be viewed as a process rather than a project to manage and conserve the forest resources with a focus on strengthening forest-based livelihoods.
7. Involvement of communities in the forest management process should be mandated. It is not that the communities do not have the scientific knowledge to manage the forest. Simple sylvicultaral techniques should be developed so that the communities themselves can manage the forest without the help of the forest department. These tools and techniques could be developed in consultation with communities and based on their existing knowledge.
8. Weak participation of women has remained as key concern. No adequate effort has been made to involve them in the forest management process though most states have mandated their participation. Efforts may be made to organize meetings among the women members and the issues in JFM may be discussed. Their opinions should be documented and accordingly the management plan should be developed.

References

Corbridge S. and S. Jewitt, 1997. From forest struggles to forest citizens? Joint forest management in the unquit woods of India's Jharkhand in Environment and Planning A, Volume 29, pp. 2145-64.

Sekhar, Madhushree, 2004. Indigenous institutions and forest conservation: user-group self-nitiatives in India, working paper 140, Bangalore, Institute for Social and Economic Change.

National Forest Policy 1988.

State forest report, 2002-03 Office of the PCCF, Directorate of Forests, Government of West Bengal.

2004, Root to Canopy, Winrock International India, New Delhi.

Rabindranath N.H. and P. Sudha, 2004, JFM in India, University Press (India).

Poffenberger, M. and B., MacGlean, 1998: Village Voices, Forest Choices: JFM in India, OUP, New Delhi.

Victor Michael, *et al.*, 1997: Community forestry at cross roads: Reflections and future directions in the development of community forestry, proceedings of an international seminar, RECOFTC, Bangkok.

Sarin, Madhu, 2001. Disempowerment in the name of 'participatory' forestry?— Village forests joint management in Uttarakhand, India: Forests, Trees and People, Newsletter No. 44.

Kashwan, Prakash, 2003. Conflicts in Joint Forest Management Cases from south Rajasthan: Issue 4, Volume 2, Community Forestry, Bhubaneswar, RCDC.

2000, Bihar forest resolution, department of forest and environment, Government of Bihar.

1992, Joint Forest Management: Concept and Opportunities, Proceedings of the National Workshop at Surajkund, New Delhi, SPWD.

2003, State of Forest Report, Dehradun, Forest Survey of India.

Prabhu, P., 2001. The greening of Haldapada- Shisne, The Humanscape, Vol. VIII, Issue XI, December.

Saigal, S., 2001. Forests and their people, *The Humanscape*, Vol. VIII, Issue XI December.

Yadav, G. *et al.*, 1997. Progress in Community Forestry in India. Community forestry at cross roads: Reflections and future directions in the development of community forestry, proceedings of an international seminar, RECOFTC, Bangkok.

Singh, R., 2004. A study of status of joint forest management in Haryana, *The Indian Forester*, Dehradun, Forest Survey of India.

Upadhyaya, S., 2003. JFM in India: Some legal concerns in *EPW*, August 2003.

2001, Conflict and conflict management in joint forest management, Ahmedabad, VIKSAT.

Sundar, N. and *et al.*, 2001. Branching out, Joint Forest Management in India, New Delhi, Oxford University Press.

Lele, S. and *et al.*, 2005. Joint Forest Planning and Management in the Eastern Plains Region of Karnataka, Bangalore, Centre for Interdisciplinary Studies in Environment & Development.

Sarap, Kailas, 2004. Participatory Forest Management in Orissa: A Review of Polices and Implementation, Working paper No. 2, Overseas Development Group, UK and Sambalpur University, Orissa.

Chatterjee, Mitali. Women in Joint Forest Management: a case study from West Bengal, Technical Paper 4, Kolkata, IBRAD.

Vikram, Kirti, 1999. 'Janjatiya Jiwan me Van ka Mahatwa' in *Bulletin of Bihar Tribal Research Institute*, Vol. XXXVIII, May, pp. 1-4, Ranchi: BTWRI.

Vidyarthi, Dr. L.P., 1991. 'Forest Economy' in the *Bulletin of Bihar Tribal Research Institute*, Vol. XXXI, Nos. 1 and 2, pp. 52-70, Ranchi: BTWRI.

Sinha, S.P., 1991. Tribals and Forests, Ranchi: Bihar Tribal Welfare Research Institute.

Sharma, P. Dash, 1997. 'Tribals and forest management' in *Yojana*, Vol. 41, No. 8, pp. 51-53. New Delhi: Publication Division, Ministry of I & B.

Louis, Prakash, 2000. 'Marginalisation of tribals in Jharkhand' in EPW, Nov. 18, pp. 1487-91, Mumbai: Sameeksha Trust.

Roy, N.K. and *et al.* 1982. 'Leadership patterns in Tribal and Non-tribal Villages of Santhal Paraganas' in *Bulletin of Bihar Tribal Research Institute*, Vol. XXIV, Nos. 1 and 2, pp. 45-54, Ranchi: BTWRI.

Chhattisgarh State Forest Policies, 2001.

Indian Forest Act, 1927.

JFM Circulars of the Orissa Forest Department, 1993 and 1996.

JFM Circulars of the Bihar Forest Department, 1990 and 2000.

JFM Circulars of the Jharkhand Forest Department, 2001.

JFM Circulars of the Chhattisgarh Forest Department, 2000 and 2002.

PCCF, 1992. A decade of forestry in Orissa, Bhubaneswar: Forest Department, Government of Orissa.

PCCF, 1999. Orissa Forest-1999, Bhubaneswar: Office of PCCF, Govt. of Orissa.

JFM circulars of Madhya Pradesh Forest Department.

Salient Features of State JFM Resolutions

Issues	*Orissa*	*West Bengal*	*Chhattisgarh*	*Jharkhand*	*Bihar*
Year of issue	1993	1989	2001	2001	1990
Amendments, if any	1994, 1996	1996	2002	—	2000
Legal status of the committee	Resolution	Resolution	Resolution	Resolution	Resolution
Forest category	Degraded and dense forests	High and protected forests	Low and high forest	Reserve and protected forests	Reserve and protected forests
Type of committee	Van Samrakshyana Samiti (VSS)	Forest protection committee and eco-development committee	Forest protection committee and Village Forest Committee	Village Forest Management and Conservation Committees and Eco-Development Committees	Village Forest Management and Conservation Committees and Eco-Development Committees
Eligible participants	Two members from each household (a male and a female)	Head of the household is the member of the general body	Two members from each household (a male and a female)	Each member of the Gram Sabha will be the member of GB	Each member of the Gram Sabha will be the member of GB
FD representation	Forester	Beat officer/deputy range officer	Beat guard/forester	Vanpal	Vanpal
Tenure of management committee	2 years	2 years	2 years	2 years	2 years

Benefit Sharing as per the Revised Guidelines

Orissa	*Chhattisgarh*	*West Bengal*	*Bihar*	*Jharkhand*
In Orissa, 50 per cent of the net value of the final harvest of the trees would go to the VSS directly after final harvesting.	Free Nistar facility to all members of the JFM committees. Free forest produce from intermediate mechanical thinning of the coupes that fall within the specified area under protection. Share from main felling 30% for Village Forest committees (VFC), 10% for Forest Protection Committees (FPC).	Government, through its resolution in 1987, decided to give a 25% share of usufructs of the intermediate and final yield to FPCs.	The Bihar government resolution proposes that 20% of the final harvest would go to the committees and this amount would be distributed equally among *Gram Vikas Nidhi, Van Vikas Nidhi and Working Nidhi*	According to 2001 resolution of Jharkhand forest department, 20 per cent of the net value of the final harvest of timber would go to committee. 90 per cent of the net value of the forest produce go to the committee.

Chapter 18

Non-Timber Forest Products and Rural Livelihoods with Special Reference to the Policies and Markets in Orissa

TAPAS KUMAR SARANGI

ABSTRACT

In the absence of adequate resource endowment such as land and access to service sector the majority of poor households rely on forest as well as on labour market. Forests play a crucial role in the livelihood strategies of many rural households in Orissa. Beyond subsistence, fuelwood and fodder collection, there are a wide range of forest products which poor households depend on, particularly for consumption and generation of income in the lean season.

Given the collection of forest produce, the income of the forest dwelling communities would depend on value addition to the products and their selling at reasonable prices. But, the marketing structure for sale of Non-Timber Forest Products (NTFPs) is exploitative to the forest-dwellers. The nature and structure of marketing in the tribal dominated area is quite different from other areas because the sellers, who are mostly illiterate and resource poor, have to deal with powerful buyers, who are mostly traders/money lenders. NTFPs have also low demand in local areas. In such situation they have to sell these products at low price.

This paper will analyse how various government policies and institutions during the last few decades in the state of Orissa have affected the livelihood of forest-dwelling communities. It will also suggest the policy measures that will help in improving the access to forest products and enhance the livelihood of the forest-dwelling communities.

I

INTRODUCTION

Rural poverty in India is generally considered to be linked to a lack of access to cultivable land, or to its low productivity. Changes in the collection of gathered items from common property resources such as forests go largely unnoticed, and are not even accounted for in the national accounts and Gross National Product (GNP). However, about 100 million people living in and around forests in India[1] derive their livelihood support from the collection and marketing of non-timber forest products (NTFPs). These NTFPs provide subsistence and farm inputs, such as fuel, food, medicines, fruits, manure, and fodder. The collection of NTFPs is a source of cash income, especially during the lean seasons, because of their increasing commercial importance. Thus, the issue of rights and access to NTFPs and incomes from NTFPs is basic to sustenance and livelihood for the forest-dwellers.

Socio-economic Profile of the State

The state of Orissa is located in the East of India, having a total geographical area of 15.57 million ha. The population of the state is 36.71 million (3.6% of the population of India), of which, 85% are rural and 15% are urban population (Population Census, 2001). The average population density is 236 persons per sq. km. The tribal population constitutes 22.2% of the total population. The livestock population is 22.7 millions constituting 4.8% of country's livestock population. The state ranks 4th among the states in terms of area under forest cover. The present day state of Orissa has a complex historical tradition because it has been formed over time by taking areas from other states with different administrative and institutional arrangements including forest institutions. The present state of Orissa is an amalgamation of different parts coming from Bengal, Bihar, Madras Presidency and Central Province.

The state can be divided into four distinct physiographic regions. These are North Plateau, Eastern Ghat, Central Tableland and Coastal Plains. Forest is mainly found in first three regions of the state, as the tribal population is high within the regions. The economy and livelihood of the state and its people is predominantly agriculturally based, involving an estimated 75% of the working population. However, 47.1% of population in the state is below the poverty line as per 61st round of National Sample Survey, 2004-05. Of the total poor,

90% live in rural areas, and the intensity of poverty is particularly high among the tribal population located in forest-fringe villages. In view of this there is a need to understand the forest and forest related issues in order to understand the livelihood of people dependent on forest.

The structure of the paper is as follows: besides the *introductory* section; the *second* section discusses the extent of dependence of rural poor on forest and forest-based resources. The *third* section analyse the management of NTFPs and the market mechanism for its sale. The *fourth* section provides the problems in NTFPs trades in Orissa and the policy reform. *Last* section ends with a conclusion.

II

DEPENDENCE ON FOREST

In the absence of adequate resource endowment such as land and access to service sector the majority of poor households rely on forest and on labour market. Forests play a crucial role in the livelihood strategies of many rural households in Orissa (Sarap & Sarangi, 2009). Beyond subsistence, fuelwood and fodder collection, there are a wide range of forest products which poor households depend on, particularly for consumption and generation of income in the lean season. The dependence on income from NTFPs has been shown to be inversely related to the size of landholdings in Orissa as well as in India (Fernandes & Menon, 1987, Sarap, 2007). It is labour-intensive, but the returns from this are quite low. Sarap's (2007) study has found that the percentage of household's total income coming from forest-related activities varied from 40% to 50% in the case of poor households. But, absolute income from the forest is low and below the poverty line income for these groups. Similarly a study conducted by the Indian Institute of Forest Management in 1996 (see MoEF 1998) found that the contribution of forests to the economy, for tribals was very high. The average tribal family drew about half of its annual income from forests and 13% from cattle (see also Singh 1997).[2]

Women play an important role in the collection of NTFP (see Khare & Rao, 1993, Mallik 2000, Sarap 2007). Total women labour engaged in the collection of forest produce in Orissa is as high as 300 million woman days per year. Throughout India, collection of kendu leaf generates part time employment for 7.5 million people—a majority of them tribal women (Arnold, 1995) while in Orissa, 1.8 million women are involved in this, collecting 45 thousands tones (Rs. 450 million) of leaves per annum. Despite higher participation of women in the collection of NTFPs the price realised from selling the product is very low.

Importance of NTFPs for the Livelihood of Forest-Dwellers

Non-timber forest products (NTFPs) have been the lifeblood of the forest dwellers especially the tribal living nearby the forest. These products provide food and cash income to the poor tribal during the lean seasons. Yet policy-makers have overlooked the potential of NTFPs in combating rural poverty and food insecurity, and state policy on NTFPs has mostly favored private business interests: private leaseholders, traders, moneylenders as well as its own interest. The private business houses were the monopoly traders of major NTFPs till 2000. As a result the primary collectors were worse-off because of the low payments received by them when they exchanged their products with the traders. Even the low income derived from the sale was irregular due to erratic procurement of these products by the buyers.

Definition of NTFPs

The state differentiates between Non-Timber Forest Products (NTFPs) and Minor Forest Products (MFPs). This is because the Constitution of India stipulates that ownership over the MFP has to be transferred to *Panchayats.* The state has identified 85 items as NTFPs. Out of these, 68 items have been transferred to *Panchayats.* The rest have been divided into nationalised NTFP (bamboo, *sal* seed and *kendu* leave) and 'lease barred' items. The products under the lease barred items are mostly gums, barks and leaves that are either banned or allowed selective extraction by primary collectors.

This classification came up only after the provision of Panchayat Extension to Scheduled Area (PESA) act came into being in 1996. In response to PESA the state has come out with legislation whereby *Gram Sabhas* have been given ownership rights over NTFP in their area of jurisdiction. In operational terms, it means that traders who want to operate in any area have to become registered with the respective *Panchayat* and pay a fee. The traders are supposed to pay the prices fixed by a district level committee.

After the promulgation of Orissa *Gram Panchayats* (Minor Forest Produce Administration) Rules in November 2002, the responsibility of fixing the minimum procurement price (MPP) has been vested with the *Panchayat Samity.* The MPP so fixed by *Panchayat Samity* is to be rectified by *Gram Sabha* and the *Gram Panchayat* has been empowered to modify the MPP if needed. According to the new rules, any person who is interested to deal in forest produce can deposit a required amount of money and register himself as a dealer. The *panchayat* has the power to cancel his registration if it finds that the dealer buys forest products at less than the fixed price approved by it. Further, each dealer has to give a statement about the amount of products bought by him in the *panchayat* areas to the *panchayat* office and the Range Office of the Forest Department. But in practice traders do not follow this. It is noteworthy that there is no restriction on movement of produce inside the state. Overall it is clear that the *panchayats* have been provided with the responsibility and authority of managing non-nationalised NTFPs, but lacks the capacity to exercise these powers, and in many cases they are not aware of their power.

Nationalised NTFPs

Sal seeds, *kendu* leaf and bamboo are the nationalised NTFP items in the state. *Kendu* leave trade was nationalised in 1973 and Sal seed during 1983, with a view ostensibly to ensuring a fair price to the gatherers, and also to enhancing government revenue. The contradiction between these objectives have largely been resolved in favour of the later: the policy environment relating to NTFP trade was characterised by revenue maximisation by the state. Furthermore, apart from the above three nationalised forest products, trading rights for several marketable NTFPs were given to private houses as monopoly leases up to 2000. In such a situation the fate of forest products and livelihood of people dependent on these products, were in the hands of private parties and industries.

The price fixation of the NTFPs is mainly based on minimum wages. For instance, as per the relevant Acts [OFP (CT) Act, 1981, and OKL (CT) Act, 1962] the price fixed for the NTFPs are mainly based on consideration of minimum wages. The Orissa Forest Products (Control of trade) Act, 1981, section (7) states that while fixing the price of specified forest products, regards may be paid to among other things, "general level of wages for unskilled labour prevalent in the units and the provisions of the minimum wage Act, 11 of 1948". However the prices fixed by state have little relevance in the absence of mechanisms to ensure that these prices are paid. The monopoly leaseholders depend on the local sub-agents/traders in varying degrees for procurement of NTFPs. Because of low bargaining power of the primary collectors *vis-à-vis* these traders, the former rarely get the state administered prices.

The overall impact of the polices and laws were depression of prices received by the primary collectors for NTFPs especially due to monopoly leases and high royalty fixed by the Forest Department, with a resultant deprivation of their livelihood. On the other hand the state generates significant revenue from its trade (*kendu* leave and *sal* seed) that is based on the hard work of the primary collectors. It is desirable that the primary collectors should get a share in the profit from the operations of NTFPs. The state has a provision for channelising of profits from *kendu* leaf operations to the primary collectors, but the system of distribution is faulty. For instance, a provision exists that at least 50% of the profit earned from Kendu Leaf (KL) operations has to be distributed to *panchayat* bodies in the KL growing subdivisions. But so far only *ad-hoc* grants have been given. Further, instead of sharing the profit within the *kendu* leave growing/collecting areas the profit is distributed widely, even to non-*kendu* leaf areas. Further the funds given to the *panchayats* under the KL grants are utilised for a variety of purposes including payment of salaries to the staff. As a result, the primary collectors hardly get much benefit from the transfer of profit to *panchayats*. Moreover, the percentage share of revenue passed on to the primary collectors of KL as wages is the lowest in the case of Orissa at around 20%. (see Vasundhara, 1997) Clearly, although NTFPs form an integral part of the livelihood of the forest dependent communities, state policies related to these items have been generally pro-rich and trader-oriented until 2000. As a result the livelihood conditions of the poor, dependent on these produce, have been very precarious. Likewise the post-2000 changes in the policies of NTFPs have not much improved the livelihood condition of the forest-dwellers.

III

NTFPS MANAGEMENT VIS-A-VIS MARKET MECHANISM

In the process of commercialisation of NTFPs, a number of agents, namely: middlemen, businessmen and traders, government agencies, etc. enter in to market network. Also, the very characteristics of NTFPs influence the market behaviour, mode of exchange and prices differently in different situations.

In Orissa, while non-traditional NTFPs such as; barks, lac, medicinal plants, resins, oilseeds are exchanged for salt, tobacco and dry fish in remote tribal regions NTFPs of greater exchange value are bartered for cloth, umbrella and other luxury goods. Lac, resin and honey are by and large bartered to peddlers. Oil seeds, barks, resin, gum, leaves, fibers, canes and other similar products are sold in large quantities to generate cash income to tribal households.

Nationalised and commercially significant NTFPs, such as; *kendu* leaves, *sal* seeds, bamboos restrict trade, and also limit number of legal buyers. It chokes free flow of goods, and delays payment to gatherers. Thus, it reduces the income that the forest-dwellers might get and impoverishes them. It creates inherently exploitative alternative markets, such as: State monopolies, Private monopolies and illegal trade channels.

The structures of marketing channels vary depending upon number of agencies involved and nature of products.

Primary gatherer ® village merchant (either an agent or sub-agent) ® wholesaler ® processor consumer. Under such an arrangement, the primary gatherer ends up receiving a small share in what the final consumer pays.

Marketing of NTFPs exhibits a wide range of variations in terms of market structure, marketing channels, price and scope for processing. Most of the NTFPs markets are essentially local, and exhibit seasonal behavioural pattern—such as, honey market in autumn, *kusum* seed market in pre-monsoon season, tamarind market in summer, and broom stick market in early spring.

Better quality products attract higher demand, and better prices in any market. But, the quality of primary NTFPs is influenced by post-harvest handling, processing and storage conditions. Admittedly, consumer markets need sustainable and continues product availability, reliable and predictable supply, and stable quality products.

A study (Mallik & Panigrahi, 1998) suggests that. While many vendors sell NTFPs for making extra income, others are supported by a network of merchants and several levels of buyers and sellers. Local traders and merchants are the main intermediaries. They buy NTFPs cheaply from the primary gatherers, and sell them to exporters/ processors or their agents at exorbitant prices.

Lack of timely dissemination of information about the support prices, market avenues, processing units for value addition, etc. indeed increase vulnerability of primary gatherers owing to "distress sales". In the absence of appropriate link between input sector and post-production sector, the gatherers, the cultivators and resource owners of NTFPs fail to secure a fair share of processing and value addition (Chandrasekaran, 1998).

NTFP trade and markets are highly disorganised. Government agencies, private middlemen stand between the primary collectors and the manufacturing units/whole sellers/outside dealers as intermediaries. But, the primary collectors in the disposal process are in close contact with the direct consumers in the local weekly markets as well as village '*baats*'.[3] The major buyers of their collections are private businessmen, traders, government agencies and consumers. The trader very often does not pay in cash, and insist on barter,

Markets for NTFPs are by and large informal and unstructured. As a result, the primary gatherers and NTFPs dependent population suffer from various exploitative practices in the hierarchical structure of market network in the tribal areas. Their exploitation is manifested in low price, credit-liked trade and by way of cheating in the measurement.

The price variation in the bordering states as well as within the districts is a matter of grate concern. Evidently, the monopoly buyers within the state pay a lower price to NTFPs as compared to the alternative markets and also their counterparts in bordering states. This transaction is supposedly illegal, but operates in everybody's knowledge.

TERMS OF TRADE AND MODES OF EXPLOITATION

Various modes of exploitation and deprivation arise owing to situations, where exchange takes place between illiterate, poverty-stricken, ignorant, impoverished and unorganised tribal forest dwellers and a group of organised vested interests, traders/business men. In the absence of an effective, vibrant and procurer-friendly institutions a number of non-tribal intermediaries, namely; middlemen, businessmen, traders seem to have infiltrated in to tribal hinterland in guise of traders, shopkeepers and medicine men to take the advantage of the poverty, ignorance, spendthriftness of the tribal people.

The mode of NTFP trade exhibits great variation by type, region, season, etc. Barter is a common mode. Traders also make advance payments to primary collectors, and later buy goods at very low rates and sell them in cities for huge profits. These modes have set up exploitative elements due to non-payment of prices fixed for NTFPs. Traders also function as money lenders, and buy NTFPs towards repayment of debt or interest. While private traders and middlemen buy NTFPs through agents and sub-agents at the primary level, the government agencies procure specified items; such as; Tamarind, Hill broom, *Mahua* flower, *Sal* seed, *Kendu* leaves, etc. directly from the primary collectors at the local collection centers.

Among the modes of exploitation in trading activities, differential prices, grading of the products, limited processing, creation of situations towards more indebtedness, means of distress sales, metric system of weight and measure, etc. are very important. In situations, where weaker sections are prone to sell as much they can to meet their pressing consumption needs, the exploitative elements become more active, to exploit the situation.

Though market is the most powerful channel of communication particularly in the tribal region, the NTFPs indeed face "buyers' market". In such a situation, the middlemen indeed largely benefit from the commercialisation process in terms of appropriating a greater share of value. Thus, relationship between primary collectors and middlemen (in a sense) is symbolic.

Much of the miseries of tribal and other forest-dependent communities are primarily due to lack of access to forests to collect NTFPs. Even if collection is not prohibited from the revenue and protected forests, the right to process some NTFPs and sell the products freely in the markets has not been granted. Market intermediaries including private traders form a dominant link between the primary gatherer and the final consumer.

The intermediaries are capable of maintaining a stronghold in the marketing network due to their ability to meet immediate needs of the primary gatherers. They offer quick and timely credit, make quick payment and also have a good network of procurement at the door step of the producers.

Poor communication and transportation facilities, highly segregated markets and unequal bargaining powers between buyers and sellers make the field more profitable for middlemen (FAO, 1995). Thus, middlemen can and often get involved in unfair activities and exploit the producers' weak bargaining power due to latter's ignorance of the market factors, and thereby retain a disproportionate share of producers' earnings.

IV

PROBLEMS WITH NTFP TRADE IN ORISSA

Given the collection of forest produce, the income of the forest dwelling communities would depend on value addition to the products and their selling at reasonable prices. But, the marketing structure for sale of NTFPs is exploitative to the tribals. The nature and structure of marketing in the tribal dominated area is quite different from other areas because the sellers, who are mostly illiterate and resource poor, have to deal with powerful buyers, who are mostly traders/money lenders. NTFPs have also low demand in local areas. In such situation they have to sell these products at low price (Sarap 2005).

Various reasons could be attributed for realising low prices by the sellers. Interlocking of credit and output market force the gatherers to sell their produce to the moneylenders, only at predetermined prices. Further, the limited surplus erodes their bargaining capacity and makes them more vulnerable to exploitation. Another reason for low prices for produce is lack of value addition at the village level. Even price realised by selling of nationalised forest products is very low and the sellers don't receive the price in time. The state policy on NTFPs has mostly favoured private business interests till 2000.

NTFPs under Public Sector Monopoly

The state government nationalised the trade of *Kendu* leaf and

Bamboo during the year 1973 and *Sal* seed in the year 1983. In most of the cases, the Tribal Development Cooperatives Corporation (TDCC) or the Orissa Forest Development Corporation (OFDC) appointed agents formally or informally (Government of India, 1988) to purchase NTFPs from village traders. This puts the forest produce gatherers at the mercy of two different sets of people, the agent as well as the government department officials. Whatever payment that gatherers received had to be routed through both of them. In a study by Fernandes *et al.*, 1987, it was found that government agencies had not managed to eliminate middlemen in majority of the villages in the purchase of NTFP. On the other hand, the same middlemen, who until recently exploited the tribals as moneylenders and merchants, continued their work in the garb of agents of government bodies (Das, 1998).

The state institutions (OFDC, TDCC) are unable to serve the primary gatherers of forest products because they are sick and no adequate funds to buy the products from the sellers. Faced with this situation, they wish to pursue a completely risk-free policy. In the few commodities that the TDCC traded (e.g. hill brooms), purchase transactions were first finalised; these selling prices were, down-marked to fix the procurement prices for the gatherers. Because of the middlemen involvement on the actual prices received by the gatherers could be lower still. More generally, the state institutions opted to limit their role by becoming rentiers (Saxena, 2003).

Monopolies reduce the number of legal buyers, chokes the free-flow of goods, and delays payment to gatherers, as government agencies find it difficult to make prompt payments. This results in contractors entering from the back door, but they must now operate with higher margins required to cover uncertain and delayed payments by government agencies, as well as to make the police and other authorities ignore their illegal activities. This all reduces gatherers' collections and incomes (*ibid.*).

Private Monopolies

The public sector's disappointing performance also led the state to take monopoly powers away from them in favour of private parties. Thus from 1985 onwards, government of Orissa encouraged private parties to acquire monopoly rights over forest produce. The largest beneficiary was Utkal Forest Products Ltd. (UFP), which was given long-term lease for 29 items for ten years in 1989 (*Ibid.*). Its control was even extended to the designated forest products growing on private lands and non-forest government lands. This was, despite emphasis in law as laid down in the Orissa Forest Code and Orissa Forest Produce (Control of Trade) Act, 1983, to encourage the Tribal/Labour Cooperative/*Gram Panchayats* as procurement agents for NTFPs. However, upto March 2000, there was no involvement of grassroots-level *Gram Panchayat* in NTFP trade.

Case of Kendu Leaf and Sal Seed

In Orissa, the share of NTFP revenue (particularly revenue from *kendu* leaf) in total forest revenue has increased from 43% in 1985-86 to 89.3% (share of *kendu* leaf is 85.1%) in 2001-02. About a million pluckers are engaged during the season to pluck *kendu* leaves, which last about 45 days in summer, at a rate fixed by the government. However, the wages paid to *kendu* leaf pluckers is much lower in comparison to the profits earned by the state. For every rupee paid to the pluckers, the State earns three rupees (in one year, i.e. 1989-90 it went upto more than Rs. 10). The wages paid to the pluckers are abysmally low and not in keeping with the amount of labour put by them in procuring the leaves (Saxena, 2003).

Sal seed is another important NTFP in Orissa whose trade, like *kendu* leaf, is often controlled by the State. The price paid to the primary gatherer has been around Rs 3 per kg and as daily collection is not more than 6–8 kg per day, a person can earn only about 20-25 rupees per day, which is just 40–50% of the minimum prescribed wage (*ibid.*). Although NTFPs form an integral part of the livelihood of the forest dependent communities, state policies related to these items have been generally pro-rich and trader-oriented up to 2000 (see Vasundhara, 1998).

Policy Reform for NTFP Marketing

Due to concerted efforts by civil society, the state NTFP policy was changed in March 2000. The new policy seeks to give primacy to welfare of forest dependant poor over revenue objectives of the state. It also seeks to deregulate NTFP trade and encourages competition for NTFP procurement by conferring rights over 68 NTFP items to *gram panchayat* as opposed to the earlier policies of monopoly leasing.

However in operational terms, it means that traders who want to operate in any area have to register with the respective *panchayat* and pay a fee fixed by a *panchayat samitte* level. In many cases the district level committees generally declare the price well after the procurement time. As a result, the collectors of the major NTFP items have to sell their produce to the traders at lower than the prices fixed by the public agencies. This is due to the fact that the latter do not have direct access to the markets and they are not involved in processing of the products for end use (Sarap and Sarangi, 2010).

Overall the *panchayats* have been provided with the responsibility and authority of managing non-nationalised NTFP, but lack the capacity to exercise these powers, and in many cases they are not aware of their powers. Clearly the recent changes in the policies of NTFPs have not improved the livelihood condition much (see Sarap, 2005).

V

CONCLUSION

The overall impact of the polices and laws were depression of prices received by the primary collectors for NTFPs especially due to monopoly leases and high royalty fixed by the Forest Department, with a resultant deprivation of their livelihood. On the other hand the state generates significant revenue from its trade (*kendu* leave and *sal* seed) that is based on the hard work of the primary collectors.

Clearly, although NTFPs form an integral part of the livelihood of the forest dependent communities, state policies related to these items have been generally pro-rich and trader-oriented until 2000. As a result the livelihood conditions of the poor, dependent on these produce, have been very precarious. Likewise the post-2000 changes in the policies of NTFPs have not much improved the livelihood condition of the forest dwellers.

The NTFP Policy has given many responsibilities to GPs in terms of monitoring and regulating the NTFP trade. This is a newfound role of the *Panchayats*. Given their earlier experiences they have little knowledge of NTFP market and trade. Thus measures should be taken urgently to enhance their capacity to regulate and monitor the trade so that they can discharge their responsibilities and the primary gatherers benefit. Their involvement in the price fixating system can be a first step towards this. Similarly, proper coordination and cooperation between the *Gram Panchayat*, Forest Department and other concerned departments involved in the process need to be stressed.

Processing is another area that needs to be looked into. If markets can be provided for simple processed items which can be done in households, then subsidies for effective training for processing can help gatherers value add and improve income. For example, broom grass can be bound into broomsticks with simple training by women and men in their own houses. The same can be said for products like tamarind, which can be processed and packed as a household/cottage industry. The market is quite extensive for these items, and household producers can have the choice of either selling in the open market, or through government outlets, depending on the pricing.

This is also true in the case of bamboo. The art of bamboo processing is a fast dying art in the tribal regions of Orissa, due to the wrong policies of the Government, which has denied access to the local artisans. However, a sizeable demand for bamboo products still exists, as the tribal economy and livelihood has a variety of uses for it. A two-pronged effort needs to be made here to regenerate bamboo forests, along with support for once again reviving the art of bamboo weaving. This would help several tribal communities to have a better income.

Rather than be a monopoly buyer of NTFPs or try to regulate price through administrative mechanisms, government should adopt market-friendly policies, facilitate private trade, and act as a watchdog rather than eliminate the trade. It should encourage local bulking, storage and processing, and bring large buyers in touch with gatherers, so as to reduce the number of layers of intermediaries. Government should encourage the formation of self-help groups (SHGs) among the forest-dwellers so that such groups are able to bargain better with the trade. Finally, a more effective implementation of credit-oriented and poverty alleviation programmes will help the poor in recovering from debt bondage, which is the single most important factor for their dependence on traders and depresses the price that forest dwellers are able to negotiate with them.

Notes and References

1. India's population in 2001 was about one billion.
2. See also Singh (1997) for similar findings for forest fringe villages located in the district of Sambalpur, Mayurbhanj and Ganjam in Orissa.
3. 'Haat' is a local term refers to village market.

References

Arnold, J.E.M. (1995). Socio-economic benefits and issues in non-wood forest product use, Report of the International Expert Consultation of Non-Wood Forest Products, Rome: Food and Agriculture Organisation.

Chandrasekhar, (1998). Role of NWFPs in Sustainable Forest Management, *Forest Usufructs*, Vol. 1 (Nos. 1 & 2), Dehradun.

Das, V. (1998). 'Human Rights, Inhuman Wrongs: Plight of Tribals in Orissa', *Economic and Political Weekly*, March 14-20, 33 (11): 571.

FAO (1995). *Report of the International Expert Consultation on NTFP*, Rome.

Fernandes, W., Kulkarni, S. and Menon, G. (1987). *Forest, Environment and Tribal Economy: Deforestation, Improverishment and Marginalisation in Orissa*, New Delhi: Indian Social Institute.

Government of India (1988). *Commodity Studies* (in six Volumes), New Delhi: Progressive Agro-Industrial Consultants, Ministry of Welfare, Government of India.

Khare, A. and Rao, A.V.R. (1993). 'Products of Social Forestry: Issues, Strategies and Priorities', *Wastelands News*, 6 (4): 7-17.

Mallik, R.M. (2000). "Sustainable Management of Non-timber forest products in Orissa: Some issues and options", India Journal of Agriculture Economics, July-September, 55 (3): 384-97.

Mallik, R.M. and N. Panigrahi (1998), Non-Timber Forest Produce collection: Benefits and Management in Orissa, The Ford Foundation, New Delhi.

MoEF (Ministry of Environment and Forests of India) (1998), 'Report of the Expert Committee on Conferring Ownership Rights of NTFPs on *Panchayats*', New Delhi: Government of India (unpublished).

Sarap, K. (2005). The Forest-Livelihood Context in Orissa, Phase-II, Project Report submitted to Centre for Economic and Social Studies (CESS), Hyderabad.

Sarap, K. (2007). "Forest and Livelihood in Orissa", in Oliver Springate Baginski and Piers Blaikie (Eds.), *Forests, People and Power—The Political Ecology of Reform in South Asia*, London: Earthscan.

Sarap, K and Sarangi, T.K. (2009). Malfunctioning of Forest Institutions in Orissa, *Economic and Political Weekly*, 44 (37): 18-22.

Sarap, K. and Sarangi T.K. (2010), "An Analysis of Forest Institutions in Orissa", *Journal of Social and Economic Development*, Volume 12, No 2, July-December, pp. 193-210.

Saxena, N.C. (2003). "Livelihood Diversification and Non-Timber Forest Products in Orissa: Wider Lessons on the Scope for Policy Change?", Working Paper No. 223, London, U.K: Overseas Development Institute.

Singh, R.V. (1997). "Evolution of Forest Tenures in India: Implications for Sustainable Forest Management (BC 1500-1997 AD)", A Ph.D. Thesis, Vancouver, Canada: The University of British Columbia. Unpublished.

Vasundhara (1998). "Non-timber Forest Products and Rural Livelihoods, with special Focus on Existing Policies and Market Constraints", London, U.K: Department for International Development.

Part III

Recent Environmental Problems

Chapter 19

Air Pollution Load in Bhadravathi, South India

SIRAJUDDIN M. HORAGINAMANI, HINA KOUSAR, M. RAVICHANDRAN, S.J.VEERESH, K.G. GIRISH AND S.R. PRAVITHA

ABSTRACT

Bhadravathi town is heartland of Karnataka state, is a famous industrial town not only of Karnataka but South India also, with many major industries like iron and steel, paper and pulp and sugar, with many medium and small-scale industries. The expansion of industrial growth and other anthropogenic activities in the town posses a serious threat to air environment. In the present study, an effort has been made to know the status of ambient air quality in terms of major air pollutants such as SPM, SO_2, and NO_2 sampled at three different sampling areas representing, residential, commercial and sensitive of Bhadravathi town. It has been observed that the concentrations of SPM concentration are high and beyond the standard given by NAAQS in all three sampling areas throughout the sampling period. Whereas the concentrations of SO_2 are well within the prescribed limits in all three sampling areas throughout the sampling of six month. The NO_2 concentrations are high and beyond the standards in residential as well as commercial areas, but exceeded in Sensitive areas and are well within the limits.

Keywords: SPM, SO_2, NO_2, Residenial area, Commercial area, Sensitive area.

INTRODUCTION

The phenomenon of environmental degradation and disruption is well known in developing countries due to expansion of industrial growth and many more pollution generating anthropogenic activities. The contaminants continuously interact with the environment to cause toxicity, disease, aesthetic distress, physiological disturbance and over and above all environmental decay (Boubel *et al.*, 1992; Agrwal *et al.*, 1991). Stack emissions from various industries emits large amounts of ash (Sanjay and Agrwal, 1991) and polluted gasses causes severe harm to living and non-living surrounding the industries (Stern *et al.*, 1992). Ambient air quality at various cities of towns throughout the world has been studied by several researchers. Of variety known air pollutants, Suspended particulate matter (SPM), Sulphur dioxide (SO_2) and Nitrogen dioxide (NO_2) are considered major air pollutants. In India a network of air quality monitoring has been established by NEERI (National Environmental Engineering Research Institute) in cities where its zonal laboratories are based and the pollution level at each station has been documented (NEERI, 1980, 1983, 1988).But no study on ambient air quality of Bhadravathi is found in recent literature. Hence, this present study has undertaken.

MATERIAL AND METHODS

Bhadravathi (13°50' N and 75°42' E, 578.9 m above sea level) heartland of Karnataka state, situated on the banks of holy River Bhadra and spread over an area of 67 Km^2 with total population of 1,60,662 as per 2001 census. It is an industrial town, the availability of iron ore at Kemmangudi and perennial water supply Bhadra reservoir facilitated the location of large-scale industries namely the Vishweshwaraya Iron and Steel Limited (VISL) Factory and Mysore Paper Mills Limited (MPM) in the Bhadravathi town during 1918 and 1936 respectively. Apart from large industries, various medium and small-scale industries are found the vicinity of the township. It is one of the most important industrial towns of Karnataka state. Because of its pollution status recently, central pollution control board identified Bhadravathi as one among 24 critical problem areas of India.

MATERIALS AND METHODS

1. Sampling Site

The selection of sampling sites for the present study was based on the location, population, regional background and other various factors. Accordingly, the study area has been classified as follows:

(a) Residential Area,
(b) Commercial area, and
(c) Sensitive area.

(a) Residential Area

For the present study Hosmane extension was selected as the residential area, since it was characterized by a thick population with schools, petty shops and houses and less commercial activities.

(b) Commercial Area

In the present study, Rangappa circle was chosen as the commercial area because it is the hub of commercial activities and heavy traffic.

(c) Sensitive Area

Sir M. Vishweshwaraya Arts and Commerce College campus was selected as the sensitive area. This area is characterized by several educational institutions, hospitals and parks.

2. Meteorological Parameters

The meteorological parameters like wind velocity, light intensity, humidity, rainfall, etc., were analyzed by means of meteorological instruments—Anemometer, Psychrometer, Thermometer, Illuminometer, Flat Plate vane and Rain Gauge developed by National Environmental Engineering Research Institute (NEERI). The meteorological station was set-up at sensitive area. Readings were taken in a day at an interval of 1 hour and average of that has been taken.

Doubled distilled water was used for preparing all the solutions and estimation. Ambient air quality was monitored for major air pollutants viz. Suspended Particulate Matter (SPM), Sulphur dioxide (SO_2) and Nitrogen dioxide (NO_2), with the help of High volume sampler (Envirotech APM-410). SO_2 and NO_2 were absorbed in Sodium tetrachloromercurate and Sodium hydroxide solution. Analysis of absorbing solutions by West and Gaeke method, and Griess-Saltzman method, respectively. SPM was collected on pre-weighed glass fiber filter (Whatman). Filter paper was again weighed after sampling and the difference in weight were used to calculate concentrations of SPM in respective areas and expressed as $\mu g/m^3$ of air. The monitoring was done for 24 hours. This research work was carried out between December 2004 to May 2005. Three sampling areas, namely, Hosamane (Residential), Rangappa Circle (Commercial) and Vishweshwaraya College campus (Sensitive), to assess the status of ambient air quality of Bhadravathi town.

For the collection of suspended particulate matter from ambient air, high volume air sample is used. The SPM was collected on glass microfiber filter paper (Whatman GF/A 20.3 × 25.4 cm). The mass concentration of SPM is calculated as per the methods mentioned by Sandhu and Gehlan, 1992 and Rao *et al.*, 1988. West and Gaeke's method [IS:5182 (Part II)-1969] pararosaline method was followed for the intermittent determination of sulphur dioxide at experimental locations. Chemicals used were mercury chloride, sodium chloride, Pararosaline hydrochloride, formaldehyde, sodium meta bisulphite, starch and iodine. Jacob and Hochheiser's method [IS: 5182 (Part VI)-1975] method was used for the determination of NO_2. Chemicals used were Sodium hydroxide, Sulphanilamide, N (1-naphtyl) ethylene diamene dihydrochloride, hydrogen peroxide.

RESULTS AND DISCUSSION

River Bhadra which is the lifeline for the residents of Bhadravathi town has been reduced to a dumping drain thanks to the indiscriminate discharge of effluents from the above mentioned industries. The ambient air quality analysis revealed startling results.

The main source for suspended particulate matter is VISP, where dust particles enter through mineral crushing area. Pollutants such as SO_2 and NO_x are released during the manufacture of cast iron and production of steel. The meteorological data of Bhadravathi showed slightly warm temperature due to the presence of these pollutants in somewhat high concentrations (Table 1). The temperature ranged between 27 to 35°C during the course of investigation. The wind direction was oriented more towards south-west, wherein lies a part of the city on the down stream.

TABLE 1

Monthly Mean Values of Meteorological Data (January-April, 2005)

Parameters	*January*	*February*	*March*	*April*
Wind velocity (km/h)	19.72	18.09	17.82	14.58
Wind direction	NE	N	SW	SW
Temperature (°C)	27	31.33	32.6	35
Relative humidity (%)	83	74	72.33	58
Rainfall (cm/day)	Nil	Nil	Nil	Nil
Light intensity (lux)	>5000	>5000	>5000	>5000

The present study reveals that the concentration of SPM varied from 189-1785 mg/m³ during the course of entire investigation. It was also found that the concentration of SPM was high during summer, reaching it's peak of 1785 mg/m³ during April and was comparatively low during January and February. Station-wise SPM concentration was highest in commercial area and the reason attributed for this is its proximity to VISP and a high vehicular density (Table 2). The statistical analysis reveals that SPM is positively correlated with meteorological factors like temperature and negatively correlated with factors like relative humidity and wind velocity (Tables 5, 6 and 7). These observations are in accordance with Chandrashekaran *et al.* (1996), Meenambal and Akil (1999) and Bhaskaran *et al.* (2002).

TABLE 2

Monthly Average Concentration (microgram/m³) of Air Pollutants in Residential Area (Hosamane)

Month	*SPM*	*SO_2*	*NO_2*
December	105.00	65.36	69.00
January	107.00	66.33	71.00
February	110.00	68.00	71.33
March	167.00	75.33	71.00
April	145.00	77.66	76.33
May	122.00	74.99	74.87

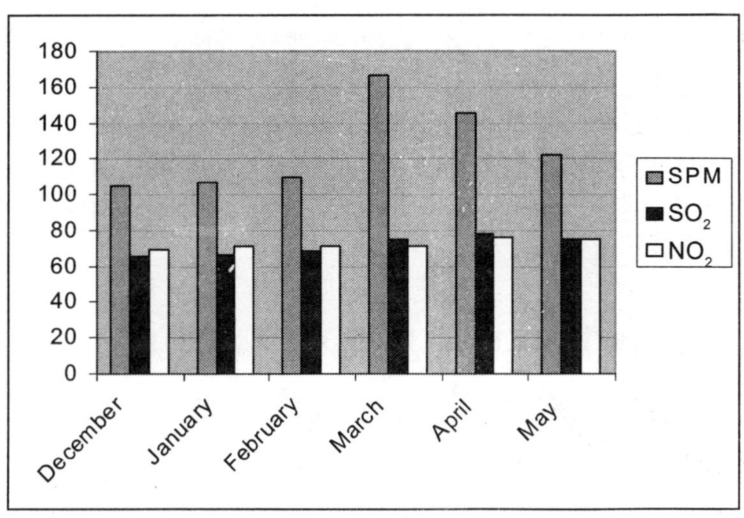

The average SO_2 concentration varied from a minimum of 66.5 mg/m^3 to a maximum of 107 mg/m^3 at residential area and sensitive area respectively, in the month of January and April (Table 3). The concentration of SO_2 positively correlated with temperature and negatively correlated with wind velocity and relative humidity (Tables 5, 6 and 7). These observations are in line with Bhaskaran *et al.* (2002), Seema Gujral *et al.* (2002) and Sanjay Tiwari *et al.* (1995). High concentration of SO_2 in this area is due to the fact that the areas are located within the industrial townships of these industries.

TABLE 3

Monthly Average Concentration (microgram/m^3) of Air Pollutants in Commercial Area (Rangappa Circle)

Month	*SPM*	*SO_2*	*NO_2*
December	1622.00	101.00	102.24
January	1626.00	103.00	104.70
February	1631.00	104.00	106.00
March	1684.00	106.00	107.70
April	1780.00	111.00	118.70
May	1785.00	111.45	119.86

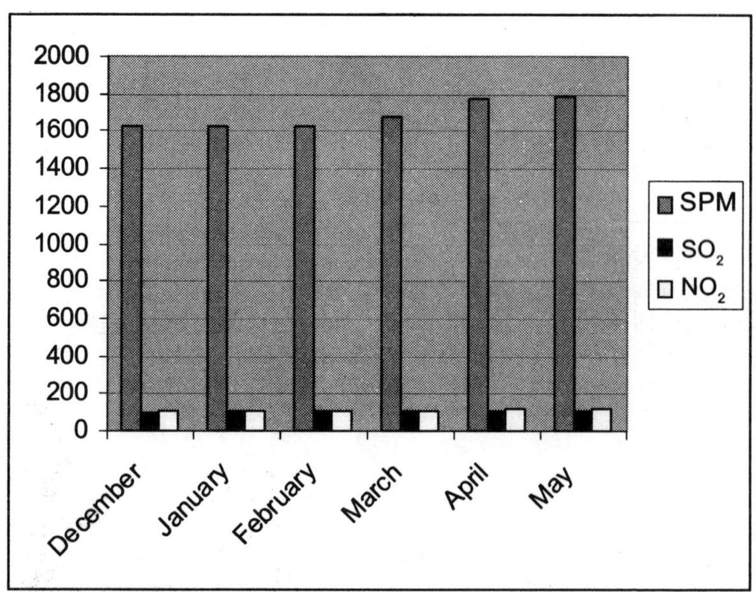

TABLE 4
Monthly Average Concentration (microgram/m^3) of Air Pollutants in Sensitive Area (Sir M. Vishweshwaraya College Campus)

Month	*SPM*	*SO_2*	*NO_2*
December	280.00	100.00	90.00
January	282.00	103.30	92.66
February	283.00	104.66	92.66
March	299.70	107.00	94.33
April	262.00	108.24	104.00
May	288.00	110.00	105.22

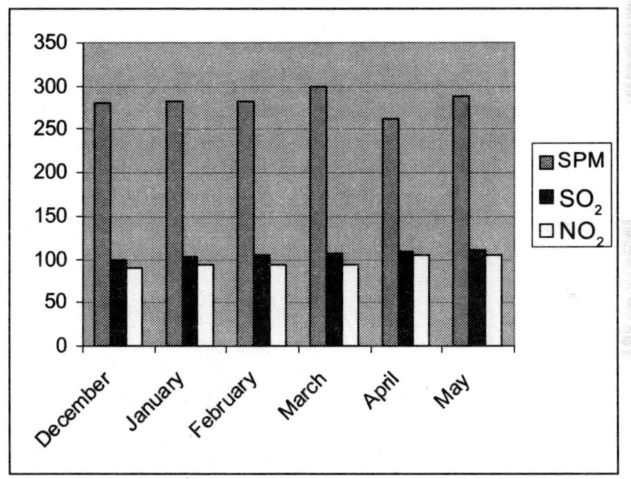

The NO_x concentration varied from a minimum of 71 to 118.7 mg/m^3. The highest concentration was recorded in commercial area during April and was lowest in January (Table 4). High concentration of NO_x may be due to high vehicular density. NO_x positively correlated with meteorological parameters like temperature and negatively correlated with wind velocity and relative humidity. These observations corroborate with the findings of Bhaskaran *et al.* (2002); Gujral *et al.* (2002) and Tiwari *et al.* (1995).

CONCLUSION

From the foregoing observations it can be concluded that SO_2 and NO_x are within the prescribed standards in residential and commercial

TABLE 5
Correlation of Meteorological Parameters v/s Air Pollutants in Residential Area

		SPM	*SO_2*	*NO_x*	*Wind velocity*	*Temperature*
SO_2	r	0.938*				
	P	0.031				
NO_x	r	0.891*	0.689*			
	P	0.054	0.155			
Wind velocity	r	-0.940*	0.853*	-0.928*		
	P	0.030	0.073	0.036		
Temperature	r	0.848*	0.895*	0.711*	-0.922*	
	P	0.076	0.052	0.145	0.039	
Relative humidity	r	-0.933*	-0.869*	-0.900*	0.998*	-0.946*
	P	0.034	0.066	0.050	0.001	0.027

*Values are significant at 0.05.

TABLE 6
Correlation of Meteorological Parameters v/s Air Pollutants in Commercial Area

		SPM	*SO_2*	*NO_x*	*Wind velocity*	*Temperature*
SO_2	r	0.996*				
	P	0.002				
NO_x	r	0.981*	0.986*			
	P	0.009	0.007			
Wind velocity	r	-0.954*	-0.977*	-0.971*		
	P	0.023	0.012	0.015		
Temperature	r	0.840*	0.875*	0.814*	-0.922*	
	P	0.080	0.062	0.093	0.039	
Relative humidity	r	-0.944*	-0.969*	-0.954*	0.998*	-0.946*
	P	0.028	0.016	0.023	0.001	0.027

*Values are significant at 0.05.

areas while SPM concentration is well above the limits. However, in sensitive area SPM, SO_2 and NO_x concentrations are beyond the standards limits.

TABLE 7
Correlation of Meteorological Parameters v/s Air Pollutants in Sensitive Area

		SPM	*SO_2*	*NO_x*	*Wind velocity*	*Temperature*
SO_2	r	0.937*				
	P	0.032				
NO_x	r	1.000*	0.940*			
	P	0.000	0.030			
Wind velocity	r	-0.950*	-0.929*	-0.945*		
	P	0.025	0.036	0.027		
Temperature	r	0.773*	0.731*	0.762*	-0.922*	
	P	0.114	0.135	0.119	0.039	
Relative humidity	r	-0.928*	-0.905*	-0.923*	0.998*	-0.946*
	P	0.036	0.047	0.039	0.001	0.027

*Values are significant at 0.05.

REFERENCES

Agarwal *et al.*, 1995, "A systematic study on meteorological and ambient air quality assessment at the industrial city Kota", *IJEP,* 15(10):776-82.

Boubel, R.W., *et al.*, 1992, "Fundamentals of air pollution" (3rd edition). Academic Press Inc., USA. 99-162.

Divya Gurtu, Meenal Vaidya and Gajghate D.G., 2001, "World Scenario of Particulate Matter NO_2 and SO_2 : A Review", *IJEP,* 21(8): 683-95.

Garg, S.S., 2002, "Seasonal Fluctuations in Ambient Air Quality of Chitrakoot Region, Satna", *IJEP,* 22(12): 1390-94.

Mahuya Das Gupta, Adak, S. and Purohit, K.M., 2002, "Study of Ambient Air Quality Based on SO_2 and NO_X Pollution in Mandiakudar", *IJEP,* 22(12): 1375-78.

Mohanty, S.K., 1999, "Ambient Air Quality Status in Koraput", *IJEP,* 19(3): 193-97.

Rao, M.N. and Rao, H.V.N., 1989, "Air Pollution", IInd Edn,Tata Mc Graw-Hill Publishing Co. Ltd., New Delhi, p. 17.

Sanjay, Tiwari, Ashu Rani and Agarwal, S.K., 1995, "A Systematic Study of Meteorological and Ambient Air Quality Assessment at Industrial City Kota", *IJEP,* 15(10): 776-80.

Seema Gujaral, Vikas Sharma, and Ashu Rani, 2001, "Assessment of Dispersion of Ambient Suspended Particulate Matter Meteorological Conditions near Kota Thermal Power Station", *IJEP,* 20(3): 238-249. www.cpcb.nic.in

Chapter 20

Effect of Air Pollution on Chlorophyll, Ascorbic Acid Content and Epiphytic Lichens Diversity

With Special Reference to Mangifera Indica and Pongamia Pinnata Species

NAVEEN, D., E.T. PUTTAIAH AND B.E. BASAVARAJAPPA

ABSTRACT

Ambient air quality monitoring was carried out for the two consecutive years from 2006 to 2008 in order to know the concentration of suspended particular matter, sulfur dioxide and oxides of nitrogen at Site-1 (Industrial area of the town) and at Site-2 (Kuvempu University Campus). In order to know the effect of air pollution on the surrounded tree species, analysis of chlorophyll and ascorbic acid in the leaves of *Mangifera indica* and *Pongamia pinnata* was carried out. In the meanwhile, the epiphytic lichens monitoring was done using European guideline. Results indicated that concentration of SPM (232.30 to 236.56 ìgm/m^3) was higher among the three analyzed air pollutants, although all the three air pollutants found well within the threshold limit at both the sites. The concentration of chlorophyll and ascorbic content were showed varied response from site-1 to site-2. Lichen species belonging to nine genera were identified on the tree barks of *Mangifera indica* and *Pongamia pinnata* at site-2 in the meantime sensitive lichen species showed their absence at site-1. The varied response of chlorophyll, ascorbic acid content and lichen species may directly attributed to air pollution scenario in Bhadravathi town.

Keywords: Air quality, Chlorophyll, Ascorbic acid, Lichens Diversity Value, Epiphytic lichen

INTRODUCTION

A major challenge facing India is to attain a proper balance between economic growth and environmental quality, of which air pollution is an important aspect. The quality of air in the urban areas is deteriorating due to rapid urbanization and industrial development, which creates a number of problems. Air pollution may be defined as any atmospheric condition in which substances are present at concentrations enough their normal ambient levels to produce a measurable adverse effects on men, animals, vegetation or materials.

During the last few decades increased industrial emissions, urbanization and heavy tonnage vehicular activity in Bhadravathi town have resulted the changes in air quality. Lichens are among the most valuable biomonitors of atmospheric pollution (Upreti *et. al.*, 2006). They have certain characteristics which meet several requirements of the ideal biological monitor. The first observation on sensitivity of lichens to air pollution dates back to 19th Century (Nylander, 1886). Since then large number of investigations in various countries have been carried out (Carreras, *et. al.*, 1998, Bargagli *et. al.*, 2002, Jayashree Rout *et. al.*, 2010). Lichens are often and effectively used as monitors of pollution. Lichens have certain characteristics that make them ideal biomonitoring organisms (Upreti *et.al.*, 2008).

Moreover, the monitoring of variation of biochemical parameters such as chlorophyll and ascorbic acid in the plants is also play an important role in bioindication of air quality. As such, the researchers who were concentrated mostly around large point sources of pollutants have started focusing on air pollution effects on vegetation in India (Agarwal, M., 2000). Plants are considered as an indicators and as well as mitigator of pollution (Mukherjee, 1993). Certain organisms by virtue of their adoptability to prevailing environment continue to exist or otherwise if sensitive become eliminated. Literature pertinent to response of trees to specific pollutant shows affluence. Such information is helpful in monitoring specific pollutants and in its mitigation.

Thus, the need for monitoring the vegetation responses to air pollution has increased than ever before, especially in the urban and industrial areas. Plants express the effects of the dose of air pollution integrating climatological, geological, biological and other environmental factors into a response. This may allow direct interpretation of the effect of air pollution exerts on the environment (Posthumus, 1980).

MATERIAL AND METHODS

Description of Study Area

Bhadravathi is an industrial town situated at 13°492 463 N to 75°422 223 Eÿþ in Shimoga District of Karnataka state, India. It is situated at a distance of about 255 km from the state capital Bangalore. Study area is notorious for its emissions from the two large scale industries (Visweshwaraya Iron and Steel Plant and Mysore Paper Mills Limited) in addition to numerous small scale industries of the town. Air quality of Bhadravathi town was compared with that of control area (Kuvempu University Campus).

Determination of Ambient Air Quality

Ambient air quality monitoring was carried out by following standard methods of the National Ambient Air Quality Monitoring (NAAQM). Air sampling was carried out using APM-410 and APM-411 high volume air sampler. The sampling frequency was 24 hours, twice a week at uniform intervals and for a period of two consecutive years (July 2006 to June 2008).

Specification of High Volume Air Sampler used for air monitoring

Flow Rate	-	0.9-1.2m^3/min
Recommended Filter	-	Whatman Filter Paper (25 cms)
Sampling Time	-	Normally 8 hours, 24 hours (max)
Power Requirement	-	220 V, single phase, 50 Hz, A.C. built in voltage stabilizer with automatic shut-off.

Determination of Suspended Particulate Matter

Filter inspection: The light table surface was cleaned with a methanol soaked wiper and allowed it to dry. Filter was handed with flowed hands to prevent contamination. Before placing each filter on the filter chamber of instrument it was examined for pinholes on the light table.

Suspended Particulate Matter was measured by weight/volume and the mass/quantity of SPM. In each case it was determined by weighing the filter paper before and after sampling with proper equilibrium each time. The monthly mean of SPM for each of the stations were obtained separately using the values of all the samplings of the respective month.

Determination of Sulphur Dioxide (Modified West and Gaeke Method)

The ambient air was bubbled through the aqueous solution of

Potassium tetrachloromercurate (TCM) and the SO_2 in air forms a dichlorosulphitomercurate complex, which resists oxidation by the oxygen in the air. This complex was stable to strong oxidants such as ozone and oxides of nitrogen and therefore, the absorber solution is stored for a couple of hours prior to analysis. This complex was made to react with pararosanilline and formaldehyde to form the intensely colored pararosaniline methylsulphonic acid. The absorbance of the solution was measured by means of spectrophotometer (Systronics 367).

Determination of Oxides of Nitrogen

The gas was collected in the absorber of the air sampler and the mixture was analyzed through the sodium arsenite method (Jacob and Hochheiser, 1958).

Ambient NO_x was collected by bubbling air through a solution of sodium hydroxide and sodium arsenite. The concentration of nitrite ion (NO^-_2) produced during sampling is determined colorimetrically by reacting the nitrite ion with phosphoric acid, sulfanilamide and N-(1-naphthyl)-ethylenediamine di-hydrochloride (NEDA) and measuring the absorbance of highly colored azo-dye at 540 nm using a spectrophotometer (Systronics 367).

Sampling of Plants

The leaves of *M.indica* and *P.pinnata* were collected during the air quality monitoring and care was taken that the sample trees were of almost same age and belong to more or less same girth and height class for a given species. Full grown leaves of the twigs at a reasonable height above the ground level and at the outer periphery of the canopy were collected for the analysis. The sampling was done thrice in a year.

Estimation of Ascorbic Acid

Ascorbic acid or vitamin C is an antiscorbutic. Ascorbic acid contents were determined by the method of Aberg (1958) and Sadasivam and Manickam (1997).

Estimation of Chlorophyll

The concentration of chlorophyll in the leaf samples of selected trees were determined by the method of Arnon (1949) and Sadashivam and Manickam (1997).

Surveying Lichen Diversity

A monitoring quadrate consisting of four independent quadrate segments of five 10 × 10 cm squares each as shown in Figure 1 is attached vertically to the trunk so that the lower edge of each segment is 1 m above highest point of the ground.

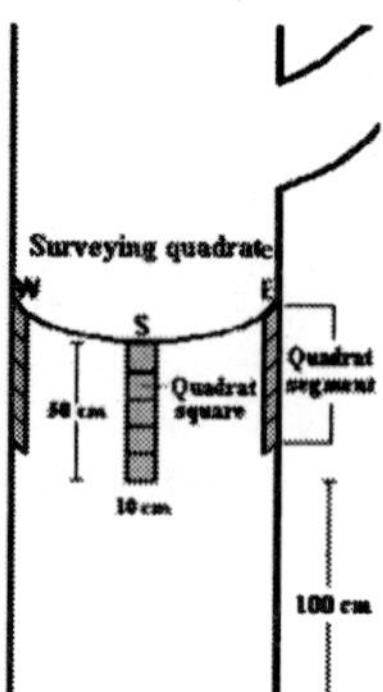

Figure 1 : Recording Quadrate Composed of Four Quadrate Segments each with Five Squares

Lichens were identified according to the methods described in the published literature and using European Guidelines (Asta, 2002). LDVs were calculated.

Statistical analysis of the obtained data was done using SPSS software v. 12.0.

RESULTS AND DISCUSSION

From the two years continuous monitoring it is evident that all the pollutants are within the threshold limit but the concentration of SPM was found to be more in concentration than SO_2 and NO_x. The SPM concentration was ranged from 10.00 ± 3.74 at control site during 2006-07 to 236.56 ± 83.22 ìgm/m^3 during 2006-07 at industrial area. On the other hand, the concentration of SO_2 ranged from 0.033 ± 0.06 ìgm/m^3 at control site to 13.62 ± 6.09 ìgm/m^3 during 2006-07 at industrial area. While, NO_x ranged from 0.28 ± 0.31 ìgm/m^3 at control site to 19.69 ± 7.88 ìgm/m^3 during 2006-07 at industrial area.

The chlorophyll and ascorbic acid parameters of the leaves of tree species at industrial area showed the significant variation from control site (Table 2). The percentage reduction of ascorbic acid was noticed to be 46.15% in *P.pinnata* and 44.90% in *M.indica*. A reduction in ascorbic acid contents in the leaves of the tested plant species at the industrial area indicated that, higher the ascorbic acid content in the leaves greater will be the tolerance of those plants to air pollution. Tinku, G. and Ambarish, M. (2003) in Burdwan found low ascorbic acid content in plant species *Ficus benghalensis* resulting in rating it least tolerant species among the thirteen tested plant species. All values of air pollutants were expressed as the mean of 96 trials. On the other hand, the biochemical parameters tested were expressed as the mean of 6 trials.

TABLE 1
Annual Concentrations of SPM, SO_2 and NO_x (ìg/m^3) Showing Significant Differences among the Industrial Area and Control Site

Sites	*Control*	*Around industrial area*	*F-value*	*P-value*
SPM				
2006-07	20.08 ± 11.91	236.56 ± 83.22	122.16	0.0001
2007-08	20.15± 16.14	232.30 ± 81.63	115.85	0.0001
SO_2				
2006-07	0.07± 0.09	13.23 ± 4.75	134.56	0.0001
2007-08	0.033 ± 0.06	13.62 ± 6.09	133.87	0.0001
NO_x				
2006-07	0.35 ± 0.36	19.69 ± 7.88	136.21	0.0001
2007-08	0.28 ± 0.31	19.15 ± 6.88	126.01	0.0001

± Std. Dev

Further, obtained data pertaining to the air pollutants and biochemical parameters of tested plants were subjected to one multifactorial analysis of variance (ANOVA) in order to determine which means were significantly different from which others ($p < 0.001$).

Ascorbic acid reduction associated with the air pollution exposure was explained by Keller and Schwager (1977), Singh and Rao (1983), Varshney and Varshney (1984), and Agrawal *et al.* (1991). Similarly, Agrawal *et al.* (1991) found correlation between ambient air pollutants concentration and decrease in the levels of ascorbic acid along with the chlorophyll.

Pollution load dependent increase in ascorbic acid content of all the species may be due to the more rate of production of reactive oxygen species (ROS) such as SO_3^-, HSO_3^-, OH-, O_2^-, etc. during photo oxidation of SO_3^- to SO_4^- where sulfites are generated from SO_2 absorbed. The free radical production under SO_2 exposure would increase the free radical scavengers, such as ascorbic acid, super oxide dismutase, peroxidase etc. (Dwivedi, A.K. and Tripathi, B.D., 2007) based on dosage and physiological status of plant. The reductions in ascorbic acid may be attributed to its consumption for scavenging cytotoxic free radicals generated during the chain reaction after the absorption of pollutants into the foliage.

The chlorophyll contents in the leaf extracts of the trees also showed varied response from sampling site-1 to sampling site-2. The

TABLE 2

Variation of Ascorbic Acid and Chlorophyll Contents (Mean ± SD) in Tested Plants

Tested Plants	*Ascorbic Acid*		*Chlorophyll-a*		*Chlorophyll-b*		*Total Chlorophyll*	
	Industrial Area	*Control Site*	*Industrial Area*	*Control Site*	*Industrial Area*	*Control Site*	*Industrial Area*	*Control Site*
(1)	*(2)*	*(3)*	*(4)*	*(5)*	*(6)*	*(7)*	*(8)*	*(9)*
M.indica	2.92 ± 0.02	5.30 ± 0.22	1.29 ± 0.02	1.41 ± 0.02	0.39 ± 0.06	0.59 ± 0.06	1.68 ± 0.001	1.98 ± 0.01
P.pinnata	1.89 ± 0.02	3.51 ± 0.05	0.98 ± 0.07	1.30 ±0.06	0.41 ± 0.01	0.60 ± 0.02	1.39 ± 0.03	1.90 ± 0.10

± Std. Dev

maximum reduction of chlorophyll a was noticed in both the trees; *M.indica* (28.51%) and *P.pinnata* (24.61%).

On the other hand the reduction of chlorophyll b was noticed in *P.pinnata* (32.78%), *M.indica* (31.57%) was higher than chlorophyll a. From the results it is evident that, the chlorophyll reduction in the leaves of tested trees is directly attributed to the air pollution effect in the industrial area. Dust may cause chlorosis and death of leaf tissue by the combination of a thick crust and alkaline toxicity produced in wet weather (Tariq J. *et al.*, 2008).

The obtained results are attributed to the findings of various other researchers all over the country Reduction of chlorophyll may be due to the increase of chlorophyllase enzyme activities, which in turn affects the chlorophyll concentration in plants (Mandal and Mukherji, 2000). SO_2 plays an important role in the reduction of chlorophyll content (Rao and Dubey, 1985; Mandloi and Dubey, 1988). Although the tree species showed significant variation in the biochemical parameters, the extent up to which plant species were affected varied from species to species and station to station. Air pollutants make their entrance into the tissues through the stomata and cause partial denaturation of the chloroplast and decreases pigment contents in the cells of polluted leaves (Tripathi, A.K. and Mukesh, G., 2007).

M.indica and *P.pinnata* showed 80% each reduction in chlorophyll at Korba Power Plants, M.P., India (Williams *et al.*, 1995). Samal and Santra (2002) at Kalyani city, Bengal noticed 5.66%, 12.08% reduction of chlorophyll content in *M.indica* and *P.longifolia* respectively. The possible secondary additive factors could not affect the morphology of tested plants, as the soil type and weather of the neighbouring areas are very much similar (Bhadravathi and Shankaraghatta).

Lichens Sensitivity to Air Pollution

The study also reveals that the diversity of lichens is affected from the air pollution scenario in the study area. The sampling site-2 have showed significantly lesser LDV from that of control site. In all, a total of 9 genera of epiphytic lichens were identified: Pyxine, Chrysothrix, Parmotrema, Teloschistes, Dirinaria, Graphis, Ramalina, Lecanora, and Heterodermia (Figure 2). Their occurrence is listed in Tables 3 and 4.

In comparison, all 9 genera of epiphytic lichens were present in the control site, but lichens belonging to the genera Ramalina and Teloschistes were absent on M.indica in the industrial area. Species of genera Teloschistes, Ramalina and Heterodermia were absent on the bark of P.pinnata while only six genera were found. These results suggest that the control site had high diversity of lichens with

TABLE 3
Variation in Lichen Diversity (mean ± SD) at Control Site

Lichen	*M.indica*				*P.pinnata*			
	North	*East*	*South*	*West*	*North*	*East*	*South*	*West*
(1)	*(2)*	*(3)*	*(4)*	*(5)*	*(6)*	*(7)*	*(8)*	*(9)*
Pyxine	08 ± 0.81	10 ± 1.41	02 ± 0.00	01 ± 0.81	01 ± 0.81	02 ± 0.00	08 ± 1.41	04 ± 3.55
Chrysothrix	02 ± 0.81	03 ± 0.81	—	—	01 ± 0.00	—	—	—
Parmotrema	—	01 ± 0.81	—	—	03 ± 0.00	02 ± .00	01 ± 0.81	04 ± 1.41
Teloschistes	—	—	—	—	—	—	—	—
Dirinaria	01 ± 0.81	01 ± 0.81	—	—	—	01 ± 0.81	02 ± 0.81	—
Graphis	01 ± 0.00	05 ± 0.81	01 ± 0.81	02 ± 0.81	01 ± 0.00	01 ± 2.15	01 ± 0.81	06 ± 0.81
Ramalina	—	—	—	—	—	—	—	—
Lecanora	01 ± 0.81	02 ± 0.81	—	—	01 ± 0.81	—	02 ± 0.81	—
Heterodermia	01 ± 0.81	—	—	—	—	—	—	—
Sum of Frequencies	14 ± 2.45	22 ± 3.21	3 ± 0.71	3 ± 1.12	7 ± 0.98	6 ± 1.51	14 ± 2.52	14 ± 2.56

± Std. Dev

TABLE 4

Variation in Lichen Diversity (mean ± SD) at Industrial Area

Lichen	*M.indica*				*P.pinnata*			
	North	*East*	*South*	*West*	*North*	*East*	*South*	*West*
(1)	*(2)*	*(3)*	*(4)*	*(5)*	*(6)*	*(7)*	*(8)*	*(9)*
Pyxine	16 ± 4.96	04 ± 0.81	04 ± 0.81	05 ± 0.81	04 ± 0.00	12 ± 1.41	02 ± 0.00	02 ± 0.00
Chrysothrix	03 ± 0.00	04 ± 0.00	02 ± 0.81	—	01 ± 0.00	05 ± 0.00	02 ± 0.00	08 ± 1.63
Parmotrema	03 ± 0.00	—	—	02 ± 0.81	06 ± 0.81	03 ± 0.00	04 ± 0.00	—
Teloschistes	02 ± 0.81	—	—	07 ± 2.12	02 ± 0.00	02 ± 0.00	—	01 ± 0.00
Dirinaria	04 ±0.00	02 ± 1.63	06 ± 0.81	—	03 ± 0.00	09 ± 0.00	03 ± 1.41	06 ± 0.00
Graphis	03 ± 0.00	06 ± 0.00	—	04 ± 1.63	—	02 ± 0.00	04 ± 0.00	04 ± 1.63
Ramalina	08 ± 1.63	—	04 ± 0.81	—	05 ± 0.81	07 ± 0.00	01 ± 0.00	—
Lecanora	03 ± 0.81	—	01 ± 0.00	05 ± 0.81	06 ± 1.63	06 ± 0.81	06 ± 0.95	01 ± 0.81
Heterodermia	03 ± 0.00	09 ± 2.16	06 ± 0.81	02 ± 00	02 ± 0.00	07 ± 0.00	01 ± 1.41	05 ± 0.81
Sum of Frequencies	43 ± 4.55	25 ± 3.19	23 ± 2.45	25 ± 2.63	29 ± 2.15	53± 3.22	23 ± 1.96	27 ± 2.84

± Std. Dev

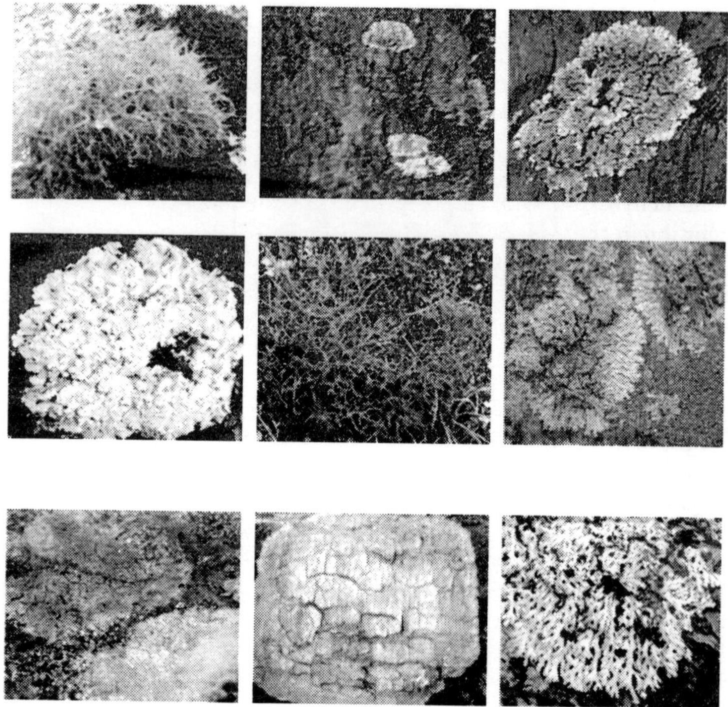

Figure 2 : Lichen Genera Identified during the Study Period in Bhadravathi Town

frequencies of 118 ± 7.23 and 132 ± 5.12 on the barks of M.indica and P.pinnata, respectively. The genus Pyxine (29 ± 5.8) dominated while Parmotrema (5 ± 0) showed the least diversity on M.indica. Dirinaria sp. (21 ± 2.8) was dominant on P.pinnata.

At the industrial site Pyxine sp. (21 ± 4.4) was most common on the bark of M.indica. The total sum of frequencies of lichen genera was 42 ± 3.5, indicating lichen diversity. Pyxine sp. (15 ± 3.1) dominated on P.pinnata. The industrial area had and LDV of 41.7, about three times lesser than the control site (122).

The varied response directly attributed to air pollution scenario in the study area. Since lichens lack roots, surface absorption of rainfall is the only means of obtaining vital nutrients which are dissolved in rainwater. Lichens lack protective surfaces that can selectively block out elements including pollutants that are dissolved in rainwater. Lichens act like sponges, taking in everything that is dissolved in the rainwater, and retaining it. The variation in response could be directly attributed to the emissions from two large industries and from a number of small-scale industries in the study area. This observation supports what was noted in earlier studies on lichens at Pauri City, Uttaranchal, India by Vertica Shukla and Upreti (2007). A qualitative survey of the epiphytic lichens in the surroundings of Ulan Bator in October 2007 also showed similar trends (British Lichen Society, 2008). Gombert *et al.* (2006) found lichens and tobacco plants as complementary biomonitors of air pollution in the Grenoble area (Isere, southeast France).

Lichen species belonging to genus Fixin dominated at industrial area including control site and thus it is rated as most tolerant genus among all the nine genera of lichens identified on the tree barks of the two *M.indica and P.pinnata*. While, the lichen species of genus Ramalena were most sensitive among the nine genera at study area since, the species of genus were infrequent at industrial area where the air pollution was noticed to be more. A number of species were no longer found and he also found significant decrease in species diversity of Ramalina species (*Science Daily*, December 26, 2002). Lichen diversity counts can be taken as estimates of environmental quality: high values correspond to good situations, while low values indicate poor air quality (Asta *et al.*, 2002). Such measures are chiefly depending from the two main reactions of epiphytic lichen communities to air pollution by phytotoxic gases, especially SO_2 and NO_x: a reduction in the number of species and in their frequency (Nimis, 1999).

CONCLUSION

Obtained results regarding to the effect of air pollution on vegetation showed significant correlation with the concentration of air pollutants at respective sampling sites. It is suggestible that the resistant plants manifest in the investigation can be employed in abatement of air pollution. They can be grown in and around the industries and road side to reduce the pollutants in air. Secondary benefits can be derived from them, as some of them have medicinal importance and yield edible fruits. *P.pinnata* seeds will be made use in the production of bio-fuel as well.

Moreover, a relationship existed been between the lichen community existing at sampling sites and the degree of air pollution. The absence of naturally appearing lichens in severely polluted areas limits the spatial differences of polluted areas. Zones based on epiphytic lichen vegetation provide a better indication of air pollution intensity than distribution maps of particular species. Hence, documentation of the lichen species at the study area will be of greater importance in the future.

REFERENCES

Aberg, B. (1958) Ascorbic acid. In: Burstorm, H. (ed.) *Encyclopedia of Plant Physiology*, Springer-Verlag, Berlin, 6:479-497.

Agarwal Madhoolika (2000) Research on air pollution: Effects on vegetation in India: A review, *The Botanica*, 50: 75-83.

A.K. Tripathi and Mukesh Gautam (2007) Biochemical parameters of plants as indicators of air pollution, *Journal of Environmental Biology*, 28(1) 127-32.

Arnon, D.I. (1949) Copper enzymes in isolated chloroplasts polyphenoloxidases in Beta vulgaris. Plant Physiology, 24:1-15.

Asta, J., Erhardt, W., Ferretti, M., Fornasier, F., Kirschbaum, U., Nimis, P.L., Purvis, W., Pirintsos, S.A., Scheidegger, C., Van Haluwyn, C., Wirth, V. (2002), Mapping lichen diversity as an indicator of environmental quality, monitoring with lichens—monitoring lichens. NATO Science Series IV (Vol. 7), Kluwer Academic Publishers, Dordrecht, pp. 273-79.

Bargagli, R., Monaci, F., Borghini, F., Bravi, F. and Agnorelli, C. (2002) Mosses and lichens as biomonitors of trace metals. A comparison study on *Hypnum cupressiforme* and *Parmelia caperata* in a former mining district in Italy. *Environmental Pollution*, 116, 279–87.

Carreras, H.A., G.L. Gudino and M.L. Pignata (1998) Comparative biomonitoring of atmospheric quality in five zones of Cordoba City (Argentina) employing the transplanted Usnea sp.,Environmental pollution. 103:317-25.

Dwivedi, A.K. and B.D. Tripathi (2007) Pollution tolerance and distribution pattern of plants in surrounding areas of coal fired industries. Journal of Environmental Biology, 28(1):257-63.

Environment (2009) Canada's Online News Magazine Issue, 97, December.

Gombert, S., Asta, J., Seaward, M.R.D. (2006) Title of paper. Ecological Indicators 6, 429-43.

Jacob, M.B. and Hochheiser, S. (1958), Continuous sampling and ultra-micro determination of nitrogen dioxide in air. Anal. Chem., 30: 426.

Jayashree Rout, Urvashi Dubey and D.K. Upreti (2010), A Comparative Study of Total.

Chlorophyll Content and Chlorophyll Degradation of Some Lichens in Disturbed and Undisturbed Sites of Along town, West Siang District, Arunachal Pradesh, Assam.

University Journal of Science & Technology. Vol. 6, Number I 46-51.

Mandal, M. and Mukherji, S. (2000), Changes in chlorophyll content, chlorophllase activity, Hill reaction, photosynthetic CO_2 uptake, sugar and starch content in five dicotyledonous plants exposed to automobile exhaust pollution. Journal of Environmental Biology, 21(1), 37-41.

Mandloi, B.L. and Dubey, P.S. (1988), The industrial emission and plant response at Pithanpur (M.P). International Journal of Ecological and Environmental Science, 14, 75-99.

Mukherjee, A. (1993), Plants as indicator and mitigators of pollution. Indian Sci. Cruiser, 7: 11-17.

Nylander, W. (1886), Less lichen du Jardin de Luxemberg, Bulletin de la Societe Botanique de France 13: 364-72.

Posthumus, A.C. (1980), Monitoring levels and effects of air borne pollutants on vegetation. In use of biological indicators and other methods: National and International programmes. Paper presented to the Symposium on the effect of air borne pollution on vegetation. Warsaw, Polland. United Nations Economic Commission for Europe.

Rao, M.V. and Dubey, P.S. (1985), Plant response against SO_2 in field conditions. Asian Environment 10:1-9.

Rao, M.V. and Dubey, P.S. (1990), Biochemical aspects (antioxidants) for development of tolerance in plants growing at different low levels of ambient air pollutants. Environ. Pollut. 64:55-66.

Rao, M.V. and Dubey, P.S. (1990), Biochemical aspects (antioxidants) for development of tolerance in plants growing at different low levels of ambient air pollutants. Environ. Pollut. 64:55-66.

Sadasivam, S. and Manickam, A. (1997), Biochemical Methods. 2nd Edition, New Age International (P) Ltd. Coimbatore, India. p. 256.

Samal, A.C. and Santra, S.C. (2002), Air quality of Kalyani township (Nadia, West Bangal) and its impact on surrounding vegetation. Indian J. Environ. HLTH. 44 (1): 71-76.

Singh, S.N. and Rao, D.N. (1983) Evaluation of plants for their tolerance to air pollution. In proceedings of Symposium on Air pollution control Vol. I. Indian association for Air pollution control, New Delhi, 218-24.

Tariq Javed, Shahzad, M.A. Basra and Irfan Afzal (2008) Model Farming "Crop air pollution assessment methodology", Pakissan.com

Upreti, D.K., Nayaka S. (2003) Review of lichenology in India during 2001-03. *Brit. Lich. Soc.* Bull 93, 37-39.

Upreti, D.K., Shukla V. and Nayaka S. (2006) Heavy metal accumulation in lichens of Dehradun City,Uttaranchal, India, *Indian Journal of Environmental Sciences,* pp. 165-69.

Varshney, S.R.K. and C.K. Varshney (1984) Effects of SO_2 on ascorbic acid in crop plants Environ, Pollut., 35: 285-90.

West, P.W. and Gaeke, G.C. (1956) Fixation of sulfur dioxide as sulfitomercurate III and subsequent colorimetric determination, Anal. Chem., 28 : 1816.

Williams, A.J., Benerjee, S.K. and Gupta, B.N. (1996) Plant biochemical.

Chapter 21

Management of the Tropical Oil Seed Plant Jatropha Curcas

ANANTHANAG, B., BASAVARAJ, S. AND PUTTAIAH, E.T.

ABSTRACT

Jatropha curcas is a multipurpose plant with many attributes and considerable potential. It is a tropical plant that can be grown in low to high rainfall areas and can be used to reclaim land, as a hedge and/or as a commercial crop. Thus, growing it could provide employment, improve the environment and enhance the quality of rural life. The establishment, management and productivity of *Jatropha* under various climatic conditions. The development of *Jatropha curcas* as a possible energy crop in India an alternative fuel produced from domestic, renewable resources and it can be blended at any level with petroleum diesel to create a biodiesel blend and can be used in compression ignition (diesel) engines with no major modifications. The oil obtained from the seeds of *Jatropha Curcas* holds promise as fuel used as alternative for diesel. These have a variety of uses, but the economic exploitation of these plants has remained neglected for long time. Survey of *Jatropha Curcas* plant species at the study area was conducted during April 2009-Oct. 2010 to know the density of the species and impact of morphological characteristics on the yield efficiency of the species.

Keywords: Biofuels, *Jatropha Curcas*, Economy, Petroleum Products, Rural Employment

INTRODUCTION

Indian economy is one of the fastest growing economies in the world and the energy demand is growing at state if 4.6% every year. (IEA 2001; Planning Commission, 2003). The transport sector has witnessed a rapid surge in energy demand, diesel, petrol, and compressed natural; gas are main fuels used on the transport sector. In 2007/08. 70% of the country's crude oil requirement was met by importing 18.36 billion dollar worth of crude oil . as this rate, the country would require 5.8 million barrels oil/day by 2030, of which more than 94% will be met through oil imports (EIA, 2001). Extreme dependence on petroleum as a primary energy source entails considerable risks, both from energy security as well as the environmental point of view. In this context, biodiesel emerges as an important alternative transport fuel. Globally, among all biofuels, is already being produced at a fair scale. Global biodiesel production during 2005 is estimated to be 41 million kilolitres. Of which about 70% is used as fuel (Shete, 2005). Majority of the production and consumption takes place in Brazil and USA.

Biodiesel has many advantages over petroleum diesel. Biodiesel operates in compression ignition engines just like petroleum diesel. It requires no engine modification and increases engine life. It is biodegradable and non-toxic. Biofuel have a prominent role in environmental protection and poverty reduction, presently energy security can be achieved by shift towards renewable. Biofuel play a key role in this regard. According to an estimate Asian countries would drive the future energy demand, but the insufficient policy guidelines have hampered the adaptation of biodiesel as a viable alternative (Ram Mohan. *et. al.*, 2006).

The renewable liquid fuel produced from biological raw material is a good substitute for petroleum diesel. It is gaining worldwide acceptance as an environment-friendly solution to the energy problem. It is an accepted option for achieving energy security, reduction in imports, rural employment, and for improving the agricultural economy. Biodiesel results in substantial reduction of unburnt hydrocarbons, carbon monoxide and particulate mater. The IOCL (Indian Oil Corporation Ltd.) reported that the maintenance cost of vehicles run on biodiesel has no sulphur and very little aromatic hydrocarbons, and has about 10% built in oxygen that helps it to burn freely. A higher cetane number improves the combustion.

All countries of the world, including those with surplus energy, are banking upon vegetable oil as an alternative source of energy by way of biodiesel. Developing countries cannot afford to utilize edible vegetable oil or even used vegetable oil. However, many of these countries, like India, have large tracts of wastelands and a tropical climate. The oil extracted from plants such as *Jatropha, Pongamia*, etc., can be processed onto biofuel. In fact, *Jatropha* is very hardy plant that grows well in semi-arid conditions and is not browsed by cattle or attacked by pests. The process involved in producing biofuel from these plants can generate employment especially in the rural area. (*Source:* www.biospectrumindia.com).

Study Area

Shimoga, a place known for its scenic beauty, flush green lush forests, eye-catching waterfalls, cool climate is situated in the Malnad region bounded by Sahyadri Ghats at an mean elevation of 640 AMSL in the western part of Karnataka. Shimoga is situated between 13027' and 14o39' north latitude and 74038' and 7604' east longitude. The district is spread over an area of 1058,000 hectares with a forest area of 327,000 hectares. The eastern part of district comes under the semi-malnad zone with plain topography and occasional chains of hills covered with semi-deciduous vegetation.

Shimoga district is rich in flora and fauna, the dense forest and green shrub jungles are main producers of sandalwood, rosewood, teak and other exotic timber. The other important trees found around the district with rich yields. Shimoga farmers are feeding the people of Shimoga and other districts of Karnataka by producing good quality paddy, coconut, Ragi, Pepper, Areca and Sugarcane.

Four sampling sites were selected from all over the Shimoga district. The sampling sites are located on a study map (Figure 1).

Seed Sources

For the present study seeds were collected from the four regions that is Sagar, Thirthalli, Sorobha and Bhadravathi area of Shimoga. The Physiologic, Physiographic and climatic conditions of the places where seeds are collected are given in Table 1.

TABLE 1

Study Sites	*Soil type*	*Annual rainfall (mm)*	*Temperature (°C)*
Sagar	Red Soil	1200	20-30
Thirthalli	Loamy soil	1100	20-30
Sorabha	Red and Black	955	24-32
Bhadravathi	Black soil	700	26-34

Note : * Average Temperature during April 2009-Oct. 2010.

DESCRIPTION OF TESTED PLANTS

There are many tree species which bear seeds rich in oil having properties of excellent fuel and which can be possesses into diesel substitute. In Karnataka the important commercial non-edible oil yielding plants are, *Pongamia pinnata, Jatropha curcas, Azadirachta*

Figure 1 : Location Map of Study Area Showing Sampling Stations

indica, Madhuca longifolia, Schlichera oleosa, Garcinia indica, Calophyllum inophyllum and many more. The oil obtained from such seeds is chiefly used for manufacture of soaps, candles, paints, varnishes, linoleum, and lighting for medicinal purposes. Among the various oil yielding plants *Jatropha curcas* were selected for the study. The detailed description of the plants are as follows :

Jatropha curcas

Family	-	Euphorbiaceae
English	-	Purging Nut, Physic nut

Hindi - Bagberenda
Kannada - Adalu Haralu, Kaadu Haralu

The study was conducted by selecting 4 sites of Sagar, Thirthalli, Bhadravathi and Sorabha area in Shimoga district (see Figure 1). The survey was conducted in the months of April 2009-Oct. 2010. In each location, mature pods from healthy seeds were collected. The pods were sun-dried and split longitudinally for separating the seeds. From each location 50 seeds were selected at random and their length, breadth and thickness were measured. Seed weight was determined with a sample of 60 seeds. Number of pods and seeds per kg and recovery percentage were also calculated and the moisture was estimated by a low constant temperature method. The seedlings were dried in a hot air oven at 90°C for 24 hours and the dry weights were determined. The data were analyzed for standard deviation.

RESULTS AND DISCUSSION

Seed weight was highest in Sagar seeds followed by Thirthalli and Sorabha while the lowest was in seeds of Bhadravathi. Seed length, breadth and thickness were also higher in Sagar and lower in Bhadravathi seeds. Later, the generated data pertaining to girth, height and canopy were correlated with yield. The height and yield is significant in seeds of site-1 and 2, but it is insignificant in seeds of site-3 and 4. The number of seeds per kg was highest in Sagar, lowest in Bhadravathi. The differences in weight, length, breadth, thickness and number per kg of the seeds were substantial among the different seed lots. Similar seed morphological observations have been made by Athaya (1985). Moisture content varied significantly among the seeds and it varied from 1.2% to 0.7% in Sagar to Bhadravathi (Table 2). The highest germination percentage and vigor index were observed in seeds collected from Sagar while the lowest was in Bhadravathi seeds. Large seeds may have more food required for vigorous early growth of seedlings. The seeds in the present study showed wide variation in seed

TABLE 2

Variation in Physical Characteristics of *Jatropha Curcas* Seeds

Location	*Seed length (mm)*	*Seed breadth (mm)*	*Seed Thickness (mm)*
Sagar	1.88	1.4	1,2
Thirthalli	1.76	1.2	1.0
Sorabha	1.66	1.0	0.9
Bhadravathi	1.53	0.8	0.7

germination. This type of variation is depends on the topography and soil type of the study area. The highest yield was noticed in Sagar site, because the rainfall of this area is higher than the other sites. Bhadravathi site having lowest seeds weight, because this study site receiving less amount of rainfall compare to other study sites. It is concluded that this *Jatropha curcas* is grow well in rainfall areas than that of drier zones.

References

Allen, O.N. and Allen, E.K. 1981. The Leguminosae The University of Wiscons in press 812 p.

Anon, 1986. The useful plants of India. Publications and Information's Directorate, C.S.I.R., New Delhi, India.

Athaya, C.D. 1985. Ecological studies of some forest tree seeds. Seed morphology. *Indian. J. For.*, 8:33-36.

C.S.I.R, 1948-1976. the Wealth of India, 11 Vols, New-Delhi.

C.S.I.R, 1976. Pongamea pinnata Pierre. Wealth of India, Raw- materials, 8:206-211.

Dutta, A.C., 1970. Botany for Degree students. Oxford University Press, London. 240 p.

Kackar, N.L. Solanki, K.R. and Jindal, S.K. (1986). Variation in fruit and seed characters of Prosopis cineraria (L) Mac bride in Thar Desert. *Indian J. For.*, 9:113-115.

Ponnammal, N., M.C. Arjuna, T. Gunamani and K.A. Antony. 1993. Germination and seedling growth of Azadaricta Indica A., *Juss. J.Tree Sci.*, 12:65-68.

R.A. Khimani, B.N. Satodiya and R.G. Yadav, 2003. Cultivation Aspects of *Jatropha* : An overview. Proceeding of the National work shop on Jatropha and other oil seeds held at Pune, India.

Ram Mohan, M.P., Thomas Phillippe G.T. and Shiju M.V. (2006). Biofuels laws in Asia: instruments for energy access, security, Environmental protection and rural empowerment, *Asian Biotechnology and Developmental review*, Vol. 8(2), pp. 51-75.

Shete, N., 2005; Biofuel development unit; bio resource technology an international conference on Biofuels, 2012; vision to reality, New Delhi, October 17-18, 2005, organized by TERI.

Standards and Warranties, National Biodiesel Board Web Page, http://www.biodiesel.org/resources/fuelfactsheets/standards_and_warranties.shtm, Accessed on November 3, 2003.

Chapter 22

Eco-system Services for Fuel Energy Consumption and Rural Poor Livelihood

The Evidence from NSSO 54th Round

A. KANNAN AND DHULASI BIRUNDHA VARADHARAJAN

ABSTRACT

The Common Property Resources (CPRs) play a crucial role in sustaining the livelihood of the rural community, particularly the rural poor. It is the vital component of community assets of the people of the country, especially in the dry areas. In recent times, social scientists have focused attention towards access to CPRs, declination of CPRs and management of CPRs. There has been a gradual decline in CPRs in the village over the last 30 years, mainly because of privatization. The conflict over CPRs is increase in times of crisis. It is therefore, necessary to understand the nature of dependence on CPRs and it role on rural economy. In this connection, the present analysis, attempts to analyze the critical issue of fuel energy consumption in rural India with the help of 54th round survey of NSSO.

Keywords: Common property resources, agro-climatic zones, agro-climatic landscapes, National Sample Survey (54th Round)

PRELUDE

Broadly speaking, common property resources refer to all such resources accessible to the whole community and to which no individual has exclusive property rights. The rights and practices determining the access to these resources are generally conventional.

The CPRs have traditionally been a source of economic sustenance of local people in general and rural poor in particular. They have played an important resource supplementing role in the private property-based farming system. It is also the main source of biomass fuel for the rural population. The use of wood and biomass energy is still increasing. Wood energy is used by households, industries, commercial enterprises and institutions, not only in rural but also in urban areas. For the poor in developing countries, for urban as well as rural, wood is usually the prime source of energy for cooking food and for keeping it warm. In these countries, it is estimated that 86 percent of all the wood consumed annually is used as fuel.

With increasing rural population and the low probability of large shifts to conventional fuels, such as kerosene and LPG, consumption of wood energy will *continue to increase*. Fuel woods are mostly gathered for own use. But in many places, particularly in urban areas, fuel wood and charcoal have become tradable goods. Fuel woods are bought by poor urban households and rural households of almost all income levels. Even where dependence is not as high, CPRs function as an irreplaceable safety net for the poor. When farm and financial assets are scant, the commons can provide secondary income and sources of food and fuel for basic survival. In reality, CPRs can generate important self-employment opportunities, and often serve as an important and flexible source of secondary income for poor households. In some areas, common lands are converted to private parcels as a form of land reform or decentralization, or to stimulate development. Common property resources is also been leased out to private enterprises in the form of fishing or timber concessions. In either case, the poor may lose access of these resources. This paper primarily aims to study the collection pattern of fuel wood and its value of the landless and landholding households of CPRs in India based on the data from National Sample Survey Organization (NSSO) of 54th round, 1998.

MATERIALS AND METHODS

The present analysis is based on a straightforward reading of the NSSO data of 54th round, 1998 on common property resources in India has been used for this research paper to carry out zone-wise analysis. This paper is made an attempt of comparative assessment of landless and landholding households in this round. In order for such a comparison to be useful, it requires categorizing these zones. The NSSO survey covered 27 states and 5 union territories with 78990 sample households in rural areas spread over 5115 survey villages and 31323 households selected from 1745 urban blocks. A zone-wise classification of different states and union territories based on geographical locations (Bhatt, 1997) was made with a view to get a comprehensive picture of dependence of landless and landholding households of CPRs in India. The classification makes it easy to understand the spatial variations in access to CPRs for fuel in India. The outcomes thereof could be useful for policy-makers to identify the priority areas for protection of CPRs; and also give a clear knowledge on this ground.

REVIEW OF RELATED STUDIES

Numbers of earlier studies have focused mainly on the collection of fuel wood and its value from CPRs and also its significant contribution to rural households in particular and all households in general. Here few of studies were discussed in worth-mentioning the necessity of fuel wood to the current context of both national and international status. The amount of fuelwood collection varied to a great extent between rich and poor, and between proximate and far-off users. Contribution of fuel-wood from the community forestry was higher for lower income group and proximate households. Rich households collected almost ten times less fuel-wood from the community forestry than the poor households, whereas far-off households did so almost four times less than the proximate users. Lower caste group collected more fuel-wood than upper caste groups. Both qualitative and quantitative analyses in this put forward that varying degree of household's fuelwood dependence on the forests is mainly determined by their socio-economic conditions. More specifically, poor depends directly on the forest as their pattern of fuelwood collection is very high (Sapkota, *et. al.*, 2008).

Based on quantitative assessment of households in various income sources of Northeastern Bengal Bhattacharya (2003) found that more than 0.7 million rural people engaged in fuel wood trade. In the study area, around 36 percent are migrants who depend on fuel wood collection do this trade which is the major source of household income (see Bhattacharya, 2003).

Another study reviewed the consumption aspect of fuel and found that traditional fuel such as fuelwood, crop residue and dung cake plays a dominant factor in the domestic energy use in rural India and accounts for about 90 percent of the total. Fuelwood alone contributes for about 60 percent of the total fuel in rural areas. In urban areas, the consumption pattern is changing due to amplified availability of commercial fuel like LPG, kerosene, and electricity. During 1983-99, the consumption of traditional fuel declined from 49 percent to 24 percent and LPG connection to households increased from 10 percent to 44 percent (Pandey, 2002).

In relation to energy consumption, a study says that 70 percent of Zimbabwe's energy consumption is derived from wood energy. These households continue to consume fuelwood for energy because of its obvious status as free goods. About 70 percent of the country's total population live in common areas and depend directly on forests for firewood, construction timber, food and fodder, etc. The free access to common property condition in these forests lends itself to over-exploitation. The average annual consumption of fuelwood in non-urban areas ranges from 8.24 million tones in 1990 to just over 10 million tons in 1996 (Mabugu, *et.al.* 1998).

Chen (1991) study on 59 poor households in Ahmedabad district of Gujarat focused that the poor households collect over 70 percent of their fuel and 55 percent of their fodder requirements from CPRs and these resources are more essential to the poor than the better-off. Jodha (1990) has pointed out that the CPRs have a greater role to the rural to the rural economy and thereby ensuring the survival of poor.

Fire wood is also the major form of energy to household cooking and accounts for 60 percent of the total. In the two lowest income groups, firewood accounts for some 80 percent of total household consumption. However, firewood is also used in substantial quantities in higher income households. Thus, middle and above averaged income households use wood for as much as 36 to 45 percent of total energy. Only the highest income households, with at least twice the average income, use wood in amounts approaching 10 percent or less of their total energy (Macauley, 1989)

Arnold *et al.*, (1978) study shows that approximately one-third of the world's population depends upon wood for energy. Supplying more wood fuel can be only one of the solutions toward meeting the continuing energy needs of those now dependent upon wood for this purpose. On the other hand, it is clear that hundreds of millions of people will continue to be so poor that they will have no choice but to receive their household energy from wood and other organic fuels. The potential for expanding the supply of wood fuels is in fact substantial.

RESULTS AND DISCUSSION

As discussed earlier, the Common Property Resources (CPRs) form a crucial part of environmental resources. There are a large number of issues involved that pertain to CPRs. Fuel wood collection from the CPRs by the poor is one such, which received poor attention from the social scientists, despite the fact that CPRs provide life sustenance to rural household in general and rural poor in particular. In recent years, most of the village commons were degraded owing to free access situation with a weak property rights relation, lack of institutional arrangements and break down of *Panchayat Raj* management systems. Since the CPRs are being used collectively and freely, a state of no custodian prevails, which leads to overexploitation, culminating in a position of degraded condition and in this manner places a further stress on the rural economy particularly the economy of the rural poor. With the result, the area under CPR is on the decline, the CPR dependants are enforced to face problems. In addition, environmental degradation caused considerable hardships to fuel wood collectors in rural areas.

Fuel wood is used as a primary source for cooking, which accounts for at least 60 percent of end-use energy consumption of households. In Cambodia, the statistics show that 92 percent of all households in Cambodia used firewood for cooking. While in India cooking accounted for about 62 percent of all urban household's end-use energy consumption. Available data also showed that this proportion increases for low income households, reaching, for instance, 79 percent for households with annual income less than 500 rupees. Whereas in Nepal, it shows, among the household end-use energy consumption met by biomass fuels, cooking accounted for 72 percent. In Pakistan it is accounted for 78.5 percent of household end-use energy consumption and in Philippines, it is accounted for about 90 percent of the end-use energy in households for cooking and water heating purposes (Lefevre, *et. al*. 1997, see http://www.fao.org/docrep/w7519e/w7519e06.htm).

Number of earlier studies argued that the landholding households depend less on commons for fuel but they benefited more in terms of value. In this context, it is worth probing the forces at work in discerning the access to all forms of CPRs across all sections of rural population.

In case of landless households, the fuel wood usage in Western zone recorded higher (808) than the all India level (709). With regard to landholders, it was 781 households in Northeastern zone as compared with all India level of 559 households. Even the landless across zones showed higher proportion in Western, Northeastern and Southern zones. Interestingly, landless households engaged more in usage of fuel wood for their life sustenance than landholders.

With regard to collection of fuel wood, the landless households engaged more, as much as 749 households per one thousands in Western zone when compared to all India level. This trend is prevalent in four out of six zones. The landholders are also involved in collection of fuel wood but lesser than landless group. Except Northeastern zone, all the

TABLE 1
Number per 1000 of Households Using and Collecting Fuel Wood Across Agro-Climatic Zones

Name of the zone	*Uses*			*Collection*		
	Landless	*Landholders*	*All*	*Landless*	*Landholders*	*All*
Southern zone (5)	757	660	710	602	392	500
Northern zone (5)	589	471	508	444	258	320
Central zone (2)	690	602	632	568	376	446
Western zone (2)	808	611	699	749	417	568
Eastern zone (3)	689	538	608	610	350	469
Northeastern zone (8)	783	781	787	655	602	624
All India (25)	709	559	623	597	337	448

Note : Number (per 1000) of households using and collecting fuel wood by landless and landholding households.

Source : Computed from NSSO data 54th round, 2000.

other zones followed a similar level of fuel wood collection by the category of landholders. In the case of landless households usage and collectors were high, while in the landholder's case, usage is more than the collectors. In other words, gap between users and collectors is less in landless group whereas the gap between the above said groups is high in landholders. Pasha (1992) studies in 14 villages of Karnataka has found

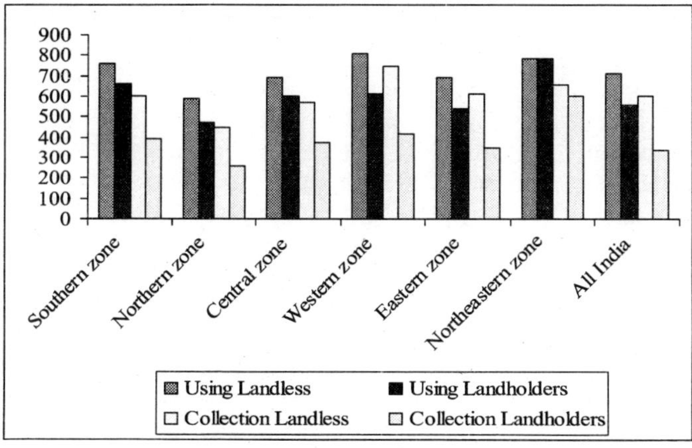

Figure 1 : Number per 1000 of Households Using and Collecting Fuel Wood Across Agro-Climatic Zones

that CPRs are contributing about 63 and 72 per cent fuel wood and fodder consumption respectively in three villages of Karnataka. It was estimated that both poor and non-poor undertake fuel wood collection by 77.2 per cent and 71.3 per cent respectively. Dadibhavi (1998) focused on fuel wood requirements and stated that people rely on CPRs where 90 percent of the poor and 75 per cent of the non-poor households used fuel wood and dung cake for cooking.

The Figure 1 illustrates the prevalence of users and collectors of fuel wood from common property resources, both landless and landholding households per thousand households across zones.

Data set out in Table 2 explains that the quantity of fuel wood collection carried out by the landless households was higher than that of landholding households. All India average of this showed 610 quintals, whereas the Northern zone collected three and half times higher than that of the Indian average with respect to landless households. But in the case of landholders' collection of fuel wood, northeastern zone is higher compared to Indian average. Southern zone stood the least in the quantity of fuel wood collection by the landless households. But in Southern zone, many alternative sources of fuel wood such as kerosene, biogas and other marketed energy resources are available, therefore, the dependence on CPR is also relatively lower.

TABLE 2

Average Quantity of Collection and Value of Collection of Fuel Wood from CPRs during 365 days in Agro-Climatic Zones

Name of the zone	*Collection in Quantity*			*Value in Rs.*		
	Landless	*Landholders*	*All*	*Landless*	*Landholders*	*All*
Southern zone (5)	578	400	492	433	299	368
Northern zone (5)	792	429	551	610	344	437
Central zone (2)	614	468	519	503	427	453
Western zone (2)	663	372	503	557	351	443
Eastern zone (3)	690	470	571	444	326	380
Northeastern zone (8)	2141	2737	2709	1126	1749	1706
All India (25)	610	419	500	475	342	399

Note : Average quantity of collection and value of collection of fuel wood from CPRs during 365 days by landless and landholding households.

Source : Computed from NSSO data 54th round, 2000.

In terms of value of fuel wood collection by the landless and landholding households are also varied across zones. For instance, northeastern zone had the highest value of fuel wood collection by the landless and landholding households than other five zones in India. The overall data depicts that southern zone have a very lowest proportion for both quantity and value of fuel collection from the common property resources on comparison to all India average. The low availability of rich forest resources in southern zone may be the causative factor for the above mentioned trend.

A study carried out by Bhattacharya in North Bengal shows that more than 200,000 people enter 11.879 km^2 of forests in North Bengal every day, collecting up to 120 kg., fuelwood per week per household on a regular basis. Fuel wood business was found to be the main source of income for 10 percent of rural households in three districts of northwestern Bengal and it also accounts for about 45 percent of their cash earnings. Around 72,000 tones of fuel wood traded in the primary and secondary markets of northwestern Bengal every week represent the 45 percent monetized portion of the total fuel wood supply in the study area (see Bhattacharya, 2003).

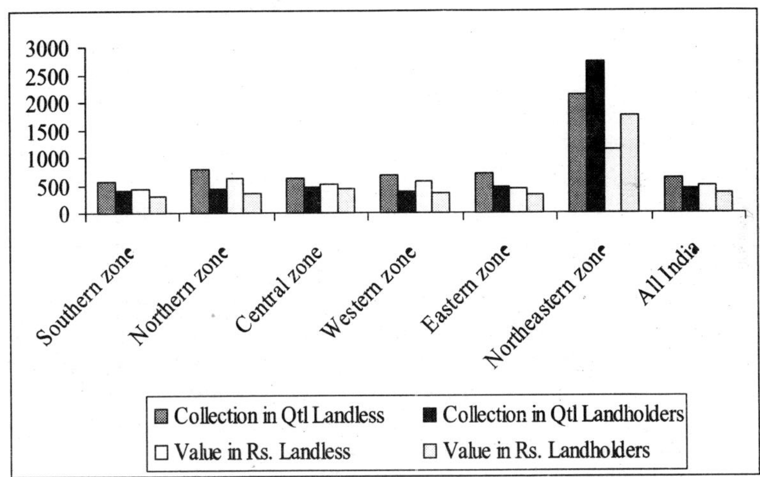

Figure 2 : Percentage of Households Collecting Fuel Wood and Value (Rs.) in Agro-Climatic Zones

Singh *et al.* (1996), found in Siwalik region of Punjab that CPRs contribute 27.3 per cent of the total gross income of the landless and 22 per cent of the cultivating households.

Most of landholding households are engaged in the sale of fuel wood, while the landless households are relatively low in fuel wood sale.

TABLE 3
Average Quantity of Sale and Value of Fuel Wood Collected From CPRs

Name of the zone	*Sale in Qtl.*			*Value in Rs.*		
	Landless	*Landholders*	*All*	*Landless*	*Landholders*	*All*
Southern zone (5)	8	21	14	7	29	17
Northern zone (5)	21	13	15	27	14	17
Central zone (2)	57	16	33	44	18	29
Western zone (2)	4	1	2	5	1	3
Eastern zone (3)	24	61	44	17	43	31
Northeastern zone (8)	104	136	139	73	91	92
All India (25)	24	25	24	19	22	21

Note : Collection of Fuel wood during 365 days by landless and landholding households.

Source : Computed from NSSO data 54th round, 2000.

All India picture explained that the sale of fuel wood with respect to landless is about 24 quintals, whereas for landholders, it was about 25 quintals. Remarkably northeastern zone and central zone alone registered higher sale of fuel wood by the landless households. It is to be pointed out that in the Northern and Eastern zones, the landholders sold higher quantity of fuel wood than the landless households. The Western zone is the least seller of fuel wood both by the landless and landholders.

In terms of monetary value of the landholders, higher values of fuel wood from the CPRs were sold when compared with the landless households. Similarly, the value of fuel wood collection and sale were the lowest in the Southern zone among both landless and landholding households. While in landless households ranged between 73 in northeastern zone and to 7 in southern zone.

Despite the fact, the landholders in northeastern and western zone are present between 91 and 1 respectively. With result of the above argument, the landholders registered higher proportion of quantity collection and sale of fuel wood than the landless households on an average across all the zones. Pasha (1992) stated that in 14 villages of Karnataka, CPRs income including the values of fodder and fuel wood, and the imputed values of grazing accounts for around 10 and 6.2 per cent of the gross income of the poor and non-poor households, respectively.

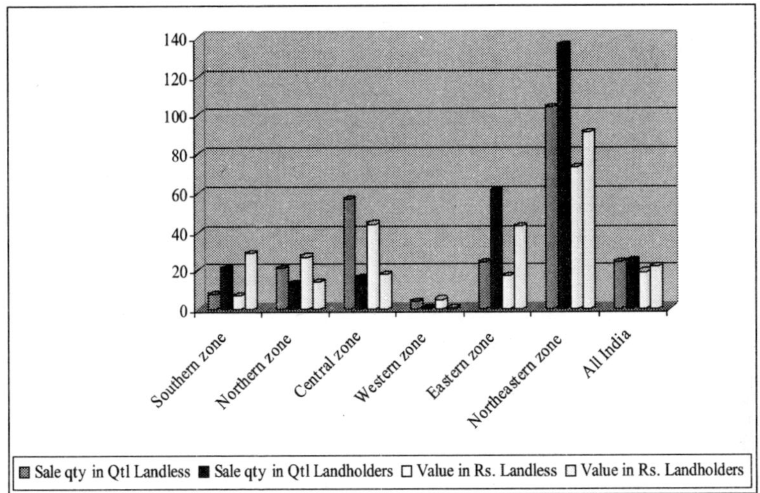

Figure 3 : Average Quantity of Sale and Value of Fuel Wood Collected From CPRs

Correlation between Quantity of Fuel Wood Collection by Landless and Landholding Households

Having illustrated the state of CPRs across zones in the foregoing sections, it is important to highlight the correlation between landless and landholders with respect to the value of fuel (both collection and sale value) and related variables to add supplementary insights. Care is taken to find out the correlation coefficient values between landless and landholding households and to collection, value of collection, sale and value of sale by the landless, landholders and all households, etc.

A crucial variable like quantity of fuel collection by all households with quantity of collection, its value, quantity of sale and value by the landless, landholding and all households at one percent and five percent level of significance are observed. The comparison made with all households' collection, value of fuel and total value of fuel collection is positively correlated at one percent level of significance. But it slightly differs in percentage terms that landholding households received higher value of 99.9 percent than that of 97.7 percent of landless households. The same trend is prevalent in quantity of collection by the landless and landholding households with one percent level of significance.

Quantity and value of fuel sale by landless and landholding households are positively correlated at five percent level of significance. But the size of sale and value of fuel wood in all households are

TABLE 4

Relationship between Collection of Fuel Wood Quantity in Quintal and Other Independent variables by Landless and Landholding Households

Independent variable	*Dependent variable*		
Total value of fuel by LHs	.999**	—	
Total value of fuel by LLs	.977**	.969**	—
Quantity of fuel collection by LLs	.993**	.990**	.982**
Quantity of fuel collection by LHs	.998**	.999**	.966**
Quantity of fuel collection by all	.998**	.998**	.968**
Value of fuel collection by LLs	.975**	.965**	.999**
Value of fuel collection by LHs	1.000**	1.000**	.973**
Value of fuel collection by All	1.000**	.998**	.979**
Quantity of fuel sale by LLs	.889*	.900*	.862*
Quantity of fuel sale by LHs	.903*	.914*	.825*
Quantity of fuel sale by all	.953**	.963**	.895*
Value of fuel sale by LLs	.857*	.864*	.862*
Value of fuel sale by LHs	.878*	.892*	.785
Value of fuel sale by All	.946**	.957**	.889*
	Total value of fuel by all	Total value of fuel by LHs	Total value of fuel by LLs

* Significant at 5% level and ** Significant at 1% level.
LLs: Landless households, LHs: Landholding households.
Source : Computed from NSSO data 54th round, 2000.

positively associated with total value of fuel value from the common property resources. Another variable ie the total value of fuel by landholding households with collection, sale and its value of landholding and all households are correlated positively at one percent level of significance. In total, the value of fuel collection by the landless households is positively associated with collection and its value of landless and all households at one percent level of significance. While, in quantity of sale and its value is positively associated at five percent level of significance. With result from proceeding, discussion and association analysis it is clear that landholding households are received higher value from CPRs than landless households. The dependent on CPRs are higher in landless households but highest beneficiaries' form the commons who are landholding households in all the zones of across India.

For instance, Jodha (1995) found from his analysis covering 80 Villages in 20 districts in six dry tropical regions of India that over 80 to 100 percent of the poor households are depending on CPRs for the purpose of fuel wood collection and animal grazing. Common property resources in different areas play a crucial part to the poor households in terms of fuel supplies, which contribute 66 to 84 per cent of fuel and animal grazing ranking from 69 to 84 per cent. In addition to this, Iyangar (1989) underlined that twenty-five villages in Kanchch district of Gujarat study revealed that cattle grazing and fuel wood collections have been identified as the important contributions of common lands. It has been revealed that 45-76 per cent of the households using common lands for grazing purpose, and 35–46 per cent for fuel wood collection.

CONCLUSION

In summary, the data from the 54th round is clearly evident that the importance of CPRs across agro-climatic landscapes and its particular concerns need to be addressed in each of these landscapes. Common property resources (CPRs) play a vital role for the survival of the local masses. It yields a numerous benefits to the society in the form of fuel, fodder, grazing, species and agricultural tools and implements. But unfortunately, the availability of CPRs has been coming down slowly. The proportion of CPRs has declined both quantitatively and qualitatively. But the dependence on the commons has not declined. Firewood gathering has been an important subsidiary occupation of the rural poor in general and landless labourers in particular. Rural households form the major beneficiaries of CPRs. It is observed that dependence on CPRs for fuel is higher in landless households, while collections of fuel wood were high in landholding households. About 98 percent of the landless and 88 percent of the cultivator households bring fuel wood from the forest to meet their fuel requirements (Singh *et. al.*, 1996). The value of collection and sale of fuel wood from the CPRs resources are also remarkably high in landholding households than the landless households across the zone of India. The association of related variables expressed that positively correlated at one and five percent level of significance. From the analysis, it is clearly evident that the landholding households were higher beneficiaries than the landless households. Other zone also contributed a lot but no marked variation between households' categories.

REFERENCES

Bhattacharya, P. (2003), "Fuel wood and displaced people: a case study from northwestern Bangal, India", Forest Energy Forum-FOPW FAO, Viale delle Terme di Caracalla, 00100 Rome, Italy, :http://www.fao.org/forestry/ fog/fopw/ energy/ energy-e.stm

Bhatt, S.C., (1997). The Encyclopaedic District Gazetteers of India. Gyan Publishing House, New Delhi.

Chen, M. (1991). Coping with Seasonality and Drought, Sage Publications, New Delhi.

Government Report: A Note On Common Property Resources in India: NSSO 54th Round (January-June 1998). National Sample Survey Organization Ministry of Statistics and Programme Implementation Government of India, *Sarvekshana*, Vol. 24 (1), 84th Issue July-September, 2000.

Dadibhavi, R.V. (1998). CPRs and their Contributions: Evidence from Karnataka. Indian Economy after 50 years of Independence Experiences and Challenges. (Eds.) Debendra. K. Das. Deep and Deep Publications. F-159, Rajouri Garden, New Delhi, pp. 355-68.

Iyengar, S. (1989). "Common Property Land Resources in Gujarat Some Findings about Their Size, Status and Use" *Economic and Political Weekly*, Vol. 24(25), June, pp. 67-77.

Jodha, N.S. (1995). Common Property Resources and the Environmental Context: Role of Bio-physical versus Social Stresses. *Economic and Political Weekly*. Vol. 30.(51), Dec. 23. pp. 3278-83.

Pasha Syed Ajmal. (1992). Common Property Resources and the Rural Poor; A Micro Level Analysis. *Economic and Political Weekly*. Vol. 27(46), November-4, pp. 2499-2503.

Lefevre, T.; Todoc, J.L.; Timilsina, J.R. (1997). *"The Role of Wood Energy in Asia"* FOPW/97/2, Food and Agriculture Organization of the United Nations Rome, Italy, http://www.fao.org/docrep/w7519e/w7519e06.htm.

Pandey, D.N. (2002). "Fuelwood studies in India: Myth and reality", Center for International Forestry Research (CIFOR), Bogor Barat 16680, Indonesia.

Macauley, Molly K., Naimuddin, M., Agarwal, P.C. and Dunkerley, J. (1989). "Fuelwood Use in Urban Areas: A Case Study of Rajpur, India, *The Energy Journal*, Vol. 10(3) p. 157-180. http://econpapers.repec.org/RePEc:aen:journl:1989v10-03-a10.

Arnold, J.E.M. and Jules Jongma (1978). "Fuelwood and charcoal in developing countries", Paper presented at 8th World Forestry Congress in September 1978 in Djakarta. *http://www.fao.org/docrep/l2015e/l2015e01.htm*

Mabugu, R.G.R. Milne and B. Campbell (1998). "Incorporating Fuelwood Production and Consumption into the National Accounts : A Case Study for Zimbabwe", Planning and Statistics Branch, Policy and Planning Division, Forestry Department, Food and Agriculture Organization of the United Nations, Viale delle Terme di Caracalla, 00100 Rome, Italy.

Sapkota. I.P. and Per Christer Odénl (2008). "Household Characteristics and Dependency on Community Forests in Terai of Nepal", *International Journal of Social Forestry*, Vol. No. 1(2) 2008, pp. 123-44.

Chapter 23

Millennium Development Goals and Pollution Exposure from Household Fuel in Tribal Society of Rural Dehradun

RAJIV PANDEY, ATIN TYAGI AND SOMNATH HAZRA

ABSTRACT

The paper discusses health and pollution asymmetry due to use of fuelwood by Jaunsary tribal, Dehradun, India and its impact on Millennium Development Goals (MDGs), particularly for Goal 3, 5 and 7. To quantify, associated parameters including pollution due to burning were explored based through comprehensive survey of 46 randomly selected households.

Result shows that all population was dependent on fuelwood and its collection was women centric. Burning was generally performed in poor ventilated kitchens and leads to releases of hazardous gases such as SO_2, NO_2 and CO. The concentration is above to prescribed standard limit, which has implications on women health. The paper outlines the exposure due to fuwelwood burning, health and therefore implications on achieving MDG's and recommendations a few policy measures.

INTRODUCTION

The Millennium Development Declaration with the objective to promote a comprehensive approach and a coordinated strategy, tackling many problems simultaneously across a broad front to make the right to development a reality for everyone by the year 2015, all member-states of the United Nations affirmed that they would "... spare no effort to free our fellow men, women and children from the abject and dehumanizing conditions of extreme poverty (United Nations, 2000). To achieve this, eight time-bound and measurable goals were defined (United Nations, 2005):

Goal 1 — eradicate extreme poverty and hunger
Goal 2 — achieve universal primary education
Goal 3 — promote gender equality and empower women
Goal 4 — reduce child mortality
Goal 5 — improve maternal health
Goal 6 — combat HIV/AIDS, malaria and other diseases
Goal 7 — ensure environmental sustainability
Goal 8 — develop a global partnership for development.

Millennium Development Goal (MDG) 7 aims to ensure environmental sustainability, based on the notion that human survival and prosperity critically depend on the sensible use of natural resources and the protection of complex ecosystems. Worldwide, 2.4 billion people continue to depend on biomass fuels (wood, dung, agricultural residues) for their basic household energy needs [International Energy Agency (IEA) and Organisation for Economic Cooperation and Development (OECD) 2004]. Moreover, the situation of India, where majority of population lives in the rural areas, the energy use pattern is characterized by a greater reliance on traditional fuels particularly fuel wood which accounts for more than 75% users (NSSO, 2002), i.e. around 3.5 billion people (WRI, 1999). In rural India, 90% of the primary energy use is bio-mass, of which wood accounts for 56%, crop residues for 16%, and dung-cakes for 21% (FAO, 1999).

Moreover, the incomplete combustion of biomass, a common phenomenon, releases complex mixture of organic compounds, which include suspended particulate matter, Carbon monoxide (CO), Sulfur dioxide (SO_2), Nitrogen dioxide (NO_2), polycyclic organic material (POM), poly aromatic hydrocarbons (PAH), formaldehyde, etc. (Edwards, *et. al.*, 2004). Daily averages of pollutant level emitted indoors often exceed current WHO guidelines and acceptable levels. Women and children are most vulnerable as they spend more time indoors and are exposed to the smoke (Smith, 1987). Low combustion efficiency of solid fuels in simple devices like traditional chulha leads to significant diversion of fuel carbon to products of incomplete combustion (PIC; typically 5-20%). Such PIC includes not only the major health damaging pollutants but also a range of greenhouse gases (Smith, *et al.*, 2000).

Suspended matter consists of dust, fumes, mist and smoke and when breathed in, lodges in our lung tissues and cause lung damage and respiratory problems. CO interferes with the delivery of oxygen throughout the body and combines with hemoglobin in blood and with high concentrations, it can cause unconsciousness and death too. Similarly, NO_2 irritates the mucous membranes in the eye, nose, and throat and causes shortness of breath with exposure of high concentrations. Further, SO_2 is responsible for diseases of the lung and its various disorders.

In India most common diseases with indoor air pollution (IAP) is probably Acute Respiratory Infections (ARI) (Parikh, *et al.*, 1999). In recent years, new evidences suggest that IAP in developing countries may also increase risk of other important child and adult health problems, such as low birth weight, perinatal mortality (stillbirths and deaths in the first week of life), asthma, and middle ear infection in children, tuberculosis, nasopharyngeal and laryngeal cancer, and cataract in adults (Bruce, *et al.*, 2000). Laxmi, *et al.* (2003), report that health impacts of the bio-fuels use are quite high for adults. Approximately 6 million people suffer from respiratory to eye-related problems.

The reduction or complete elimination of indoor air pollution can be an important contribution to improve health of household members. In Asia alone, > 460,000 deaths due to chronic respiratory disease among women could be prevented every year. Easier access to other fuels and higher fuel efficiency are likely to bring about additional health improvements to specially pregnant women and mothers. For example, some studies suggest that carrying heavy loads (e.g., biomass fuels) may be associated with an increased risk of prolapse (Pandey, 1997).

However, it is not possible to understand the problem until and unless, proper data is available in general and spatially in particular. The present study has been undertaken in the households of rural tribal of Jaunsar Bawar region of Dehradun (UK) to assess the health and level of indoor air pollution on the daily life of rural tribal of Jaunsar Bawar in terms of released hazardous gases from fuel wood burning as the cooking energy for them is mainly fuel wood (Pandey, 2007).

Methods

The present work is based on a comprehensive survey conducted in the rural areas of Jaunsar Bawar to address the objectives under study. The information required for the study are socio-economic characteristics of the households, fuel wood consumption pattern, cooking behavior of the households, kitchen structure, *chulhas*, exposure to hazardous gases (NO_2, SO_2, CO and LEL,- i.e. Least

Explosive Level gases) and health impacts. Individual households were approached and information pertaining to the study was collected. The pollution monitoring was done in each household with before burning of chulha in kitchen and outside; after 10 minutes, 30 minutes in kitchen and after 60 minutes in kitchen and living rooms.

Balance has been made between the number of households and available resources. Information pertaining to IAP and associated health hazards were collected from 20 randomly selected households spread in 8 randomly selected villages. In this area number of families residing in each village varied from 10 to 30. Therefore, 2-3 households were randomly selected for the study from each selected village. Larger sample size could not be considered due to the similar fuelwood utilization patterns, types of chulhas and kitchen followed by same cooking habits besides the environmental conditions. However, for health assessment, besides this, 46 households were also surveyed.

Indoor Air Quality (IAQ) measurement was collected from the selected households. IAQ assessments were conducted by determining concentration of emitted gases from fuelwood burning. Multi-gas Analyzer for SO_2 and NO_2, CO Analyzer for Carbon monoxide and Gas Alert Micro 5 IR for CO_2 and LEL measurement were used. Multiple measurements based on time intervals on the concentration of gases were taken before and after chulha burns. These measurements are considered as the indices of exposure. The first exposure is personal exposure to hazardous gases while cooking (first 10 minutes), second is for others through measuring it in kitchens as a surrogate for personal exposure to others (first 30 minutes) and third is for others through measuring it in rooms as a surrogate for personal exposure to others.

RESULT AND DISCUSSION

Quality of Life of the Jaunsar People

From the cursory examination of the Table 1, it is evident that the Jaunsar families, in general, are big in size with a maximum of 30 members in some cases. The average age of the head of the house was observed to be nearly 45 years. Although their education status ranged up to the post-graduation level education qualification, its averaged value came out to be 4.61 in number of years of education fairly showing poor education standard of the head of house. Majority of Jaunsaries were observed to be Farmers. Proportion of irrigated land was very small with an average of 1.85 Bighas (Table 1). Though the geographical area in Jaunsar supports good forests, in some cases people have to travel long distances for fuel wood collection and spend considerable amount of time for the same. In most of cases, it is primarily women's sensitive.

Moreover, all households of the region use fuel wood as cooking energy. This formed the basis for estimating the burden of disease attributable to indoor air pollution from solid fuel use, therefore, responsible for the poor quality of life. As, per WHO 2002, in developing countries with high-mortality, indoor air pollution was responsible for 3.7% of the overall disease burden, making it the most important risk factor after malnutrition, unsafe sex, and lack of safe water and adequate sanitation.

TABLE 1

Descriptive Statistics of Quality of Life

Parameters	*Mean ± SE*	*Min.*	*Max.*
Family size	11.37 ± 0.99	5	30
Total land holding (in Bighas)	13.85 ± 2.18	1	60
Irrigated land (in Bighas)	1.85 ± 0.82	0	28
Monthly income per household (Rs.)	3984.80 ± 567.00	500	20000
Monthly expenditure per household (Rs.)	3726.10 ± 509.30	800	20000
Education of head of house (years)	4.61 ± 0.69	0	17
No. of rooms in a house	1.24 ± 0.11	1	12
Age of head of house	44.72± 2.02	25	70
Education of head of house (no. of years)	4.61 ± 0.69	0	17
Fuel wood collected at a time (Kg.)	18.26 ±1.81	6	80
Time spent for fuel wood collection (Hours)	4.33 ± 0.26	1	8
Distance traveled for Collection (Km.)	3.11 ± 0.14	1	4

KITCHEN AND COOKING PARAMETERS

The location of kitchen generally depends on the size of house and family status as revealed by the respondents. Generally, kitchens are made of wood. It lacks proper ventilation in 50% Jaunsary kitchen. Only 20% kitchens has small window too. Only 15% chulha have chimneys for smoke emission. Moreover, overall, the kitchens of region lack proper ventilation. They cook 2 to 3 meals per day depending on the requirements and their engagements for livelihood activities. The concentrated smoke emission and burning time for chulhas depend upon the family size, morning and evening meal (Table 2). The chief cook and their associates are in general young women up to age 40. Based on the chulha location as per survey, the Jaunsary kitchen, which was in general not spacious, may be classified into four types (Figure 1), i.e. during survey, 40% families were observed to have indoor kitchen with partition, 35% families indoor kitchen without partition, 20%

families attached kitchen outside the house and the rest 5% families were observed to have separate kitchen outside the house.

TABLE 2
Descriptive Statistics of Kitchen and Cooking Parameters

Parameters	*Mean ± SE*	*Min.*	*Max.*
Chulha burning in morning (Hours)	1.82 ± 0.16	1	5
Chulha burning in evening (Hours)	2.37 ± 0.14	1	7
Chulha burning in mid day (Hours)	0.79 ± 0.15	0	3
Time for Smoke emission morning (Min)	12.02 ± 0.98	3	30
Time for Smoke emission evening (Min)	11.91 ± 0.91	3	30
Time for Smoke emission mid-day (Min)	4.46 ± 1.00	0	25
Number of pots in a chulha	2.78 ± 0.09	1	5
Cooking frequency per day (No. of Meals)	2.37 ± 0.09	0	3
Chief cook's age	39.30 ± 1.73	20	62
Regular Associate cook' s age	27.34 ± 1.83	14	32
Kitchen size (sq. ft)	26.50 ± 3.18	10	60
No. of doors in kitchen	1.24 ± 0.07	1	3
No. of ventilations	1.61 ± 0.23	0	6
No. of chimneys	0.24 ± 0.06	0	1
No. of windows	0.30 ± 0.73	0	3

Air Pollution Monitoring at House Level

The result is based on the survey of twenty households.

Sulfur Dioxide: Mean value of SO_2 concentration came out to be 1.52 ppm which is nearly 6 times more than prescribed by National Ambient Air Quality Standards (NAAQS). Even in living room after 1 hour of chulha burning it was found to be 3 times more than standard (Table 3).

Nitrogen Dioxide: In Jaunsari kitchens the level of NO_2 concentration came out to be 1.69 ppm and 1.73 ppm at ½ hour and 1 hour after chulha burning, again far more than standard (6 times and 4 times respectively) (Table 3).

Carbon Monoxide: According to NAAQS standards, CO concentration should not exceed to 20 ppm. In Jaunsary houses the mean values of CO after 30 and 60 minutes were measured 42.44 ppm and 57.69 ppm which is approx. thrice to the standard values. Besides this CO concentration inside living room was also found nearly 1.5 times more than the standard values (Table 3).

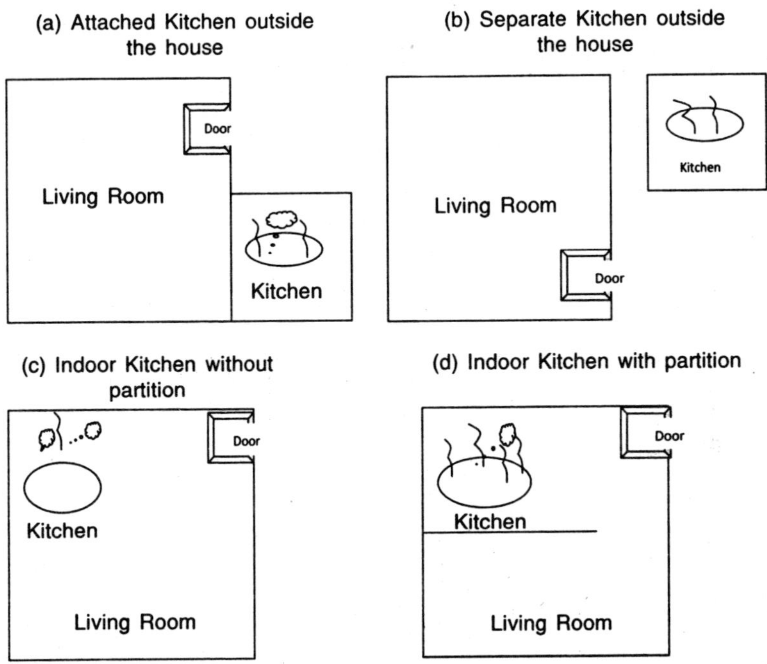

Figure 1 : The Existing Layout of Jaunsary Kitchen

Carbon Dioxide: High level of CO_2 may cause occupants to grow drowsy, get headaches, or function at lower activity levels. in large concentration it is a major cause of global warming. CO_2 concentration was also observed too high to meet the standard (Table 3).

Health Implications of Fuel Wood Burning

Jaunsaries showed poor health consciousness. 89.1% of the respondents did not contact doctors frequently. Proportion-wise various reasons for such finding are mentioned in Figure 2. The main reason was unawareness regarding the health implications, it seems. The health related issues for women were also investigated and it was found that eye related issues viz. watering and redness of eyes were very prominent among them (Figure 3). However headache, back and leg pain- related problems were also commonly observed.

CONCLUSION

The major strength of the study has been determination of actual exposure under a variety of exposure situations prevailing in the rural houses, which ultimately determine the achievement towards MDGs

TABLE 3
Spatio-temporal Polluted Gases Status in the Jaunsary Houses (in ppm)

Parameter	*SO_x Mean (Min-Max.)*	*NO_2 Mean (Min-Max.)*	*CO Mean (Min-Max.)*	*CO_2 Mean (Min-Max.)*
(1)	*(2)*	*(3)*	*(4)*	*(5)*
Outside house before chulha burns	0(0 – 0)	0.04(0.3 – 0.5)	1.68(1.00 – 4.20)	0(0 – 0)
Inside kitchen before chulha burns	0.09(0 – 0.11)	0.06(0 – 0.08)	1.74(1.00 – 4.20)	15.00(0 – 30.00)
Inside kitchen after 10 minutes of chulha burning	0.89(0 – 2.40)	1.48(0 – 3.50)	33.05(2.80 – 110.90)	877.50(300 – 1900)
Inside kitchen after 30 minutes of chulha burning	1.52(0 – 5.10)	1.69(0 – 4.00)	42.44(5.60 – 109.40)	1157.50(400 – 2600)
Inside kitchen after 60 minutes of chulha burning	2.55(0.1 – 5.00)	1.73(0.20 – 2.70)	57.69(1.80 – 129.60)	1712.50(400 – 3100)
Inside room after 60 minutes of chulha burning	0.89(0.30 – 2.50)	0.97(0 – 1.80)	33.27(4.40 –- 129.60)	1178.57(150 – 2850)

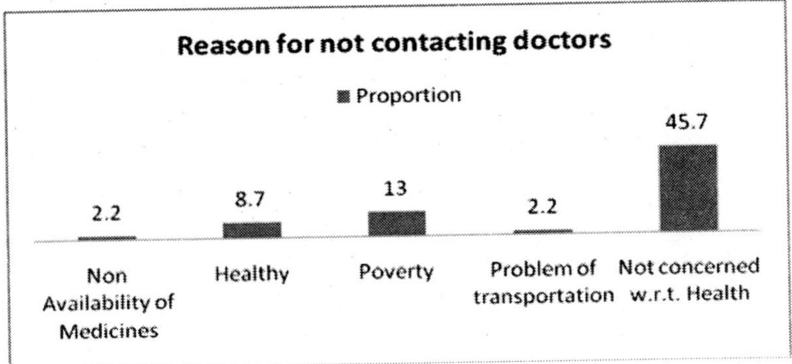

Figure 2 : Reason for not Contacting Doctors

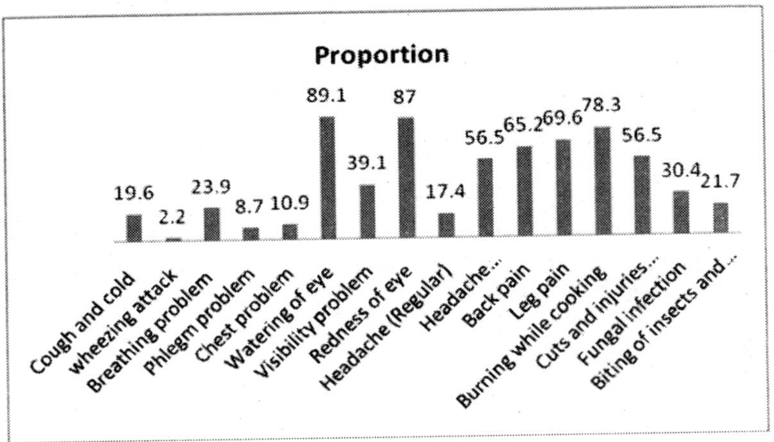

Figure 3 : Health Related Issues with Jaunsary Women

although at local level. Moreover, the small local issues are also critical section of achieving these goals. However, from an environmental point of view, widespread use of solid fuels could mistakenly be interpreted as a positive development, given that most biomass fuels—the major share of solid fuels—constitute a source of renewable energy. From a public health point of view, widespread use of solid fuels is, of course, interpreted as a negative development because of the health risks associated with indoor air pollution. Yet, solid fuel use is a poor proxy for indoor air pollution levels as the concentrations of small particles, CO, and other pollutants vary markedly between the same fuel being burnt in a traditional open fire (Eva., *et. al.*, 2006) in different types of kitchens.

By drawing the conclusion that 100% households use fuelwood, it is apparent through the indicator "the proportion of the population relying on solid fuels" as a measure of achieving MDG 7 that we are far away from the universal goals. In the study, we have also outlined about the exposure to indoor air pollution from solid fuels and which has distinct implications for achievement towards other MDGs. Therefore, development agendas and partnerships must recognize the fundamental role that household energy plays for achievement towards MDGs.

Moreover, till then, the study therefore can lay down the framework for the evaluation to further establish definitive linkages between bio-fuel smoke exposure and health. The complete rectification of the problem in totality is not possible because households can't shift the utilization of the cleaner fuels immediately instead of fuel wood. But several other aspects may be considered to control the problem. Based on the revelation and logical conclusion following points may be considered:

1. House and kitchen structures should be redesigned for the diffusion of polluted gases.
2. Chulhas should be modified as per the traditional knowledge of villagers, locally available raw materials and resources.
3. Cheap, clean and assessable substitute of fuel wood may be provided to the households with awareness about effect of fuel wood utilization.

References

Bruce, N., Perez-Padilla, R. and Albalak, R., 2000. Indoor Air Pollution in Developing Countries: A Major Environmental and Public Health Challenge for the New Millennium, *Bull. WHO* 78, 1078-92.

Edwards, R., Smith, K.R., Zhang, J., Ma Y., 2004. Implications of changes in household stoves and fuel use in China. *Energy Policy,* 32:395-411.

Eva Rehfuess, Sumi Mehta, and Annette Prüss-Üstün, 2006. Assessing Household Solid Fuel Use: Multiple Implications for the Millennium Development Goals. *Envir. Health Pres.*, 114(3)-373-78.

Laxmi, V., Parikh, J., Karmakar, S. and Dabrase, P., 2003. Household Energy, Women's Hardship and Health Impacts in Rural Rajasthan, India: Need for Sustainable Energy Solution. *Energy for Sustainable Development*, VII(1): 50-68.

Food and Agriculture Organization of the United Nations (FAO). 1999. The State of the World's Forests, 1999 (FAO, Rome).

Pandey, R., 2007. Contribution of Forestry-Role in Socio-Economic of Jaunsaries and Development of Index. Ph.D. Thesis, FRI University, Dehradun.

Pandey, M.R. 1997. Women, wood energy and health. *Wood Energy News,* 12(1):3-5.

Parikh, J., Smith, K., and Laxmi, V., 1999. Indoor Air Pollution: A reflection on gender bias. *Economic and Political Weekly*, 34(9) pp.

NSSO, 2002. Report on Village Facilities, National Sample Survey Organization, Min. of Stat. & Prog. Impl., New Delhi.

Smith, K., 1987. Bio fuels, Air Pollution and Health: A Global Review. New York: Plenum Press.

Smith, K.R., Samet, J.M., Romieu, I., Bruce, N. 2000a. Indoor air pollution in developing countries and acute lower respiratory infections in children, *Thorax*, 55:518–532.

United Nations, 2005a. Millennium Development Goals. Available: http://www.un.org/millenniumgoals/ [accessed 10 Sept., 2008].

United Nations, 2000. United Nations Millennium Declaration. Available: http://www.un.org/millennium/declaration/ares552e.pdf [accessed 10 Sept., 2008].

IEA and OECD, 2004. World Energy Outlook, 2004, Paris: International Energy Agency and Organisation for Economic Co-operation and Development.

TERI, 2000. TERI Energy Data Directory & Yearbook, 1998-99. TERI, New Delhi.

World Resources Institute, UNDP, World Bank, 1999. *1998-99 World Resources: A Guide to the Global Environment*, Oxford: Oxford University Press.

WHO, 2002. World Health Report, 2002: Reducing Risks, Promoting Healthy Life. Geneva: World Health Organization. Available: http://www.who.int/whr/2002/en/ [accessed 31 June, 2008].

Chapter 24

Solid Waste Management Problems and Solution in India

M. BALASUBRAMANIAN AND
DHULASI BIRUNDHA VARADARAJAN

ABSTRACT

Solid waste management is one of the important problems in India especially in rural areas. After globalisation five times increase in waste due to increase in population, urbanization and commercialization in urban areas in rural India. Now it is changing lifestyle and standard of living which had lead to an increase in solid waste generation. Management is big problem in rural areas for us collection, transportation; disposal and segregation even finance. Given the absence of appropriate legislation or of any monitoring mechanism on the performance of municipal authorities, the system of waste management has remained severely deficient and outdated. Inappropriate and unhygienic systems are used. At disposal sites, municipal authorities dump municipal waste, human excreta from slum settlements, industrial waste from small industrial establishments within the city, and biomedical waste without imposing any restrictions, thus provoking serious problems of health and environmental degradation. A typical solid waste management system in a developing India displays an array of problems, including low collection coverage and irregular collection services, crude open dumping and burning without air and water pollution control, the breeding of flies and vermin, and the handling and control of informal waste picking or scavenging activities. In India waste management practices and inefficient and economic instruments are very silent one. This paper attempts to problems of waste management and economic instrument for control or reduction of waste generation.

Keywords: Solid waste management, Globalization, Economic instruments.

1. INTRODUCTION

Economic, as well as social development has strong links with the environment and with sustainability. Without proper planning, design, and management, development activities result in considerable generation of waste and other environmentally objectionable materials. Rapid industrialisation and population explosion in India has led to the migration of people from villages to cities, which generate thousands of tons of Municipal solid waste daily. Generation of municipal solid waste (MSW) in India has increased significantly in the past few decades. Between 1991 and 2001, MSW generation in urban India increased from 23.86 million tons/year to more than 39 million tons/year (Sharholy *et al.* 2007). In per capita, the growth of MSW in Indian cities has been estimated to be 1-1.33%, and within next two decades, the annual waste generation is estimated to increase more than five times that of the present level (Singhal and Pandey, 2001).

The state of an economy, to a large extent, influences waste generation and municipal solid waste in particular. In developing countries, even though the per capita waste generation is low at 300 g, changes in living conditions and the influence of western 'throw away' culture results in increased solid waste generation, leading not only to environmental degradation but also a huge loss of natural resources. Improper disposal of this waste leads to the spread of communicable diseases, causes obnoxious conditions and spoils the biosphere as a whole. On the other hand, cleanliness is another factor that influences the development of any nation that is otherwise hampered owing to improper disposal of solid waste (Sudhakar Yedla and Jyothi K. Parikh, 2001).

2. PROBLEMS SOLID WASTE MANAGEMENT

Collecting, transporting, and disposing of MSW represents a large expenditure for developing country cities, waste management usually accounts for 30-50 per cent of municipal operational budgets. Despite these expenses, cities collect only 50-80 per cent of the refuse generated. In India, for instance, about 50 per cent of the refuse generated is collected, 33 per cent in Kharachi, 40 per cent in Yangon, and 50 per cent in Cairo. Disposal receives less attention, as much as 90 per cent of the MSW collected in Asian cities ends up in open dumps (Cointreau, 2008; Medina, 1997a). Solid waste management has been the most neglected area of urban development over the years and has accounted for severe health problems in urban areas all over the country. A number of cases have come to light because of mismanagement of municipal solid waste management. Due to the changing consumption pattern, quantities and characteristics of municipal solid waste vary significantly with time. This aspect is especially significant in the context of developing countries, where fast changing socio-economic conditions influence the pattern of solid waste generation significantly (David N. Beede and David E. Bloom, 1995).

According to India's constitution, SWM falls within the purview of state government. The activities are local ones and are entrusted to urban local bodies (ULBs) through state legislation. Because these activities are non-exclusive, non-rival, and essential, the responsibility for providing them lies within the public domain. Many municipal authorities in India provide SWM services very inefficiency. Old and inappropriate vehicles and tools for collection, inadequate transport, and inefficient disposal not only cause unhygienic working conditions and slow down the process but also severely affect the environment. Productivity is very low, resulting in a high unit cost of services. Collection coverage rates are only 50 to 70 per cent. The collected waste is disposed of at open dumping grounds within or outside cities, causing health hazards and environmental degradation.

Poor collection and inadequate transportation are responsible for the accumulation of MSW at every nook and corner. The management of MSW is going through a critical phase, due to the unavailability of suitable facilities to treat and dispose of the larger amount of MSW generated daily in metropolitan cities. Unscientific disposal causes and adverse impact on all components of the environment and health (Rathi, 2006; Sharholy *et al.*, 2005; Ray *et al.*, 2005; Jha *et al.*, 2003; Kansal, 2002; Kansal *et al.*, 1998; Singh and Singh, 1998; Gupta *et al.*, 1998). Due to the inadequate transport capacity and deficient workforce, collection efficiency in Indian cities has been in the range of 70-73% (Gupta *et al.* 1998: Nema, 2004). An average, one-third of the total waste remains uncollected even though the municipal bodies allocate 85-90% of their total budget for collection and transportation activities (Sharholy *et al.*, 2008). The uncollected waste mostly comprises of organic substance and in some cases human and animal excreta. It may also contain hazardous waste like industrial and medical waste. The uncollected garbage is often dumped indiscriminately resulting in clogged drains and sewerage that serves as a breeding ground for rodent vectors and insects leading to spread of diseases (Zhu *et al.* 2009).

The predominant mode of disposal has been open dumping (94%), and only 5% of the total waste is taken for composing (CPCB, 2000). The dumping fields are mostly open ground without any seepage proof lining for leachate collection involving huge risk of groundwater contamination. In fact, analysis of physicochemical characterstics of groundwater samples from tubewells and borewells near dumping sites in some major Indian cities have shown substantial concentration of heavy metals and chemicals in excess of permissible limits (Kumar *et al.* 2009). This becomes increasingly expensive and complex due to the continuous and unplanned growth of urban centers. The difficulties in providing the desired level of public service in the urban centers are often attributed to the poor financial status of the managing municipal corporation (Mor *et al.*, 2006; Siddiqui *et al.*, 2006, Raje *et al.*, 2001; MoEF, 2000; Ahsan, 1999).

Solid waste management has been the most neglected area of urban development over the years and has accounted for severe health problems in urban areas all over the country. A number of cases have come to light because of mismanagement of municipal solid waste management. The devastating pneumatic plague in Surat, Gujarat, in September 1994 is a true example of how uncontrolled garbage led to the out-break of the disease. Huge expenditure on solid waste disposal with very poor efficiency, pollution due to the burning of waste, local as well as global air pollution due to the uncollected and poorly disposed waste and dirty streets and cities failing to attract foreign investment and markets (S. Yedla and J.K. Parikh, 2001).

3. GLOBALISAITON, SOLID WASTE GENERATION

The capitalist economic system " is one that never stands still, one that is forever changing, adopting new and discarding old methods of production and distribution, opening up new territories, subjecting to its purposes societies too weak to themselves, "The exploitation of nature and labour serve as a means to the paramount ends of project-making and still more capital accumulation". Hence, the expansion and intensification of the social metabolic order of capital generates rifts in natural cycles and process, forcing a series of shifts on the parts of capital, as it expands environmental degradation. (Paul Sweezy, 1999).

Higher incomes and economic growth also tend to have an impact on the composition of wastes. Wealthier individuals consume more packaged products, which results in a higher percentage of inorganic materials—metals, plastics, glass, textiles, and so on—in the waste stream. More wastes being generated and with a higher content of organic materials could have a significant impact on human health and the environment. If those additional wastes resulting from population and economic growth are not collected, treated and disposed of properly, health and environment in Third World cities will further deteriorate. Profound differences exist between industrialized and developing countries in terms of income, standard of living, consumption patterns, institutional capacity, and capital available for urban investments (Martin Median, 2010).

Industrialized countries enjoy a relative abundance scarcity of capital and an abundance of unskilled and inexpensive labour. It makes sense for the former to devise waste management systems intensive in capital and that save in labour costs, but it often does not make sense for the latter to follow the same approach. Developing countries need low-cost, labour intensive solutions that reduce poverty particularly among the most underprivileged segments of society. A positive correlation tends to exist between a community's income and the amount of solid wastes generated. Wealthier individuals consume more than lower-income ones, which results in a higher waste generation rate for the former. The processes of accelerated population growth and urbanization translate into a greater volume of wastes generated. Globalization can promote economic growth, a desirable outcome. However, this economic growth—in addition to population increase and urbanization—will seriously strain municipal resources to deal with a booming amount of wastes. Table 1 summarizes the waste generation per capita as well as total waste generation in countries of different income levels.

TABLE 1

Waste Generation per Capita and Total Waste Generation

	Waste generation rate Per/day	*Total waste generation Million/tons/year*
Low income countries	1.3	569
Middle income countries	1.8	986
High income countries	3.1	566

Source : Sandra Cointreau (2008).

An important difference between industrialized and developing countries refers to the dissimilar amount and characteristics of waste generated. The waste generated tends to go up as income. Increases First world cities have higher waste generation rates than Third world cities. In the U.S. cities can have generation rates of over 1.2 kg/person/day, while the residents of some African cities generate less than 200 gr/ person/day. A positive relationship also tends to exist between income and waste generation rates within each city: in Mexico city, for example, low-income household generate 2.6 kg a day, middle-income households produces 2.7 kg a day, and upper income households, 3.7 kg a day. It can be seen from Table 2 that the per capita generation rate is high in some states (Gujarat, Delhi and Tamil Nadu). This may be due to the high living standards, the rapid economic growth and high level of urbanization in these states and cities. However, the per capita generation rate is observed to be low in other states (Meghalaya, Assam, Manipur and Tripura). The collection efficiency is high in the cities and

TABLE 2

Municipal Solid Waste Generation Rates in Different States in India

Sl. No.	*Name of the State*	*No. of Cities*	*Municipal population*	*Municipal solid waste t/ day*	*Per Capita generated*
1.	Andhra Pradesh	32	10,845,907	3943	0.364
2.	Assam	4	878,310	196	0.223
3.	Bihar	17	5,278,361	1479	0.28
4.	Gujarat	21	8,443,962	3805	0.451
5.	Haryana	12	2,254,353	623	0.276
6.	Himachal Pradesh	1	82,054	35	0.427
7.	Karnatka	21	8,283,498	3118	0.376
8.	Kerala	146	3,107,358	1220	0.393
9.	Madhya Pradesh	23	7,225,833	2286	0.316
10.	Maharashtra	27	22,727,186	8589	0.378
11.	Munipur	1	198,535	40	0.201
12.	Meghalaya	1	223,366	35	0.157
13.	Mizoram	1	155,240	46	0.296
14.	Orissa	7	1,766,021	646	0.366
15.	Punjab	10	3,209,903	1001	0.312
16.	Rajasthan	14	4,979,301	1768	0.355
17.	Tamil Nadu	25	10,745,773	5021	0.467
18.	Tripura	1	157,358	33	0.21
19.	Uttar Pradesh	41	14,480,479	5515	0.381
20.	West Bengal	23	13,943,445	4475	0.321
21.	Chandigarh	1	504,094	200	0.397
22.	Delhi	1	8,419,084	4000	0.475
23.	Pondichery	1	203,065	60	0.295
	Total	229	128,113,865	48,134	0.376

Source : Status of MSW generation, collection, treatment, and disposal in class-I cities (CPCB, 2000).

states, where private contractors and NGOs are employed for the collection and transportation of Municipal solid waste. Most of the cities are unable to provide waste collection services to all parts of the city. Generally, overcrowded low-income settlements do not have MSW collection and disposal services. The reason is that these settlements are often illegal and the inhabitants are unwilling or unable to pay for the services.

4. CONTROL OR REDUCTION OF WASTE GENERATION

Economic instruments[1] have drawn increasing attention in recent years as an important tool for reinforcing and implementing environmental legislation while simultaneously contributing to sustainable development. Specifically, economic instruments for solid waste management promise to both lessen the size of the solid waste problem and improve the delivery of solid waste collection and disposal services. The 1992 Rio Declaration of Environment and Development endorsed the use of economic instruments as a means to obtain a

TABLE 3

Per capita Generation, Disposal and Collection Efficiency of MSW for Indian States

State	*Per capita generation (g/cap/day)*	*Per capita disposal (g/cap/day)*	*Collection Efficiency (%)*
India (sample average)	377	273	72
Andhra Pradesh	346	247	74
Bihar	411	242	59
Gujarat	297	182	61
Haryana	326	268	82
Karnataka	292	234	80
Kerala	246	201	82
Madhya Pradesh	229	167	73
Maharashtra	450	322	72
Orissa	301	184	61
Punjab	502	354	71
Rajasthan	516	322	62
Tamil Nadu	294	216	73
Uttar Pradesh	439	341	78
West Bengal	158	117	74

Source : (Nema, 2004).

sustainable environmental improvement. This endorsement addresses a development strategy wherein there is a continuous effort to balance economic and environmental needs. In each country the conditions affecting the choice will be different. Each country will need to choose economic instruments that have a potential for improving its own particular situation.

Economists have long advocated the use of economic instruments as an alternative, or supplement, to direct regulation. Most importantly, economists argue that economic instruments can create a system for pollution reduction that achieves the same level of environmental protection for a lower overall cost (or achieves more for the same cost). Given the importance of the overall costs of environmental protection in political debate, this is a crucial advantage. Economic instruments also allow for a more hands-off regulation and decentralized decision-making, giving greater freedom to firms and plants about how to comply (Duncan Austin, 1999).

India has so long maintained a regulatory regime which has been found ineffective, and the level of pollution has been on the rise. More recently, the focus has shifted towards the adoption of market based instruments to address the pressing pollution problems. In February 1992, the Policy Statement for Abatement of Pollution issued by the Government of India stated an aim of providing industries and consumers clear signals abut the cost of using environmental and natural resource, since there was an increasing trend in environmental pollution (Mehta *et al.*, 1994).

4.1 Economic Instruments in Solid Waste Management

In the environmental economics policy literature the term *economic instrument* is generally understood to refer to policy, tool or action which has the purpose of affecting the behaviour of economic agents by changing their financial incentives in order to improve the cost-effectiveness of environmental protection efforts (pollution control and avoidance).

Economic instruments comprises all incentives/disincentives the self-interest of consumers, producers, and services providers to make environmental improvements or reduce adverse environmental consequences. These instruments may be used to address basic environmental needs, or may motives actions to address environmental protection beyond the prescribed minimum accepted standards of command-and-control regulatory approaches. The underlying logic for the use of economic instruments to decrease the human-produced stresses on the environment is that there is always a higher level of environmental performance that some polluters could achieve, but that the law cannot readily require. The make polluters go beyond regulatory requirements involves incentives. Economic instrument may provide these incentives and encourage people to extend their actions beyond the minimum standards required by law, resulting in "over-performance" as opposed to "compliance". Therefore, economic instruments have a theoretical potential to internalize negative environmental impacts in a cost-effective manner at a local, national and global scale.

Specifically, economic instruments for solid waste management can be used as a tool to:

- Reduce waste generation;
- Decrease the amount of generated waste that is hazardous;
- Segregate hazardous waste for special handling and disposal;
- Optimize recovery, reuse and recycling of waste;
- Support cost-effectiveness solid waste collection, transport, treatment and disposal systems;
- Minimize adverse environmental impact related to solid waste collection, transport, treatment and disposal systems; and
- Generate revenue to cover costs.

4.2 Taxes and subsidies

The economic instruments like tax or subsidy is advocated on the ground that it would determine the right balance between recycling, disposal and incineration. Pearce and Brisson (1993) show that various economic instruments can be used to balance the marginal costs and benefits of recycling to arrive at an optimum level of recycling. The optimum condition of recycling is given as:

$$P_R + C_D + C_{DE} = C_{SC} + C_R + C_{RE}$$

Where:

P_R = price of recycled material
C_D = marginal cost of disposal
C_{DE} = marginal environmental cost of disposal
C_{SC} = marginal cost of separate collection
C_R = marginal financial cost of recycling
C_{RE} = marginal environmental cost of recycling

The left-hand side of the above equation is the benefit of recycling. The right-hand side is the cost of recycling. Hence we get

$$MB_R = MB_C$$

Suppose for simplicity $CR_{RE} = 0$, then by rearranging we get

$$(P_R - C_{SC} - C_R) = C_D + C_{DE}$$
$$\text{Or} - \Pi_R = CD + C_{DE}$$

Thus the socially desirable recycling level occurs when the marginal loss – (Π_R) on recycling is just equal to the marginal and financial cost of disposal. The socially optimum level or recycling could be achieved by introducing (a) a recycling credit $C_{D,}$, (b) a further recycling subsidy C_{DE} or (c) a credit C_D and a tax or levy on waste disposal $C_{DE.}$

Analysis of the optimal design of the market-based instrument in the economic literature generally follows two methodologies: (a) household production framework, and (b) general equilibrium models. The household production model Freeman (1993) provides a framework for examining interactions between demand for market goods and the availability of public goods such as environmental quality. This framework is based on the assumption that there is a technical relationship among goods used by households in the implicit production of utility yielding final services.

Household waste is a by-product of household production process and therefore it can be analysed within the framework of household production function. Particularly, the consumer's decision with respect to disposal options offered. Generally, recycling involves more time inputs but involves less cost than conventional disposal subject to the condition that recycling is subsidized and there's a large savings in collection and extraction costs. A consumer chooses to recycle if the marginal benefit of recycling is higher than the cost of extra time inputs involved in it. The solution of these models would give the demand for waste disposal services and recycling efforts as function of incremental fee associated with conventional disposal and the opportunity cost of time associated with recycling.

In the general equilibrium model it is the firm who produces final goods and services and the household only consumers them. Waste here is a by-product of consumption and no waste is generated in the process of production. In these models consumers choose a composite good at given prices. A waste is considered to be a function of consumption, the utility maximizing choice also determines the amount of waste to be disposed of or recycled. Producers, on the other hand, choose the recycled good and a virgin material (that may be extracted) as inputs to produce the composite good. In this set-up the optimal instrument is obtained by comparing the social planner's first-order conditions to the market's first order conditions, Baumal and Oates (1988) and Kinnaman and Fullerton (2000).

The size of the levy is determined by the valuation of environmental costs associated with disposal and incineration. This topic has been discussed in the preceding section. However, a large literature has arisen regarding the efficiency of these economic instruments used in the waste management options. They are follows.

4.3 Unit Pricing for Garbage Collection

The most direct way to internalize the external cost of garbage disposal is to tax each bag of garbage. However in most developing and also developed countries households receive fixed price collection services, regardless of the amount they dispose of Dinan (1993) points out that such system leads to increased disposal of wastes in two ways: (i) since household does not face any extra charge for the additional unit of garbage dumps, he lacks a financial incentive to consider disposal cost in his purchasing and recycling decisions. He is endowed with unlimited dumping rights in a local land fill or transfer station at a fixed fee, Ackerman (1997), (ii) this, in turn, does not provide producers with an adequate incentive to produce goods that are less costly to dispose of Strathum *et. al.* (1995) estimated efficiency losses arising out due to the absence of marginal cost pricing of disposal services. Using data set for Portland, Oregon metropolitan areas they estimated a fairly high elasticity of demand for landfill disposal and found that the magnitude of the deadweight loss relative to the level of total expenditure grew exponentially with the increase in the difference between actual price charged long-run marginal costs.

However marginal cost pricing of garbage may be effective as long as households reduce the amount of waste generated and increase recycling. Kinnaman and Fullerton (2000) demonstrate that of household wastes are constrained by two disposals options, i.e. garbage disposal at landfill and recycling then, marginal cost pricing would tend to substitute recycling for garbage disposal. But if illegal disposal or burning features as a third alternative in the household disposal choice set, then unit pricing would encourage illicit dumping. If marginal cost pricing results increase in illegal dumping, and if the encourage illicit dumping, and if the externality costs are high, the efficiency losses from under pricing services may be smaller to bear (Page, 1977). In fact, the initial introduction of unit pricing system causes only a modest reduction in waste disposal through dumping (Ackerman, 1997). Fullerton and Kinnaman (1996) estimated that 28 per cent of the reduction in garbage resulting from pricing garbage at the curb might be as a result of illegal dumping. Jenkins (1993), Blume (1991) and Miranda and Aldy (1998) have also come out with similar findings. Using a household production framework Morris and Holthausen (1994) have shown that price increase on conventional disposal did not affect recycling. In presence of unit pricing the households found it more attractive to increase total waste reduction effort. The resources like time and purchased inputs devoted to recycling were high but they become less effective because of reduction of wastes.

Again the administrative costs of implementing the programme may exceed the social benefits. Fullerton and Kinnaman (1996) estimated the costs for Charlottesville, Virginia and found that they exceeded the social benefits from marginal cost pricing. According to Dinan (1993) a uniform tax on all types of garbage might be inefficient if materials within the waste steam produce different social costs. But differentiated tax foundation would amplify administration costs.

4.4 Virgin Material Taxes and Subsidies

In view of the above mentioned problem associated with taxing or pricing garbage economists advocate the implementation of a tax on virgin materials (or reuse subsidy in other way) to achieve an efficient allocation of resources in a scenario where garbage disposal produces external costs. It is argued that since external costs of virgin material extraction and external benefits or recycled material are not included from the profit maximum problem of the firm, too much of virgin input is extracted and too little of recycled input is used.

In support of this theoretical argument Miedima (1983) finds that a tax on virgin material setting equal to the marginal cost of disposing of any resulting waste material produces welfare gains greater than that from a subsidy on producer's use of secondary material, or a direct tax on garbage or an advanced disposal fee. Dinan (1993) accepts that virgin material fees can increase the demand for recycled materials and hence reduce disposal requirements. They can also raise prices for recycled materials and thus increase the recycling rate of the suppliers in the secondary market.

A large number of studies, on the other hand, have justified virgin material taxes on the ground that primary inputs enjoy effective subsidies compared to secondary inputs and these have an unfair comparative advantage over their recycled counter part. For instance, in United states EPA's study Federal Disincentives: A study of federal Tax subsidies and other materials posed a serious obstacle to recycling. Two other studies 'Energy Information Administration' by Department of Energy and the 'Alliance to Save Energy' also obtained estimates of huge energy subsidies. By Kinnaman and Fullerton (2000) virgin materials, infact, would earn subsidy on account of the fact that (a) income earned by the produced is taxed at the rate of capital gains instead of the corporate income tax rate, (b) mineral exploration is traditionally encouraged on public lands, and (c) freight rates charged for recycled materials are often higher than that for their virgin material counterpart.

4.5 Advance Disposal Fees

Advance Disposal Fee (ADF) is a tax on goods levied at the time of sale, based on the cost of ultimate disposal of goods. The rationale behind such fee is that when goods are exchanged the externality from disposal of these goods are not reflected in the product price. Collecting the disposal cost in advance would correctly raise product prices to reflect the true costs to the society leading to reduce consumption of goods that are waste generating in nature.

Ackerman (1997) had described the success stories of such ADF programme in Florida, where the policy succeeded in raising substantial portion of revenue ($45 million in its first year) and even forced the firms to switch over to recyclable packages. But he argued that this was not because the ADF successfully internalized the disposal externality. He argued that the price rise due to ADF was insignificant. He attributed its success to the "symbolic meaning" it conveyed. Any product that met the recycling standards was exempted from ADF and this increased its marketability as the state seal of environmental approval served as an advertisement for the product.

4.6 Deposit Refund System

Deposit Refund System (DRFs) involves the payment of a deposit when a product is purchased. The deposit is repaid when the product is returned after use. This system provides a strong financial incentive for returning products to a centralised facility to better ensure product reuse, safe disposal or recycling.

5. CONCLUSION

The informal policy of encouraging the public to separate solid waste management and market it directly to the informal network appears to be a better option. The involvement of people and private sector through NGOs could improve the efficiency management of solid waste. Public awareness should be created among masses to inculcate the health hazards of the wastes. Municipal authorities should maintain the storage facilities in such a manner that they do not create unhygienic and unsanitary conditions. Proper maintenance of the municipal corporation transportation vehicles must be conducted, and the Dumper placer should replace the old transportation vehicles in a phased manner. Currently, at the level of waste generation and collection, there is no source segregation of compostable waste from the other non-biodegradable and recyclable waste. Proper segregation would lead to better options and opportunities for scientific disposal of waste. Recyclables could be straightway transported to recycling units that in turn would pay a certain amount to the corporations, thereby adding to their income. This would help in formalizing the existing informal set up of recycling units. It could lead to several advantages such as enabling technology upgradation, better quality products, saving of valuable raw material resources of country, reducing the need for landfill space, a less energy-intensive way to produce some products and employing labour in recycling industries. Organizing the informal sector and promoting micro-enterprises are an effective way of extending affordable services. Promotion and development of recycling is a means of upgrading living and working conditions of rag pickers and other marginalized groups (M. Sharholy *et al.*, 2008).

The current regulations (Municipal Solid Waste Management Rules, 2000) are very stringent. Norms have been developed to ensure a proper Solid waste management system. Unfortunately, clearly there is a large gap between policy and implementation. The producer responsibility is to avoid having production market that cannot be handled effectively and environmentally correctly when they become waste products. The use of economic instruments would allow additional environmental improvements to be met at least cost, or would allow present standards to be met more cost-effectively.

Note and Reference

1. Hence, a technical definition for an economic instrument is a 'tool that affect estimates of the costs and benefits of alternative actions open to economic agents (OECD, 1997).

References

Ackerman, Frank (1997), Why do we recycle?, Washington D.C., Island Press.

Ahsan, N. (1999) Solid Waste Management Plan for Indian Mega cities, *Indian Journal of Environmental Protection,* 19 (2), 90-95.

Baumol, William J. and Wallace, E. Oats (1988), ***The Theory of Environmental Policy***, 2nd Ed, New York, Cambridge University Press.

Blume, Daniel R. (1991), Under What Conditions should Cities Adopt Volume-based Pricing for Residential Solid Waste Collaboration, Unpublished Manuscript, The office of management and budget office of Information and Regulatory affairs, National Resources Branch, May.

Cointreau, S. (2008), 'Methane-2-Markets Fund: The Solid Waste context of Developing countries', Washington DC; World Bank.

CPCB (2000), Status of Municipal solid waste generation, collection, treatment and disposal in Class I cities, Series: ADSORBS/31/1999-2000.

David, N. Beede and David, E. Bloom (1995), "The Economics of Municipal Solid Waste", ***The World Bank Observer***, Vol. 10, No. 2, pp. 113-50.

Dinan, Terry M. (1993), Economic Efficiency Effects of Alternative Policies for Reducing Waste Disposal, ***Journal of Environmental Economics and Management***, 25(3), November, 245-256.

Dinan Terry M. (1993), Economic Efficiency Effects of Alternative Policies for Reducing Waste Disposal, ***Journal of Environmental Economics and Management***, 25(3), November, 245-56.

Duncan Austin (1999), Economic Instruments for Pollution Control and Prevention—A Brief Overview, ***World Resource Institute.***

Freeman, III, A.M. (1993), The Management of Environmental and Resoruce Values: Theory and ***Resources for the Future.***

Gupta (1998), "Solid waste management in India: Options and opportunities", ***Resources, Conservation and Recycling***, 24: 137-54.

Jenkins, Robin R. (1993), The Economics of Solid waste Reduction, Hans, England, Edward Elgar Publishing Limited.

Jha, M.K. Sondhi O.A.K. Pansare, M. (2003), "Solid Waste Management Case Study", ***Indian Journal of Environmental Protection,*** 23 (10) 1153-1160.

Kansal A. (2002) "Solid Waste Management for India", *Indian Journal of* ***Environmental Protection***, 18 (2) 444-48.

Kansal, A. Prasad and R.K. Gupta, S. (1998), "Delhi Municipal Solid Waste and Environment—An Appraisal", ***Indian Journal of Environmental Protection***, 18 (2), 123-28.

Kinnaman, Thomas C. and Don Fullerton (2000), The Economics of Residential Solid Waste Management, in T. Tietenberg and H. Folmer (eds.) 'The International Yearbook of Environmental and Resource Economics 2000/ 2001, Cheltanham, UK and Northampton, MA, USA, Edward Elgar Publishing Ltd., 100-47.

Kumar, Sunil (2009), Assessment of the Status of Municipal solid waste management in Metro cities, state capitals, Class I cities, and class II towns in India: An insight, ***Waste Management*** 29, pp, 883-895

Medina Martin (1997a), "Scavenging on the Border: A study of the Informal Recycling sector in Laredo, Texas, and Nuevo Laredo, Mexico" Ph.D Dissertation, Yale University.

Medina Martin (2010), "Solid wastes, Poverty and the Environment in Developing Country Cities", United Nations University, Working Paper Series No. 2010/23.

Mehta, Shekhar, Sudipto Mundle, and U. Sankar (1994), Incentives and Regulation for Pollution Control. National Institute of Public Finance and Policy, New Delhi.

Miedema, Allen K. (1983), Fundamental Economic Comparisons of Solid Waste policy options, Resources and Energy, 5, 21-43.

Miranda, M.L. and J.E. Aldy (1998), Unit Pricing of Residential Municupal Solid Waste: Lesson form Nine Case Study Communities', ***Journal of Environmental Economics and Management***, 52(1), January, 79-93.

Mor, S., Ravindra, K., Visscher, A.D., Dahiya, R.P., Chandra, A. (2006), Municipal Solid Waste Characterization and its assessment for potential methane generation: A case study. *Journal of Science of the Total Environment,* 371(1), 1-10.

Morris Glenn, E. and Duncan, M. Holthausen, Jr. (1994), The Economics of Household Solid Waste Generation and Disposal, ***Journal of Environmental Economics and Management***, 26, 215-34

Nema, A.K. (2004), Collection and transport of municipal solid waste. In: Training program on solid waste management. Springer, Delhi, India.

OECD (1997), Evaluating Economic Instruments for Environmental Policy, OECD, Paris, France.

Page, T. (1997), Conservation and Economic Efficiency: An Approach to Materials Policy Baltimore, John Hpokins University Press.

Paul Sweezy (1999), "Capitalism and Environment", ***Monthly Review***, 56, No, 5

Pearce, D.W. and I. Brisson (1993), Using Economic Incentives for the Control of Municipal Solid waste', paper presented to conference of Urban Solid Waste: Economic and Environmental Perspective, Milan, April.

Raje, D.V., Weakhare, P.D., Despande, A.W., Bhide, A.D. (2001), An Approach to assess level of satisfaction of the residents in relation to SWM system, ***Journal of Waste Management and Research***, 19, 12-19

Rathi, S. (2006), "Alternative Approaches for better municipal solid waste management in Mumbai", India, ***Journal of Waste Management**, 26 (10)*, 1192-1200.

Ray, M.R., Roychoudhury, S., Mukherjee, G. Roy and S. Lahiri (2005), "Respiratory and General Health Impairments of Workers Employed in a Municipal Solid Waste Disposal at Open Landfill Site in Delhi", ***International Journal of Hygiene and Environmental Health***, 108 (4), 225-62.

Sharholy, M. Ahmad, K. Vaishya, R.C. Gupta, R.D. (2007), Municipal solid waste characteristics and management in Allahabad, India, ***Journal of Waste Management***, 27 (4), 490 -496.

Sharholy, M., Kafeel Ahmad Gauhar Mahmood and Trivedi (2008), Municipal solid waste management in Indian cities—A Review, ***Journal of Waste Management***, 28, pp. 459-67.

Sidddiqui, T.Z., Siddiqui, F.Z., Khan E. (2006), Sustainable Development Through integrated municipal solid waste management approach—A case study of Aligarh District. In: Proceedings of National Conference of Advanced in Mechanical Engineering (AIME-2006), Jamia Millia Islamia, New Delhi, India, pp. 1168-75.

Singh, S.K. and R.S. Singh (1998), A Study on Municipal Solid Waste and its Management Practices in Dhanbad—Jharia Coalifield, ***Indian Journal of Environmental Protection***, 18 (11), 850-52.

Singhal, S. and Pandey, S. (2001), solid waste management in India: status and future directions. ***TERI Information Monitor on Environment and Science***, 27, 100-108.

Strathman, James G., Anthony, M., Rufolo and Gerard, C.S. Mildner (1995), The Demand for Solid Waste Disposal, ***Land Economics***, 71 (1) February 57-64.

Sudhakar and Parikh, (2001), "Economic Evaluation of Landfill System with Gas Recovery for Municipal Solid Waste Management: A Case Study", International, ***Journal of Environmental Pollution***, 15: 433-47.

Zhu, D., Asnani, P.U., Zurbrugg, C. Anapolskey, S., Mani, S. (2009), Improving Municipal Solid Waste Management in India: A Source : Book for Policy-makers and Practitioners, ***The World Bank***.

Chapter 25

Mountainous Eco-tourism and Environmental Impacts

Case Study of Rafting Industry on Ganges in Uttarakhand, Himalayas

RAJIV PANDEY AND PARTH SARATHI MAHAPATRA

ABSTRACT

Mountainous regions particularly Himalayas, coupled with varied topography and climate provide variety of habitat for flora, fauna and aesthetic beauty. This is being elaborated with reference to white water rafting on Ganges in Tehri Garhwal, Uttarakhand based on primary and secondary data collected from stakeholders during 2008 to 2010. Supplementary data was collected by means of arranging group discussion and individual interactions through pre-tested questionnaires. It includes economic, social, culture; institutional aspects and associated impacts.

Number of tourists enjoying rafting varies with seasons and comprises 32% foreigners and 68% national. The total camps were 74 in number with 0.18 Mm^2 areas. During good seasons, 2000-4000 visitors enjoy rafting per camp depending on capacity. In general, local people favoured eco-tourism due to livelihood opportunities, however, impacts on public amenities, traffic congestion, damage to natural resources, disturbance to fauna, and invasion to unique traditional culture were also observed. Correspondingly, waste disposable and defecation at the river bank, put the Ganges in peril and pollute the water. These all were at low level at present due to low influx, however, may be devastating in future without management.

The study recommends focussed approach for environmentally sound practices with proper garbage management, external interventions for well-being of locals, strategies for visitor's management, spreading environmental awareness with active involvement of local people for promoting sustainable tourism and proper infrastructure development. However, the implementation of various strategies and outcomes of environmental review remains a challenge for planners and regulators, particularly in these remote and fragile mountainous locations.

1. INTRODUCTION

Tourism is one of the world's largest and fastest-growing industries and contributes for economic development of developing countries (OECD, 2008). The growth can be attributed to changing lifestyles in affluent societies, with high desire of leisure and more concern for the quality of the environment (Veenhoven, 1999). The need for leisure combined with the desire to enjoy serenity and pristine natural beauty and to experience different cultures attracts the overseas tourists to exotic destinations (Tapper, 2001). Tourism comprises the activities of persons travelling to and staying in places outside their usual environment for not more than one consecutive year for leisure, business and other purposes (Holden, 2000). Tourism provides domestic services and experiences of domestic resources to foreign consumers in exchange for foreign currency, thus it is essentially an export industry (Woods, *et al.*, 1991). However, increasing tourist inflow has raised expectations that tourism can be an agent of socio-economic development (Tapper, 2001) as well as create pressure on resources, and a threat to cultural and biodiversity.

Globalization coupled with advance communication technology makes the world a smaller place, therefore, a stronger tourism environment has emerged allowing people to experience and enjoy other cultures, create economic and social ties with a variety of communities, and broaden and diversify their lives. Tourism has gradually become a mass phenomenon reaching large numbers of people throughout the world. Tourism is beneficial for individuals along with contributions to national economies and in turn to global economy. Tourism: 2020 Vision, the (WTO) World Tourism Organization predicts that the tourism sector will grow by an average of 4.1% per year over the next two decades, surpassing a total of 1 billion international tourists by the year 2010. Projections for year 2020 indicate that tourist arrivals will be around 1.6 billion with business of US$ 2 trillion (Milne and Ateljevic, 2001; Neto, 2002).

Eco-tourism is part of tourism with the potential of being an important sustainable development tool. Eco-tourism involves travelling to relatively undisturbed natural areas with the specified object of studying, admiring and enjoying the scenery and its wild plants and animals, as well as any existing cultural aspects (both of the past and present) found in these areas, World Tourism Organisation (WTO) defines. Eco-tourism is one way creating economic opportunities with opportunity to conserve natural resources, educating visitors about sustainability, and benefiting local people (Soleimanpour, 2005). However, proper planning and management are critical for development of ecotourism. Fifth meeting of the Conference of Parties (COP) to the Convention of Biological Diversity (CBD) signifies eco-tourism role for educating travellers about the value of healthy environment and biological diversity.

World Conservation Union (1996) defines eco-tourism, as environmentally responsible travel and visitation to relatively undisturbed natural areas, in order to enjoy and appreciate nature (and any accompanying cultural features—both past and present) that promote conservation, has low negative visitor impact, and provides for beneficially active socio-economic involvement of local population. Ideally, eco-tourism includes the followings:

- Contributes to conservation of biodiversity.
- Sustains the well-being of local people.
- Includes an interpretation/learning experience.
- Involves responsible action on the part of tourists and the tourism industry.
- Is delivered primarily to small groups by small-scale business.
- Requires lowest possible consumption of non-renewable resources.
- Stresses local participation, ownership and business opportunities, particularly for rural people.

The sustainable tourism concept was introduced in the late 1980s based on the Brundtland Report on our common future as an outcome of Wold Commission on Environment and Development (WCED) in 1987 (Brundtland website, 1987). Sustainable tourism development is the tourism that meets the need of the current generations without compromising the ability of future generations to meet their needs (Weaver, 2001). This leads to any form of sustainable nature and community-based, non-consumptive, low environmental and cultural impact tourism to the tune of carrying capacity and economic generation of the destination, which fulfils the wishes and aspirations of all kinds of visitors in a learning environment (Soleimanpour, 2005). These coupled with the deliberations of meetings of stake holders of

International Eco-tourism Society has develop a set of principles (Wood, 2002). These are as follows:

- Minimize the negative impacts on nature and culture that can damage the destination.
- Educate the traveller on the importance of conservation
- Stress the importance of responsible business, which works co-operatively with local authorities and people to meet local needs and deliver conservation benefits
- Emphasize the need for regional tourism zoning and for visitor management plans designed for either regions or natural areas that has been defined for eco-tourism destination
- Emphasize use of environmental and social base-line studies as well as long-term monitoring programs, to access and minimize impacts
- Strive to maximize economic benefits for the host country, local business and communities, particularly people living in and adjacent to natural and protected areas
- Tourism development within the carrying capacity of available resources
- Development of infrastructure in harmony with the environment, minimizing use of fossil fuel, conserving existing vegetation and wildlife, and blending with natural and cultural environment.

In some developing countries, the improvement of socio-economic conditions of certain sections of the local population, the rise in workers' rights such as paid holidays, and expansions and improvements in the transport system have led to the growing phenomenon of domestic tourism. In addition, a rapid globalization process encouraging Western lifestyles and promoting the "Northern-style consumerism" and leisure travel has encouraged many locals (Ghimire, 2001). In India, the notion of travelling for leisure had been for years, considered too much of a luxury for the majority of the population, and therefore the focus was often in attracting foreign visitors who brought U.S. dollars (*BBC News*, 2001). "Domestic tourism was ignored because it competed for facilities and services with international tourists and these were in short supply" (Rao and Suresh, 2001).

India has a tradition of protection and conservation of natural resources by indigenous and local communities. These age-old practices guide community to live in harmony with nature (Chopra *et al.*, 1993).

India with diverse topography ranging from mountain, costal and plain region offers a lot of opportunities for eco tourism. Mountain tourism ranges from mass tourism to popular sites, the ski industry, adventure tourism (trekking, climbing, rafting), cultural tourism, eco-tourism, and pilgrimage. Himalayas, the land of high biodiversity is a favourite destination of mountain eco-tourism in India. Amidst Himalayas the hills of Uttarakhand have all the ingredients to fulfil the tourist's desire. The unexplored valley, towering peaks, flowing rivers, snow-capped mountains along with a splendid combination of flora and fauna add to the exquisiteness of Uttarakhand. Lot of opportunities exist for adventure sports such as skiing, canoeing, para-glidng, rock climbing and river rafting known as 'white water rafting' in Uttarakhand.

The unique resources of fragile mountains and constraints for the essential resources pose various challenges to sustainable eco-tourism development. Therefore, focussed approach for conservation and community integrity on sustainable basis is essential for eco-tourism. The present paper addresses the issues with a case study of river rafting industry in river Ganges on "Kaudiyala – Tapovan Ecotourism Zone" in Uttarakhand. Primary stakeholders include national and international visitors along with camp owner and workers and secondary stakeholders comprise the locals and the associated state departments. The data collected from them, form the basis of impact evaluation and management.

1.1. Impact Assessment

Impact assessment is an evaluation that is intended to determine the consequences of an intervention, in terms of outcomes. This provides opportunities to understand and anticipate the possible consequences of an industry on the local environment, human populations and communities of the area. Environmental Impact Assessment (EIA) is an efficient method for preserving natural resources and protecting the environment. It involves individual assessments of aspects of environment (e.g. population, landscape, heritage, air, climate, soil, water, fauna, and flora) likely to be significantly affected by a project (Morris and Therivel, 2001). Impact Assessment is part of the planning process and hence appraises the planner and the project proponent to the likelihood of various impacts. Various biological, physical, economic and social impacts have to be identified and measured in order to be understood and facilitate to the decision-makers. Thus, impact is the process link in the chain between evidence and decision-making and can provide a realistic appraisal of possible ramifications and suggestions for project alternatives and possible mitigation measures. The functions of impact mechanism are accountability, feedback and future vulnerability.

- Accountability provides empirical evidence of the effectiveness of past investment for driving outcomes of interest and to validate the relevance of overall strategies pursued.
- Feedback helps to inform allocation decisions by providing insights as to the comparative effectiveness of different investment options, and to generate lessons that can improve the implementation of ongoing and future research.
- Future vulnerability provides knowledge about the vulnerable areas, i.e. resources and communities with their association to climate change or environment. This also includes trend analysis with respect to time, and hence helps to develop a sustainable Environment Management Plan.

The feedback function of impact assessment is important, but considerably more challenging to evaluate, as it is affected by implicit and explicit causes and effect relationships and has methodological limitations. These challenges need to be addressed for the role of impact assessment to be more effectively realised in the future and to fulfil this purpose. It may also include a rigorous cost benefit analysis. Vulnerability is another important aspect, which generated due to the secondary processes against the realised impacts due to the project.

Tourism industry has also directly and indirectly influence the various associated resources. Tourism provides employment opportunities thus increasing the economic levels and restricts migration of local population, cultural interaction open dialogues among various socio-culture populations, provide ways for commercialisation of the local products, avenues for interchange of ideas, and develops interests on protection of natural resources. Tourism also influence usages level of resources, changes in landscapes with the creations of news infrastructures, increases waste disposals, alter ecosystems, introduce exotic species of animals and plants, affect traditional habits, raise narcotic traffic, increases prices of the amenities. Broadly, tourism activities may also contribute to climate change due to emissions from vehicular fossil fuel use and waste disposal. Therefore, if tourism is to act a sustainable development and poverty alleviation agent, proper planned policies are to be devised to ensure that the benefits are shared and spread among poor communities without making them more vulnerable (Anonymous, 1997) with minimise the adverse impact of tourism on the environment (OECD, 2008).

2. ECO-TOURISM IN HIMALAYAS

Nature, setting of natural resources and the associated communities are instrumental for development of eco-tourism in remote areas. Mountainous regions particularly Himalayas, coupled with varied topography and climate provide variety of habitat for flora, fauna and places of scenic beauty (Explore Himalayas web site, 2010). The region offers dramatic mountain scenery to be imagined anywhere on the earth and therefore a versatile eco-destination (Fukuda, 1971). This forms excellent eco-tourism opportunities for rejuvenation and enjoy to visitors by allowing fair and equitable growth of host communities. These tourism resources need to be managed properly by mobilizing the local participation in order to attract the tourists (Bhusal, 2007).

Himalayan eco-tourism includes pilgrimage tourism, mountain excursion, mountaineering and white water rafting, etc. Marked by its difficult geographic terrain and hard accessibility the region has long enjoyed self-sufficiency, which was supplemented by beneficial pilgrim economy (Pauw, 1986). However, extreme seasonality, lack of appropriate infrastructure and planning to address intricacies of natural and manmade resources in these fragile ecosystems has made mountainous tourism a major growing environmental concern. In addition, steep slopes in the Himalayas, sparse forest cover, and high seismicity multiply the environmental problems resulted to landslide, soil erosion and sedimentation thereby habitat loss of floral and faunal species. This may be aggravated due to carbon release from anthropogenic activities such as increased volume of waste materials and burning of fossil fuels due to tourism. Moreover, it also provides opportunities for income generation to the locals through development of associated primary and secondary enterprises. This may also influence local cultures and traditional life styles.

The Himalayas, stretching 3,200 kilometres along India's northern frontiers, cradle numerous rivers with slopes which drain them all year round. These abundance of mountain rivers make these locations as white water destination, with plenty of first-descent and exploratory possibilities. The development of river rafting as a leisure sport has become popular since the mid-1970s. The tourism activities along the hilly river are of various types.

- Tourism of beach is primarily includes the activities of tourist near the bank of rivers like camping, enjoying the scenic beauty and camp firing. These provide the intended objectives of tourist in terms of joy and pleasure.

- Adventure tourism involves exploration of remote areas and exotic locales and engaging in various activities such as rock climbing, trekking, river rafting, etc. and hence, may also be most dangerous among all forms of ecotourism. This provides opportunity to enjoy the wilderness against the uncertainty of life in order to enjoy.
- Rural tourism, or mountain tourism are enjoying the beauty of mountains by staying at the laps of the peak. This phenomenon is causing a change of the traditional flow country-city, to the opposite side. People are moving to rural areas that are far away from cities, and this is causing a re-organisation of the economic activity of the rural areas.

These tourism activities may put enormous pressure on the area depending as nature of tourism and lead to impacts such as soil erosion, increased land pollution (polythene use and others), natural habitat loss, increased pressure on certain endangered species and heightened vulnerability to forest fires besides proving opportunities for economic and social developments. This leads to specific evaluation of specific tourism activities.

2.1 River Rafting on River Ganges—The Spatial Focus

In India, rafting is commonly exercised on the River Ganges near Rishikesh, Uttarakhand and the Beas River in Himachal Pradesh. River rafting in Uttarakhand started during last decades of the twentieth century and has rapid growth. The initial beach camps on Ganges have been established during 1988 for camping purpose. The Rafting Industry on the Ganges was became regular phenomenon since 1994 with only four camps. In 2008-09, total operational entities were 74 with 208 numbers of rafts afloat during 2008-09. Every year after rains, sand gets deposited to make clear and distinct formation of beaches on both banks of Ganga. This area has been denoted as an eco-tourism zone namely Kaudiyala-Tapovan eco-tourism zone.

The study site is situated along the Badrinath National Highway on river Ganges, district Tehri Garhwal, Uttarakhand, between 30°042 233-30°072 343 N latitude and 78°302 133-78°192 483 E longitude; popularly known as "Kaudiyala-Tapovan Eco-tourism Zone" because annually after the rains sand enormously gets deposited on the banks of Ganges in this region forming distinct beaches. This zone includes villages such as Byasi, Kaudiyala, Singtali, Timli, Bawani, Brahmapuri, Guller, Shivpuri, Mala, Kathya, Bawani, Silan, etc. The area covers a road distance of 40 km along with a river transect of 36 km from

Rishikesh city. Along the paths, deep gorges, reverberating with the gurgle of innumerable streams are very common features as the region lies at external range of the Middle Himalaya Mountains and comprises of Shivaliks. Emerald green dense forests containing Sal (*Shorea robusta*), Kanju (*Holoptelea integrifolia*), Bakli (*Anogeissus latifolia*), Dhauri (*Lagerstroemia parviflora*), Bamboo species and tree cover both sides of the highway in gorges and hill tops. The summer temperature ranges between 9°C-30°C and winters temperature ranges between 3°C - 15°C with average annual rainfall of 70 cm. In rainy season, climate is very cool and full of greeneries. The summer is very pleasant with the harsh mountain sun balanced by the mesmerizing air flow. The winter is very harsh at times with severe cold draughts all over the Tehri Garhwal region (Travel smart website, 2010).

Camping and rafting may result in resource-use conflicts, such as competition between tourists and the local population for use of prime resources like water and energy because of the scarce supply of resources in these regions against the future huge inflow of tourists. Changes in the host community's quality of life are influenced by two major factors: the tourist-host relationship and the development of the industry itself. The tourist-host relationship can be seen in terms of social, cultural and economic changes to host societies which include changes in value systems, traditional life-styles, family relationships, individual behaviour or community structure. However, the economic activity is mobilised due to tourism, may boost the local production of resources and helps in infrastructure development without analysing the full ranges of adverse impacts. Thus, taking into account the high vulnerability of the area towards depletion of natural resources and un-sustainability, there is a need for comprehensive ecological, social, cultural and economic impact assessment of River Rafting on the Ganges. Along with, it certain uncontrolled activities like location of toilets within the submergence zone of the river beach, trekking in the forests, regular campfires and dumping of partially burnt wood into the river system due to camping and river rafting carry potential threats to the immediate environment and also the quality of river water. A diagrammatic representation of the various drivers and resultants associated with the River Rafting and Camping Industry in Himalayas has been represented below for a better understanding about the working of this industry and the associated factors (Figure 1). This includes the external and internal resources required for the rafting industry and possible outcomes. This briefly also identifies various impacts.

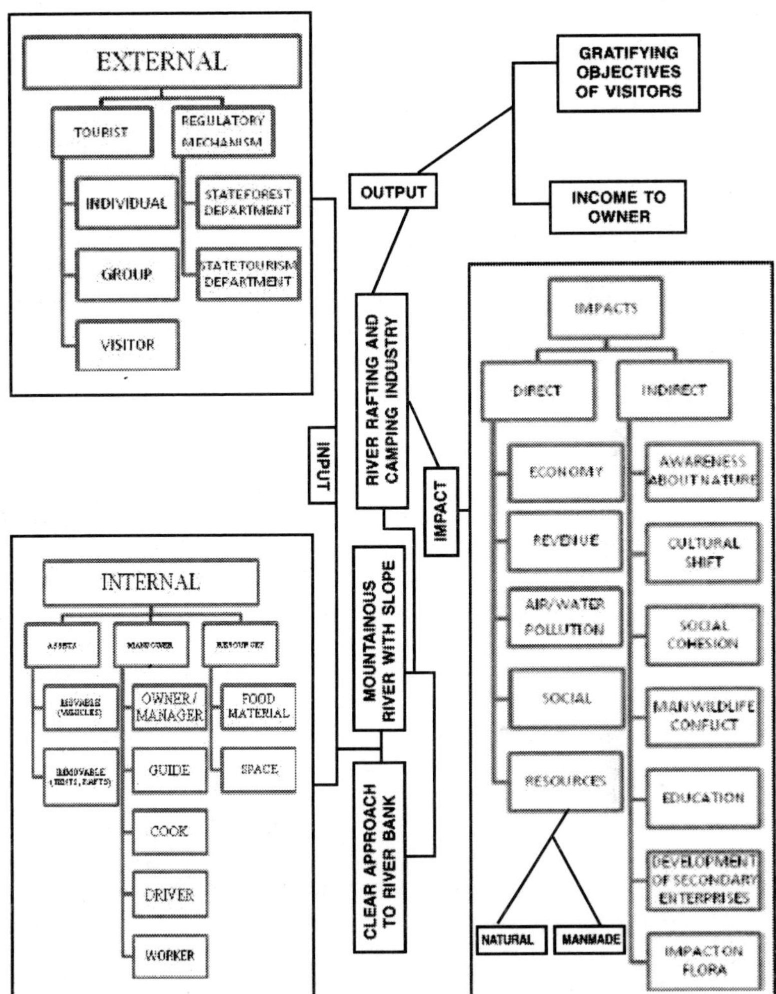

Figure 1 : Model Representation of Various Drivers and Resultant of River Rafting and Camping Industry

3. DEVELOPMENT OF RAFTING INDUSTRY IN UTTARAKHAND (1994-2008)

Rafting industry has ample scope of growth keeping in view of setting of natural resources in Uttarakhand. The rafting on Himalayan river of Uttarakhand provides excellent opportunity due to remoteness and wilderness in the association of vegetation along the river coupled with flow of river. The rafting camps are functional for 9-10 months in a year and remains closed during rainy seasons because of high wind speed and heavy water current along with unexpected and sudden river volume and flow change.

Starting with the supply side, the local market grew from 2,000 m^2 licensed area for tour operators in 1994-95 and then jumped to 1,80,000 m^2 in 2007-08 (Figure 2). This indicates that the tourism sector is profitable and yet competitive. This growth signifies that the industry is increasing with leaps and bound to cater the intended objectives of thrill, excitement and pleasure of visitors. The growth generates revenue volume (the tourism business-supply side) to camp owners and provides employment opportunity to the rural destitute in these remote locations.

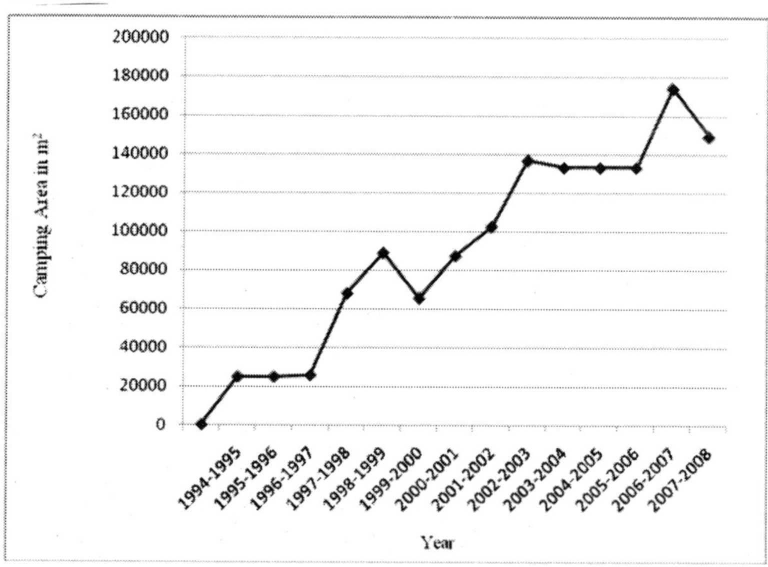

Figure 2 : Temporal Growth of Camping area on the Ganges

Source : Office of DFO, Muni-ki-Reti, Uttarakhand, 2008.

3.1. Institutional Norms and Actual Practices

Rafting, kayaking and camping on the River Ganga from Kaudiyal to Rishikesh are regulated by the State Forest and Tourism departments. The increase in volume of tourists (demand side) invited attention of regulation body to facilitate better opportunity to visitors at one side and attain sustainable development on the other side. The effectiveness of regulation mechanism may jeopardize in the absence of proper implementation. The norms with present practices are being discussed in brief (Table 1).

TABLE 1

Existing Norms and Practices

Institutional Norms	*Present Practices*
No lighting after 9 pm in the night to avoid disturbance to the surrounding wildlife.	People use various sources of illumination, though at low level and hence, rules were not followed strictly.
Music and use of crackers are prohibited in the camps to minimise disturbance in the forest area against the noise and prevent forest fire.	Tourists play music in high volume and rarely use crackers. This needs to be regulated effectively.
Toilets are permitted in the form of dry pit tanks situated 60 m away from the sand bank to prevent water pollution.	This is cause of concern as most of the toilets are not more than 10 m away from the sand banks and within submergence region. Therefore, needs strict functional guidelines.
Camp fire is only allowed on a metallic plate and not after 11 pm in night. The ash is to be disposed in municipal dustbin to prevent water pollution.	Camp fire is usually occurred beyond the time limit and ash disposal is also not proper. Strict functional guideline is essential to regulate the issue.
The use of any kind of detergent is prohibited. This restricts washing of clothes and utensils too.	Occasionally, this rule is not being properly functional. This may be addressed with awareness.
Fishing is prohibited.	Fishing is common and therefore needed proper attention.

Source : Institutional norms: Source for Institutional norms: *Based on own survey.*
Reports from Office of DFO, Muni-ki-Reti, Uttarakhand, 2008.

The comparative assessment reveals that strictures, norms should be effectively addressed to minimise the adverse impacts and for providing pathways to attain the sustainable capability.

4. IMPACTS EVALUATION METHODS FOR CASE STUDY

The basic approach for the study comprises, collection of various primary and secondary data from field surveys and government records for various possible impacts ranging from social to economic and biological issues. Villagers and local shopkeepers of nearby areas were connected for interview with respect to the social and economic descriptors. Their perception about biological issues has also been explored on spatial and temporal scale. Group deliberation has also been held to record the views of stake holders. Some youth from both genders has also been contacted and their views and perception especially on cultural shift due to interaction with the visitors has also been noted down. Series of discussion with the camp owners, managers, workers and labours of the camp has also been held to obtain unbiased information about the state of affairs.

The primary survey was conducted using a questionnaire, followed by informal and formal meetings and group discussions with the stake holders. The various domain of associated information on various aspects of the industry was included in the questionnaire to take into account of the relevant issues by means of suitable indicator. Similarly, some tailor made questions were also raised during the various discussions. This facilitates the generalisation of impacts on the society, environment, wildlife and bio-diversity. Secondary data pertaining to the rafting industry has been collected from the state machinery.

The assessment of possible impacts has been scaled in the following way depending upon the extent. Following criteria for assessment of impacts for individual components have been used to evaluate the significance of impacts (Table 2).

TABLE 2

Criteria Used in the Evaluation of Impacts

Criteria	*Significance*			
Status	*Positive*		*Neutral*	*Negative*
Extent	Site specific	Local	Regional	National
Duration	Short-term	Medium-term	Long-term	Permanent
Intensity	Low	Medium to Low	Medium to High	High
Probability	Improbable	Possible	Highly Probable	Definite
Cumulative Effects	None	Low	Medium	High
Significance	Low	Medium to Low	Medium to High	High
Significance with Mitigation	Low	Medium to Low	Medium to High	High

The criteria are defined as follows:

I. *Status of the Impact*: Negative (a 'cost'); Positive (a 'benefit') or Neutral.

II. *Extent*: Site specific (restricted to the site); Local (the site and surrounds); Regional (surrounding districts) or National.

III. *Duration*: Short-term (up to 2 months), Medium-term (3 months to 1 year), Long-term (life cycle of the project, i.e., 25 years) or Permanent.

IV. *Intensity*: The effects of the impact will be quantified as low, medium-low, medium-high or high, and the rationale for this is discussed in the evaluation of the impacts.

V. *Cumulative effects*: This refers to the degree to which an impact may combine with other project-related (or other unrelated) environmental effects. The cumulative effect is described as not applicable, low-medium, or high.

VI. *Probability of occurrence*: Improbable (unlikely), probable, highly probable, or definite (certain).

VII. *Significance before mitigation and after mitigation*: Based on a synthesis of the information contained in (I) to (VI) above, and considering mitigation measures into account an evaluation of the significance of the impacts is undertaken in terms of the following significance criteria:
- *No significance*—requires no further investigation and no mitigation or management.
- *Low significance*—the project will have negative effects but these are low, and should be totally eliminated with correct management.
- *Moderate to low significance*—requires mitigation and management to reduce impacts to acceptable levels.
- *Moderate to high significance*—requires extensive mitigation and management to reduce impacts to acceptable levels.
- *High significance*—should influence a decision about the project if the impact cannot be mitigated or managed.

5. ANALYSIS OF CASE STUDY

5.1. Economic Transaction

The camp owners pay a handsome amount of taxes to gain permits for settling camps and floating rafts on the rivers to the concerned departments for each rafting season and invest a large sum in case of purchase, repair and maintenance of the camp utilities. Based on deliberations, it was found that the tourist inflow for rafting varies from 1500 to 4000 per season, with 32% share of foreigners. This leads to profit of ₹ 15 lakhs-₹ 50 lakhs per season per camp depending of camp characteristics. This development of tourism on the demand side very much reflects the supply side (Figure 2).

This industry also deploys the locals with approximately 35% employment share, mainly for mid-supervisory and labourer positions. It is probably due to lack of sufficient skill with the local people against the demand of the industry. Therefore, measures needed to be taken for addressing the issues by providing training to locals to meet the demands of the hospitality and services management.

Furthermore indirect means of income generation options have also grown up near the camps. It mainly includes selling of refreshment items, i.e. tea, cold drinks, chips and others. These secondary enterprises can be accounted to only about 10-15% of benefit due to the tourism. Nevertheless a rare means of economic gain is achieved by selling of collected firewood for the purpose of bonfire at the rate of ₹ 100- 200 per kattha (about 20-25 Kg).

Therefore direct economic benefits from the industry are though low, however can be increased by various means in lieu of community interest. On the other hand indirect means of income earning opportunity may also be improved with high inflow.

5.2. Social and Cultural Impacts

The advent of white water rafting and camping has brought this pristine area of Uttarakhand under the lust of human desires. The region is as such under considerable anthropogenic pressure due to the increasing demands made by local populations on environmental resources. These coupled with the tourism has created pressure on resources, which may cause of social concerns, if resources are limited. However, this also provide opportunity of interactions among various socially diverse groups, and therefore sharing of cultures, traditions, etc.

5.2.1. Impact on Education

It was found though rarely that children escape their schools to enjoy the eco-tourism, though at low level. This distracts the children particularly male of unprivileged poor family from education. However, with the opportunity of income, the poor households may create opportunity for education of their children. Further, as per corporate social responsibility, the camp owners also provide opportunity to improvement of educational infrastructure.

5.2.2. Impacts on Lifestyle

Direct and indirect interaction of locals with visitors influences the lifestyle of people of these mountain communities. It was generally cosmopolitan lifestyle of tourist with a mix of western lifestyle. Senior members of the community have categorically expressed their views for the changing attitude of the adolescent males in some immoral activities with lust of rapid income earning. In a few cases smoking by the female tourist and there skimpy outfit has put some challenges into the rural lifestyle of the youth, as revealed by the elders.

Moreover the prodigious ways of the tourists fosters a covetous feeling among the young mass. Consequently, they migrate in lust of higher earnings or else feel sullen and remain in disturbed mindset. Villagers were having feelings that the lifestyle change is not beneficial to the local life and if continued in similar manner may alter the society, drastically.

5.2.3. Impacts on Amenities of Social Services

Our survey established that only 18% of the camp owners support social services mainly, small grants to schools. However, except the road, which is also yet to fully developed, other amenities are yet too shaped up.

5.2.4. Impacts on Social Cohesion

From our survey and discussion with people of different age groups, it was found that preference for traditional food was decreasing. Lust for earning money was haunting the mind of the youth thereby causing increased number out migrations. Moreover, it also provides opportunity to the locals, to come close among them, and provide a path ways for social cohesion.

5.3. Impacts on Fauna and Environment

5.3.1. Impacts on Terrestrial Fauna

With the advent of white water tourism in Uttarakhand, the fragile ecosystem of the region is threatened. The anthropogenic intrusions endanger the wildlife. In the favour of economic gains, and lack of proper regulation, wildlife was not properly taken care. A behavioural shift from diurnal to nocturnal habits is reported in case of wild life may be attributed to these industries. Even animals like leopards and bears due to the shift in habitat attacked the livestock of villagers during night. Therefore, villagers are violent and set traps to catch them.

However, close interaction and not exactly conflicts have been undertaken by 59% of the people in the camps surveyed. Though, incursions into the camps are not heard but they are not even far off distances which may thereafter pose a threat to human life. Hence, if mitigation measures are not taken instantly and left unseen then perhaps this eco-tourism zone may come amongst the hot spots of man wildlife conflict.

5.3.2. Impacts on Aquatic Fauna

Kitchen refuge (mostly leftover food) tends to be feed for some of

TABLE 3

Brief of the Impacts of River Rafting Industry in Mountainous Region

Possible Impacts of River Rafting Industry	*Prediction Criteria*							
	Status	*Extent*	*Duration*	*Intensity*	*Probability*	*Cumulative Effects*	*Signifi-cance*	*Significance with Mitigation*
(1)	*(2)*	*(3)*	*(4)*	*(5)*	*(6)*	*(7)*	*(8)*	*(9)*
Economic Impact on Locals	Positive	Site Specific	Long-term	Low	Definite	Low	—	—
Impact on Education	Positive	Site Specific	Long-term	Medium	Definite	Medium	—	—
Impact on Lifestyle	Positive	Site Specific	Permanent	Medium	Definite	Medium	—	—
Impact on Amenities of Social services	Positive	Site Specific	Permanent	Medium	Definite	Medium	—	—
Impact on Social Cohesion	Positive	Local	Long-term	Medium-High	Possible	High	Moderate	Low
Pollution	Negative	Local-Regional	Long-term	Medium-High	Possible	High	Moderate	Moderate-Low
Impact on Terrestrial Fauna	Negative	Local	Long-term	Medium-High	Possible	Low	Moderate	Moderate-Low
Impact on Aquatic Fauna	Negative	Local	Long-term	Medium	Possible	Low	Moderate	Moderate-High
Impact on Forest Areas	Negative	Local	Long-term	Medium	Possible	Medium	Moderate	Moderate-High

the aquatics. This might be a better opportunity for certain period, however in long-run; they become vulnerable to human lust and joy. Even the pollution ranging from dumping of solid waste to faecal waste may pollute their habitat and force them to undertake various adaptations.

5.4. Impacts on Environment and Climate

Most of the populaces were aggrieved with the drunken revelry of tourists and profanity towards the holy river as they; throw polythene, wrappers and various kinds of bottles into the river. Further, river is being polluted by use of detergents for washing utensils, soap and shampoo for bathing purposes.

The toilet, most of them were of dry pit type and are situated within the submergence zone dumped faecal matter during rainy season causing pollution. With the growth of industry, the pollution will keep on rising if unchecked. Similarly along the national highways, solid waste including wrappers, plastic and bottles, rugged footwear and many others were found untreated and lying on the roadside as well as on trekking trails amid forests which are also environmental concerned in near future. During the movements of tourists, it was found that trampling of vegetation and soil eventually causing damage and impacts on regeneration, which lead to loss of biodiversity.

The increased vehicular transportation in this area for tourism has contributed a lot for local air pollution as well as contributes to climate change. Noise pollution from vehicles are also now noticeable and causing distress especially to wildlife.

6. ENVIRONMENTAL MANAGEMENT PLANS

6.1 Mitigation Measures

Environmental management plan (EMP) referred to as an impact management plan, is usually part of EIA. It translates recommended mitigation and monitoring measures into specific actions that should be addressed to mitigate the impacts. The EMP will form the basis for impact management during project implementation, operation and functions. The mitigation measures for the eco-tourism industry may revolve around the settings of procedures to the effective social and economic well-being attainments.

6.2. Environment Management Plan for River Rafting

River rafting industry should intend to contribute towards better quality of life to the local people with sustainable environment. This may be achieved through the following measures:

TABLE 4

Issues and Mitigation Measures of Rafting Industry

Issues	*Descriptions*	*Facilitations*	*Mitigation Plans*
(1)	*(2)*	*(3)*	*(4)*
Concepts	Mainly focused on conceptual definition, use of specific resources	Understanding of settings and fabrics of interactions among the resources	Properly defined in communicative manner with celerity to local and tourists
Procedural Issues	Includes capacity building, the dissemination of information, the exchange of information on services, the research and study on needed areas, and the establishment of global networks	Dissemination of precise understanding of purpose of the settings of the industry	Clearly defined and precise guidelines for code and conduct of each stakeholders
Institutional Issues	Includes an *ad-hoc* informal open-ended working group on tourism, the implementation and enforcement of standards and guidelines related to tourism, the development of guiding principles for sustainable tourism development, international guidelines on tourism activity in sensitive areas, and a supporting Code of Ethics for tourism	Settings of spatial and temporal framework, instruments and guidelines; implementing of a master plan, locally integrated planning approaches, and design with nature in mind Monitoring and regulating the guidelines; Strategy setting and policy-making	Well defined activities with the scope to address the emergent needs; Trained and qualified staffs-Information sharing through various audio-visual means

Socio-cultural Issues	Includes consultation with major groups and local communities, educational initiatives and responsible tourism behaviour, tourism information awareness, sexual exploitation, inflight education videos, and art and music	Attainment of proper community response with improvement in quality of life	Specific guidelines to the tourism destination integrated with existing interaction and harmony within the community
Economic Issues	Includes the eradication of poverty, and small and medium size enterprise issues	Attainment of better employment and earning opportunity	Specific location-based capacity development trainings to locals; Settings of guiding tours
Environmental Issues	Includes voluntary initiatives, eco-efficiency, waste reduction and management, the development of indicators, and coastal zone fragility	Environmental Sustainability	Awareness among locals to create sensitivity; Strict mechanism for waste disposals; Waste treatment plans; Strict guidelines for any sort of polluting actors

A. For socio-economic development:
   - (i) Public participation for decision making for sites selection and other activities of the industry.
   - (ii) Training to youth with specific needs of rafting industry.
   - (iii) Settings of locally available raw materials for marketable products and food items.
   - (iv) Incorporation of cultural components through visits of nearby villages.
   - (v) Implementation of social corporate responsibility of rafting industry.

B. Environment and protection of flora and fauna:
   - (i) Awareness camps for environment and wildlife protection, to reduce man wildlife conflict situations to the locals
   - (ii) Development and awareness about pollution control through various extension strategy to the locals and tourists.
   - (iii) Management of garbage disposal
   - (iv) Development with the tune of specificity and fragility of mountains
   - (v) Effective co-ordination between various Government departments and the camping and rafting agencies for effective management and sustainable development of river rafting in mountainous region.
   - (vi) Building of 'Ecolodge'

C. Infrastructure and other measures:
   - (i) Developments of roads, parking and other public amenities
   - (ii) Diversion of tourists for other eco-destinations
   - (iii) Maintenance of hygiene environment through effective cleaning
   - (iv) Eco-parks developments with shops of local handicraft items

7. CONCLUSION

Eco-tourism allows the natural and pristine eco-systems and associated population to appreciate nature wilderness, while sharing with local economies. Eco-tourism has an impact on natural eco-systems, but more importantly, it offers a way to promote conservation in ecologically fragile regions (Lowman, 2004). Community participation has a key role in spreading awareness among the local villagers, tourists; etc. for sustainable form of eco-tourism (Shashi *et al.*,

2010). The socio-cultural impulse also exists amidst the development and local population accept superiority in culture of the tourists and tries to adapt the prevailing customs. This may lead to change from intricacies of the traditional local culture. Consequently, level of cross-cultural awareness, understanding and respect between visitors and local people may influence the quality of the experience and level of community tolerance and support for tourism (UNEP, 2007).

Present study provides baseline scenario for further studies to be carried out on the subject. It can be deduced that River Rafting on the banks of The Holy Ganga has become a double-edged activity due to the high inflow of tourists and lack of proper implementation of rules and regulations. Therefore, it's potential for a sustainable development is not shaping in the precise direction. Even the growth is also leading to the degradation of the environment and loss of local identity and change in traditional culture which may challenge social cohesion. It can be recommended at this juncture that mountain-specific resource characteristics are considered particularly, diversity, marginality, difficulty of access, fragility, niche and aesthetics. These characteristics are unique to mountainous regions and, as such, have specific implications for mountain recreation and tourism development. A system of recreation and tourism land-use settings may be formulated to resolve planning and management challenges associated with increasingly diverse needs of these users. Tourism planning and management in mountainous regions should consider and incorporate mountain-specific resource characteristics. Broadly, strategies for sustainability should be formulated with sufficient funding infrastructure, collaboration and cooperation among various stake-holders including response of individual business and political will.

Eco-tourism, in partnership with research, has the potential to significantly contribute for conservation and sustainable development. These may be supported with advances researches of mountain resource characteristics, mountain amenity users and mountain recreational zoning with the eco-tourism. The question of sustainability remains unanswered because many sites with nature-based tourism are relatively new and the long-term impacts have yet to be realised.

References

Anonymous (1997): Coastal Tourism—A Manual for Sustainable Development. Portfolio Marine Group, (Canbera: Environment Australia).

BBC News (2001): India Targets Domestic Tourism, October 2001, http://news.bbc.co.uk/1/hi/business/1595831.stm.

Bhusal, N.P. (2007): Chitwan national park: A prime destination of eco-tourism in central tarai region, Nepal, *The Third Pole*, 3-5; 70-75.

Brundtland website (2001): Brundtland Report webpage, Retrieved, from the World Wide Web: http://www.brundtlandnet.com/brundtlandreport.htm.

Chopra, K., Kadekodi, G.K., and Mongia, N. (1993): Environmental Impact of Projects: Planning and Policy Issues: (New Delhi: Institute of Economic Growth).

Explore Himalayas web site (2010): Himalayan Eco-tour travel, http://www.himalaya2000.com/himalayan-facts/eco-travel.html (Assessed on 1st October, 2010)

Fukada, K. (1971): The Great Himalayas, In: Shirakawa Yoshikazu (Ed.), *Himalayas*. (New York: Harry N. Abrahams. Inc.).

Ghimire, K.B. (2001): The Growth of National and Regional Tourism in Developing Countries: An Overview, In: K.B. Ghimire (ed), The Native Tourist: Mass Tourism Within Developing Countries. (London: Earthscan).

Holden, A. (2000): Environment and tourism. (London: Routledge).

Lowman, M. (2004): Eco-tourism and Its Impact on Forest Conservation : An ActionBioscience.org original article, http://www.actionbioscience.org/environment/lowman.html (Assessed on 15th October, 2010)

Milne, S. and Ateljevic, I. (2001): Tourism, Economic Development and the Global-Local Nexus: Theory Embracing Complexity, *Tourism Geographies*, 3(4); 369-93.

Morris, P., and Therivel, R. (2001): Methods of Environmental Impact Assessment, 2nd Ed. (London: Spon press).

Neto, F. (2002): Sustainable Tourism, Environmental Protection and Natural Resource Management: Paradise on Earth?, Paper to the International Colloquium on Regional Governance and Sustainable Development in Tourism-driven Economies, Cancun, Mexico, 20-22 February.

OECD. (2008): Organisation for Economic Co-Operation and Development, Policy Brief, www.oecd.org/cfe/tourism.

Pauw, E.K. (1986): Report on the Tenth Settlement of Garhwal District. (Allahabad: Govt. Press)

Rao, N. and Suresh, K.T. (2001): Domestic Tourism in India, In: K.B. Ghimire (ed), *The Native Tourist: Mass Tourism Within Developing Countries*. (London: Earthscan).

Shashi, G.B.G., Tiwari,. K., Pananjay, G.B.G., Tiwari, K. and Tiwari, S.C. (2010): A Review on the Future of Eco-tourism in the Valley of Flowers National Park: A Case Study of Garhwal Himalaya, India, *Nature and Science*, 8(4); 101-06.

Soleimanpour, H. (2005): Requirements of an international legal framework on nature-based tourism. *International Journal of Environmental Science and Technology*, 1(4); 335-44.

Tapper, R. (2001): Tourism and socio-economic development: UK tour operators. Business approaches in the context of the new international agenda, *Int J Tour Dev.*, 3; 351-66.

Travel smart website (2010): Uttrakhand: Weather, http://www.bharatonline.com/uttarakhand/tehri-garhwal/weather.html) (Assessed on 15th May, 2010).

UNEP (2007): Tourism and Mountains: A Practical Guide to Managing the Environmental and Social Impacts of Mountain Tours. (France: UNEP)

Veenhoven, R. (1999): Quality of life in individualistic society: A comparison of 43 nations in the early 1990s. *Soc Indicat Res*, 48; 157-86.

Weaver, D. (2001): Tourism in 21st century. (New York: Continuum).

Wood, M.E. (2002): Eco-tourism: Principles, Practices and Policies for Sustainability (France: UNEP, Division of Technology).

Woods, L.A., Perry, J. and Steagall, J.W. (1991): Tourism and Economic Development: The Case of Post-Independence Belize, Paper presented at the Fifth Annual Studies on Belize Conference, Belize City, Belize C.A., September 3-6.

Chapter 26

Sustainable Development and Flora-Tourism

Subrata Goswami

ABSTRACT

In this study an attempt has been made to draw a link between sustainable development and flora-tourism. Sustainable development in the context of tourism is tourism that is developed and maintained in an area in such a manner and at such a scale, that it remains viable over an indefinite period and does not degrade or alter the environment in which it exists to such a degree that it prohibits the successful development and well-being of other activities and processes. An obvious implication of this is that—"meeting the needs of the present generation without compromising the ability of the future generations to meet their own needs".

The study is based on a field survey undertaken in Pune, Bangalore and Hyderabad. The findings reveal many inadequacies in the infrastructure like transport, poor marketing structure, and market research. The natural beauty may be effectively utilized for promoting flora-tourism. This will automatically generate more employment opportunities in rural sector. Demand for eco-friendly products from flower like perfume, rose water, attar, soft drinks and medicines will be boosted up, which in turn, will act as an impetus to the growth of rural economy. The study was therefore undertaken to promote this concept of flora-tourism amongst the farmers so that they can start it as a supplementary or complementary activity on their farms. The study also proved that multimedia can be an effective tool to promote flora-tourism.

CONCEPT OF SUSTAINABLE DEVELOPMENT

Sustainability is an integrative concept because it looks at the human use and management of resources in a manner that should not destroy or disturb the habitat that is the basis of survival. The most widely accepted definition is the one given by *Brundtland Commission* in 1987 which defined sustainable development as "a process of change in which the exploitation of resources, the direction of investments, the orientation of technology development, and institutional changes are made consistent with future as well as present needs" and a "meeting the needs of the present without compromising the ability of future generations to meet their own needs". Socio-economic and environment dimensions thus become the focus of the management approach. Changes in the views of the community and its attitudes towards development are relegated to a secondary position.

In 1990, an effort was made at the international level during the *Globe 90 Conference* in *Vancouver, Canada*, to link tourism with sustainable development. In order to achieve sustainable development, environmental protection should constitute an integral part of the development process and cannot be considered in isolation from it.

The concept of sustainability has become a fundamental issue in tourism development and growth after the debate at the Rio Earth Summit in 1992. Seeing the rapid changes in tourism and the world trends, we are now examining geopolitical, socio-economic, technological and environment impacts of contemporary tourism. It was realized that tourism requires an agenda of its own, and not a part of the overall post-structural adjustment process.

The concept of sustainability when applied to tourism can be perceived and interpreted in various ways. Sustainability for attractions (both natural and man-made), infrastructure, cultures, environment, economy, etc. will have different meaning for different disciplines and the methodologies adopted also may not be the same. According to *Victor T.C. Middleton* and *Rebecea Hawkins*, "Sustainable tourism means achieving a particular combination of numbers and types of tourists, the cumulative effect of whose activities at a given destination, together with the actions of the servicing businesses, can continue into the foreseeable future without damaging the quality of environment on which the activities are based. Achieving sustainability for tourism, according to them requires that "the cumulative volume of visitor usage of a destination and the associated activities and impacts of servicing business should be managed below the threshold level at which the regenerative resources available locally become incapable of maintaining the environment.

FLORA-TOURISM

Urbanization and city life has all its merits but with a pinch of demerits too. Modern life is product of diversified thinking and activities. People therefore desire for peace and pollution free air. They are disillusioned with overcrowded resorts and cities and have nostalgia for their roots on the farm. Cities are growing at the cost of villages. Yesterday's villagers are today's urbanites. They are fed up with the flat culture and the concept of either owning a farm house on the outskirts of the cities or spending sometime at a firm house for relaxation is becoming a habitual trend. They even have a little extra cash at their hand for this purpose. The earning generation in the cities has lost touch with their roots in villages. They want to live their childhood once again with their little ones.

'Tourism and Floriculture' are far from people's minds when they go on their holidays. Yet the relationships between tourism and floriculture can be important and far reaching for local communities. Flora-tourism is therefore an emerging trend which is a mild form of sustainable tourist development and multi-activity in rural areas through which the tourist have the opportunity to get acquainted with flower producing areas, by-products of flower, aroma therapy, traditional cuisine and the daily life of the people, as well as the cultural elements and the authentic features of the area, while showing respect for the environment and tradition. Moreover, this activity brings tourist closer to nature and rural activities in which they can participate, by entertained and feel the pleasure of touring, learning and discovering. At the same time, it mobilizes the productive, cultural and developmental forces of an area, contributing in this was to the sustainable environmental, economic and social development of the rural areas.

Flora-tourism is often practiced in flower producing regions, as in the Netherlands, Spain, France and Italy. In the Netherlands, Flora-tourism is wide-spread where people spent their week end and so many farm open to the public during peak season. People can pluck flower, fruits and vegetables, ride horses, taste honey, learn about production of wine, shop in flower bouquet and much more. Every firm generally offers a unique and memorable experience package suitable for the nature lovers.

Our country has specific flower crop growing areas like Pune, Bangalore and Hyderabad. This area not only offer the flower but also their fruit products and one can experience how the mango pickles or jellies/jams, or wines are actually made. In fact, these days hi-tech floriculture units itself can be a great site for flora-tourism as many of us would like to experience how gerberas, orchids, anthuriums or carnations are cultivated in green houses. The method of packaging and storing these hi-tech flowers itself can be a memorable sight.

Flora-tourism is being developed as a valuable component of a business model to support horticulture or floriculture entities when the farm products they produce are no longer economically competitive otherwise. People are more interested in how cut flower is produced and talk with them about what goes into food/flower production. Children who visit the farms often have not seen a live duck, or rabbit, and have not picked an apple or mango right off the tree. This form of expanded flora-tourism has given birth to what are often called "entertainment farms". These farms cater to the pick-your-own crowd, offering not only regular farm products but also food, flower, open animals, train rides, picnic facilities and pick your eco-friendly products.

Flora-tourism is not something you describe but something you experience. Such places are conductive to a harmonious co-existence sought by more and more tourists on their holidays. Thus the concept of flora-tourism is a direct expansion of eco-tourism, which encourages tourists to experience agricultural like at first hand. Flora-tourism is gathering strong support from small communities as rural people have realized the benefits of sustainable development brought about by similar forms of nature travel.

Flora-tourism will certainly help farmers through additional income and dignity of life. Hence, it should be considered as help for floricultural development but they should not depend on it completely. This gives farmer an opportunity for change of work in his daily routine of hard work. It serves as a means of economic diversification for agriculture producers, along social benefits like meeting people from different cultures and walks of life. The constant instability of net farm incomes and the loss of jobs in rural areas have led to a human and financial capital drain from rural areas, with many farming families and businesses under economic stress. Those rural youth who are interested in undertaking this activity should be given working knowledge through practical and workshops. The negative mood in the rural areas has proved to be detrimental for progress by slipping away opportunities. If rural folks find this activity worthwhile, they will develop confidence in rural life. Visits of urbanites to rural areas will bring flow of income to the region automatically. However, flora-tourism is not a "quick fix" of a "get rich quick" scheme. Resources must be allocated if the recreation business is to be successful. The impacts can be greatest in locations where tourism is growing rapidly, and where tourism offers an alternative source of income to traditional cultivation of crops and allied activities.

Although people of European countries are already aware of flora-tourism, the concept is still very new in India. Farmers who have been practicing floriculture since many years are still to be motivated to divert to this new and yet supplementary business farm activity. In recent times, even agriculture graduates turn to cities and become a part of the rat race to get a high paying job. Very few of them revert back to their ancestral farms. The following study was therefore undertaken to make this class of farmers, educated and business-oriented, understand and to motivate them to start flora-tourism as a part of their farming activity.

OBJECTIVES

- To advocate the concept of flora-tourism
- To study the application of multimedia technology in promoting of flora-tourism
- To study the scope of rural employment and uplift the rural areas

SIGNIFICANCE

India is basically an agricultural country with 85% of the population engaged in agriculture as their main occupation. Agriculture contributes to 26% of the GDP. With a population of more than 110 million farmers and 6.25 lakh villages, Indian agriculture has got tremendous scope to add to the GDP if directed in the right manner. Modern techniques of floriculture have of course caused a revolution and the concept of hi-tech farming, floriculture; in general more profitability per unit cost, is trying to create a niche in this sector. But still farmers with vast land areas and insufficient funds are striving hard to increase their per capita income. However, this cannot be achieved by merely pushing large sums of government money into the rural sector with vaguely defined objectives. Sustainable rural development requires that we create modern private sector business ventures that can effectively compete with the non-farm sector. To accomplish this, rural residents must garner the entrepreneurial and business development skills and resources needed to drive the economic engine of rural India. This will require the creation of "business development organization" where rural entrepreneurs can work together and share resources and knowledge in creating rural business ventures. Successful business development involves building the capacity of rural residents to create viable business ventures. The basic outlook towards Agriculture/ Floriculture from the present is "just a way of living" should be shifted to an organized business opportunity. A relatively recent trend in the tourism industry is the partnership with recreation and floriculture.

Tourism can be instrumental in national integration, create international understanding and support local handicrafts and cultural activities. India's share in the world tourist market is less than one percent. Flora-tourism is a form of domestic tourism which can promote national integration. Flora-tourism a subset of rural tourism is about being ecological sustainable and influencing communities and tourist. It is also about influencing the tourism industry's behavior in ways that minimize environmental impacts, protect the environment and contribution to conservation.

There are many possibilities of flora-tourism activities to be carried out in India like camel rides or desert safaris as in Rajasthan or trekking in the hilly areas. Our ayurvedic treatment, body massage practices, tribal medicines, to even innovative concepts like buffalo rides owning a boat as in Kerala and other eco-adventure sports can make flora-tourism very entertaining. Since the beginning of the human civilization, flowers have been known for their soothing effects and healing abilities. Even today, many people depend on the health-giving properties of flowers in their day-to-day lives. Flowers have such 'relax' and comfort ingredients which cannot be generated in laboratories by using chemicals. Most flowers have some beneficial effects on human health (*Mukumdrao and Gaikwad, 1981*). Flower extracts help bring about the ideal harmony and peace between mind and body. While flowers nourish the body, their fragrance has a calming effect on the mind. This body-mind balance is a prime pre-requisite to good health. New and continuous relationships can hereby be established between tourists and flower producing farms. We continue to undervalue our agriculture, our heritage and our culture while rest of world is busy taking advantage of opportunities to produce and market their agricultural products.

Flora-tourism provides effective participation and income generating opportunities to local communities. Three quarters of agricultural outputs are used as inputs for other industries. It help to start own business to educated people and also helps to decrease unemployment to certain extent. Regional programmes can be conducted as tools for enhancing rural development, which will also help in changing attitude of Indian society towards agriculture. Besides educational programmes and training like organic gardening, compost making, wild herbs collection and drying, traditional food and beverage can make this a memorable experience for all ages.

RESEARCH METHODOLOGY

The emphasis in this study was on knowledge of flora-tourism and it promotion amongst the rural youth. The methodology included preparation of a multimedia based program on flora-tourism and then exposing the respondents to it. The knowledge gained by them was then accessed by a structured questionnaire. Since flora-tourism is basically based to horticulture/floriculture, it was necessary to have respondents from horticulture/floriculture backgrounds. Also as flora-tourism is a business venture, it was necessary to have respondents who are business-oriented. Hence, horticulture/floriculture graduates were selected who were pursuing their studies in post-graduate degree or diploma of flora-business management.

At first a thorough study was done on flora-tourism, its meaning and prospects in India. The researcher conducted field surveys in Pune, Bangalore and Hyderabad. During these surveys, some cut flower export units and various stakeholders of this industry were covered. Meetings were held with the officials of some financial institutions and the National Bank for Agriculture and Rural Development (NABARD). Officials of the embassies of some selected countries, including their trade counsels, were interacted with to appreciate the steps taken by their countries to promote the flora-tourism. A visit was made to a few flora-tourism farms in Pune and Bangalore. Information was also gathered from the internet. A multimedia presentation was then prepared keeping in mind all the above aspects. Discussion was also done on the topic of flora-tourism to know the concept in the minds of the respondents before showing the presentation. Questionnaire was then given to each respondent and desired data was thus collected.

FINDINGS AND OBSERVATIONS

The data thus collected was classified, tabulated and analyzed using proper statistical tools. Flora-tourism does not mean starting all together new activities but it requires marketing of the present activities to the tourists. It was found that activities like traditional composing, vermi-composing, fishing, swimming, ornamental plant nursery already existed on the farms and the respondents desired to start some new activities in this context.

76.45% of the respondents had innovative ideas for starting some or the other activity at the farm for promotion of flora-tourism. Amongst the new activities desired to start of the farm, entertainment plays showcasing the Indian art and culture was given highest preference followed by traditional cuisine and then night halt facility, and lodging in natural conditions.

The study aimed at motivating farmers/firms to start flora-tourism. Thus, it was desired to know what the opinion about flora-tourism amongst the respondents was. Following were the findings obtained:

- 93.25% of respondents confirmed that flora-tourism is a good business activity, would definitely uplift the rural areas and scope of employment. It is an activity suitable for all ages and that everyone can enjoy this form of tourism and it should be promoted, encouraged and advertised.
- 87% confirmed that flora-tourism would increase income of the farmers.
- 72.35% felt that it can be practiced throughout the year.
- 89.50% of respondents agreed that multimedia is a best technique for advertising flora-tourism.
- 81% of the firms wanted to add additional components to flora-tourism apart from those shown or explained in the presentation.
- 98% of the respondents were confident of starting the flora-tourism after watching the CD.

The study has proved that flora-tourism can be a supplementary or complementary activity. These findings can be used by the extension personnel to motivate the farmers to start this activity. It also proved that multimedia can be an effective tool for the promotion of flora-tourism. This can give a boost to flora-tourism which is still unexplored in India. The study made it clear that infrastructural facilities have to be developed at the farms for a comfortable stay of the tourists. Thus the government can take certain measures for increasing infrastructural facilities in the remote rural areas, train the farmers for this activity, and help them in marketing the same through the internet and all other modern technological facilities.

References

Lynn, C. Harrison and Winston Husbands (1996), Practicing Responsible Tourism, Toronto.

Martin Mowforth and Ian Munt, Tourism and Sustainability (1998), *New Tourism in the Third World*, London.

Mukumdrao and Gaikwad (1981), Chrysanthemum: A Flower of Commercial Value, *Indian Horticulture*, Vol. 26, No. 3, p. 11

Richard Butler and Douglas Pearce (1995), Change in Tourism: People, Place and Processes, London.

Victor, T.C. Middleton (1998), Sustainable Tourism, Oxford.

Chapter 27

Impacts of Climate Change on Indian Agriculture

An Overview

SUVRANSHU PAN

ABSTRACT

Indian agriculture has, since Independence, made rapid strides. In taking the annual food grains production from 51 million tons of the early fifties to 206 million tons at the turn of the century, it has contributed significantly in achieving self-sufficiency in food and in avoiding food shortages in our country. The pattern of growth of agriculture has, however, brought in its wake, uneven development, across regions and crops as also across different sections of farming community and is characterized by low levels of productivity and degradation of natural resources in some areas. Capital inadequacy, lack of infrastructural support and demand side constraints such as controls on movement, storage and sale of agricultural products, etc., have continued to affect the economic viability of agriculture sector. Consequently, the growth of agriculture has also tended to slacken during the nineties.

At the same time, agriculture has been shown to produce significant effects on climate change, primarily through the production and release of greenhouse gases such as carbon dioxide, methane, and nitrous oxide, but also by altering the Earth's land cover, which can change its ability to absorb or reflect heat and light, thus contributing to radiative forcing. Land use change such as deforestation and desertification, together with use of fossil fuels, are the major anthropogenic sources of carbon dioxide; agriculture itself is the major contributor to increasing methane and nitrous oxide concentrations in earth's atmosphere.

To find out actual impacts of climate change on Indian agriculture this paper discussed in five sections namely, Introduction, Climate change as such in India, Impacts of climate change on Indian agriculture, Agricultural surfaces and climate changes in other countries, and Major findings of the paper.
Keywords: Agriculture, Climate change.

I. INTRODUCTION

In India, agriculture has become a relatively unrewarding profession due to generally unfavourable price regime and low value addition, causing abandoning of farming and increasing migration from rural areas. The situation is likely to be exacerbated further in the wake of integration of agricultural trade in the global system, unless immediate corrective measures are taken. Since independence, Indian agriculture has made rapid strides. In taking the annual foodgrains production from 51 million tons of the early fifties to 206 million tons at the turn of the century, it has contributed significantly in achieving self-sufficiency in food and in avoiding food shortages in our country. The pattern of growth of agriculture has, however, brought in its wake, uneven development, across regions and crops as also across different sections of farming community and is characterized by low levels of productivity and degradation of natural resources in some areas. Capital inadequacy, lack of infrastructural support and demand side constraints such as controls on movement, storage and sale of agricultural products, etc., have continued to affect the economic viability of agriculture sector. Consequently, the growth of agriculture has also tended to slacken during the nineties.

At the same time, agriculture has been shown to produce significant effects on climate change, primarily through the production and release of greenhouse gases such as carbon dioxide, methane, and nitrous oxide, but also by altering the Earth's land cover, which can change its ability to absorb or reflect heat and light, thus contributing to radiative forcing. Land use change such as deforestation and desertification, together with use of fossil fuels, are the major anthropogenic sources of carbon dioxide; agriculture itself is the major contributor to increasing methane and nitrous oxide concentrations in earth's atmosphere.

On the light of the above background, this paper is analyzed. It has five sections namely, Introduction, Climate change as such in India, Impacts of climate change on Indian agriculture, Agricultural surfaces and climate changes in other countries, and Major findings of the paper.

II. CLIMATE CHANGE AS SUCH IN INDIA

Despite technological advances, such as improved varieties, genetically modified organisms, and irrigation systems, weather is still a key factor in agricultural productivity, as well as soil properties and natural communities. The effect of climate on agriculture is related to variabilities in local climates rather than in global climate patterns. The Earth's average surface temperature has increased by 1 degree F in just over the last century. Consequently, agronomists consider any assessment has to be individually considering each local area. On the other hand, agricultural trade has grown in recent years, and now provides significant amounts of food, on a national level to major importing countries, as well as comfortable income to exporting ones. The international aspect of trade and security in terms of food implies the need to also consider the effects of climate change on a global scale. A study published in Science suggest that, due to climate change, Southern Africa could lose more than 30% of its main crop, maize, by 2030. In South Asia losses of many regional staples, such as rice, millet and maize could top 10%.

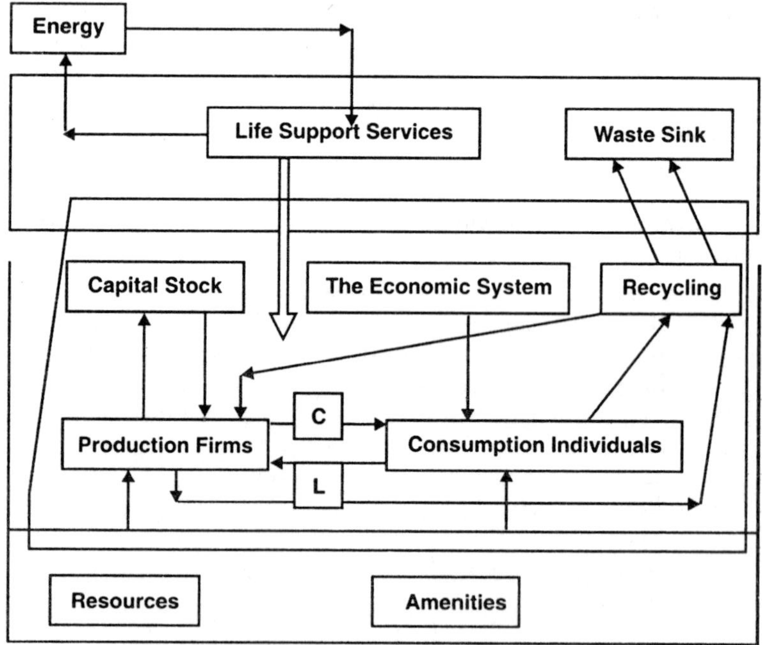

Figure 1 : The Economics of Climate Change

The 2007 *IPCC Fourth Assessment Report* concluded that the poorest countries would be hardest hit, with reductions in crop yields in most tropical and sub-tropical regions due to decreased water availability, and new or changed insect pest incidence. In Africa and Latin America many rainfed crops are near their maximum temperature tolerance, so that yields are likely to fall sharply for even small climate changes; falls in agricultural productivity of up to 30% over the 21st century are projected. Marine life and the fishing industry will also be severely affected in some places. Climate change induced by increasing greenhouse gases is likely to affect crops differently from region to region. For example, average crop yield is expected to drop down to 50% in Pakistan according to the UKMO scenario whereas corn production in Europe is expected to grow up to 25% in optimum hydrologic conditions. More favourable effects on yield tend to depend to a large extent on realization of the potentially beneficial effects of carbon dioxide on crop growth and increase of efficiency in water use. Decrease in potential yields is likely to be caused by shortening of the growing period, decrease in water availability and poor verbalization.

From Figure 1, it is found that the economics of climate change is basically depends upon two-fold system boundary. One is known as Economic System Boundary (which comprises capital stock, production firms consumption individuals and recycling) and the other is known as Environment System Boundary (which comprises the economic system, waste sink, life support services, resources and amenities). These two systems when added with energy then the economics of climate change are formed.

In the long-run, the climatic change could affect agriculture in several ways:

- *Productivity*, in terms of quantity and quality of crops
- *Agricultural practices*, through changes of water use (irrigation) and agricultural inputs such as herbicides, insecticides and fertilizers
- *Environmental effects*, in particular in relation of frequency and intensity of soil drainage (leading to nitrogen leaching), soil erosion, reduction of crop diversity
- *Rural space*, through the loss and gain of cultivated lands, land speculation, land renunciation, and hydraulic amenities.
- *Adaptation*, organisms may become more or less competitive, as well as humans may develop urgency to develop more competitive organisms, such as flood resistant or salt resistant varieties of rice.

They are large uncertainties to uncover, particularly because there is lack of information on many specific local regions, and include the uncertainties on magnitude of climate change, the effects of technological changes on productivity, global food demands, and the numerous possibilities of adaptation. Most agronomists believe that agricultural production will be mostly affected by the severity and pace of climate change, not so much by gradual trends in climate. If change is gradual, there may be enough time for biota adjustment. Rapid climate change, however, could harm agriculture in many countries, especially those that are already suffering from rather poor soil and climate conditions, because there is less time for optimum natural selection and adoption.

II.I. Projections of Climate Change

Schneider *et al.* (2007) assessed the literature on key vulnerabilities to climate change. With low to medium confidence, they concluded that for about a 1 to 3°C global mean temperature increase (by 2100, relative to the 1990-2000 average level) there would be productivity decreases for some cereals in low latitudes, and productivity increases in high latitudes. With medium confidence, global production potential was predicted to:

- increase up to around 3°C
- very likely decrease above about 3 to 4°C.

Most of the studies on global agriculture assessed by Schneider *et al.* (2007) had not incorporated a number of critical factors, including changes in extreme events, or the spread of pests and diseases. Studies had also not considered the development of specific practices or technologies to aid adaptation.

II.II. Regional Impacts of Climate Change

(a) *Africa*: Africa's geography makes it particularly vulnerable to climate change, and 70% of the population relies on rain-fed agriculture for their livelihoods. Tanzania's official report on climate change suggests that the areas that usually get two rainfalls in the year will probably get more, and those that get only one rainy season will get far less. The net result is expected to be that 33% less maize—the country's staple crop—will be grown. Alongside other factors, regional climate change—in particular, reduced precipitation—is thought to have contributed to the conflict in Darfur. The combination of decades of drought, desertification and over-population are among the causes of the conflict, because the

Baggara Arab nomads searching for water have to take their livestock further south, to land mainly occupied by farming peoples. With high confidence, IPCC (2007) concluded that climate variability and change would severely compromise agricultural production and access to food.

(b) *Asia*: With medium confidence, IPCC (2007) projected that by the mid-21st century, in East and Southeast Asia, crop yields could increase up to 20%, while in Central and South Asia, and yields could decrease by up to 30%. Taken together, the risk of hunger was projected to remain very high in several developing countries.

(c) *Australia and New Zealand*: Hennessy *et al.* (2007) assessed the literature for this region. They concluded that without further adaptation to climate change, projected impacts would likely be substantial. By 2030, production from agriculture and forestry was projected to decline over much of southern and eastern Australia, and over parts of eastern New Zealand. In New Zealand, initial benefits were projected close to major rivers and in western and southern areas. Hennessy *et al.* (2007) placed high confidence in these projections.

(d) *Europe*: With high confidence, IPCC (2007) projected that in Southern Europe, climate change would reduce crop productivity. In Central and Eastern Europe, forest productivity was expected to decline. In Northern Europe, the initial effect of climate change was projected to increase crop yields.

(e) *Latin America*: With high confidence, IPCC (2007) projected that in drier areas of Latin America, productivity of some important crops would decrease and livestock productivity decline, with adverse consequences for food security. In temperate zones, soybean yields were projected to increase.

(f) *North America*: According to a paper by Pan (2006), predicted increases in temperature and precipitation will have virtually no effect on the most important crops in the US. With high confidence, IPCC (2007) projected that over the first few decades of this century, moderate climate change would increase aggregate yields of rain-fed agriculture by 5-20%, but with important variability among regions. Major challenges were projected for crops that are near the warm end of their suitable range or which depend on highly utilized water resources.

(g) *Polar regions (Arctic and Antarctic)*: For the *Guardian* newspaper, Brown (2005) reported on how climate change had affected agriculture in Iceland. Rising temperatures had made the widespread sowing of barley possible, which had been untenable twenty years ago. Some of the warming was due to a local (possibly temporary) effect via ocean currents from the Caribbean, which had also affected fish stocks. With medium confidence, they concluded that the benefits of a less severe climate were dependent on local conditions. One of these benefits was judged to be increased agricultural and forestry opportunities.

(h) *Small islands*: In a literature assessment, Mimura *et al.* (2007) concluded, with high confidence, that subsistence and commercial agriculture would very likely be adversely affected by climate change.

III. IMPACTS OF CLIMATE CHANGE ON INDIAN AGRICULTURE

The vulnerability of Indian agriculture to climate change is well acknowledged. But what is not fully appreciated is the impact this will have on rain-fed (non-irrigated) agriculture, practiced mostly by small and marginal farmers who will suffer the most. The crops that may be hit include pulses and oilseeds, among others. These are already in short supply and are consequently high-priced. Nearly 80 million hectares, out of the countrys net sown area of around 143 million hectares, lack irrigation facilities and, hence, rely wholly on rain water for crop growth. Over 85 per cent of the pulses and coarse cereals, more than 75 per cent of the oilseeds and nearly 65 per cent of cotton are produced from such lands. The crop yields are quiet low.

The available records indicated that the predominantly rain-fed tracts experience three to four droughts every 10 years. Of these, two to three droughts are generally of moderate intensity and one is severe. Most of the rain-fed lands, moreover, are in arid and semi-arid zones where annual rainfall is meagre and prolonged dry spells are quite usual even during the monsoon season. This makes crop cultivation highly risk-prone. If the quantum of rainfall in these areas drops further or its pattern undergoes any distinct, albeit unforeseeable, change in the coming years, which seems quite likely in view of climate change, crop productivity may dwindle further, adding to the woes of rain-fed farmers.

It is found that medium-term climate change predictions have projected the likely reduction in crop yields due to climate change at between 4.5 and 9 per cent by 2039. The long-run predictions paint a scarier picture with the crop yields anticipated to fall by 25 per cent or more by 2099. This will have a detrimental effect on farmers' income and purchasing power, with obvious down-the-line repercussions. Though the rainfall records available with the India Meteorological Department (IMD) do not indicate any perceptible trend of change in overall annual monsoon rainfall in the country, noticeable changes have been observed within certain distinct regions.

At least three meteorological sub-divisions—Jharkhand, Chhattisgarh and Kerala—have shown significant decrease in seasonal rainfall though some others have recorded an uptrend in precipitation as well. Since rain-fed crops, like coarse grains, pulses and oilseeds are grown mostly during the kharif season, these are impacted by both low as well as excess rainfall. The groundnut crop in the Rayalaseema area of Andhra Pradesh in 2008 can be a case in point. It suffered substantial damage because of high as well as low rainfall at different stages of crop growth. While heavy rainfall early in the season adversely affected the development of pegs (which bear groundnut pods below the soil), the relatively drier spell at the later stage hit the development of pods.

This aside, climate change is also reflected in the increasingly fluctuating weather cycle with unpredictable cold waves, heat waves, floods and exceptionally heavy single-day downpours. The most noticeable of such events in recent years included the country-wide drought in 2002, the heat wave in Andhra Pradesh in May 2003, extremely cold winters in 2002 and 2003, and prolonged dry spell in July 2004 and January 2005 in the north, unusual floods in the Rajasthan desert in 2005, drought in the north-east in 2006, abnormal temperature in January and February in 2007, and 23 per cent rainfall deficiency in the 2009 monsoon season. All these events took a heavy toll on crop output.

Indeed, the silver-lining in this dismal scenario is the National Action Plan on Climate Change, launched in 2008, which aims at developing technologies to help rain-fed agriculture adapt to the changing climate patterns.

At least four of the eight 'national missions' started under this programme will have direct or indirect bearing on rain-fed farming. These are the missions on sustainable agriculture, water, green India and strategic knowledge. The ICAR-led national agricultural research system is also conducting research on specific projects under the umbrella programme on climate change. Apart from the use of technological advances to combat climate change, there has to be sound policy framework and strong political will to achieve this objective.

State agricultural universities and regional farm research centres, too, will have to play a role in developing local situation-specific strategies for adapting the rain-fed farming to emerging climate patterns.

III.I. Impact of Global Warming

Climate change and agriculture are interrelated processes, both of which take place on a global scale. Global warming is projected to have significant impacts on conditions affecting agriculture, including temperature, carbon dioxide, glacial run-off, precipitation and the interaction of these elements. These conditions determine the carrying capacity of the biosphere to produce enough food for the human population and domesticated animals. The overall effect of climate change on agriculture will depend on the balance of these effects. Assessment of the effects of global climate changes on agriculture might help to properly anticipate and adapt farming to maximize agricultural production. Figure 2 explain the annual greenhouse gas emissions by various sectors.

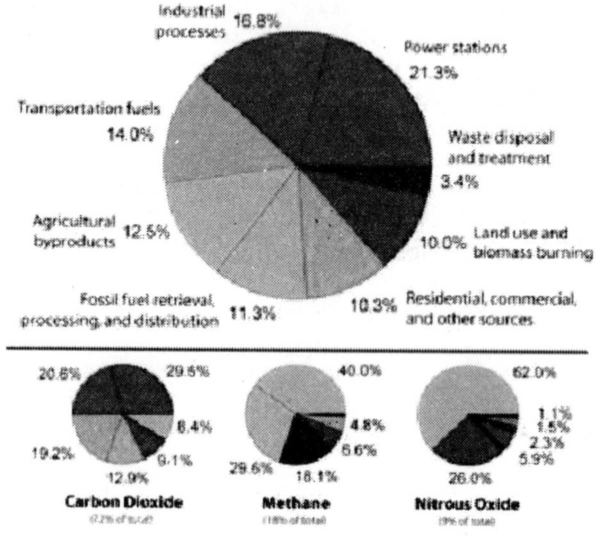

Figure 2

III.II. Shortage in Grain Production

Crops such as these sunflowers can be affected by severe drought conditions in Australia. Between 1996 and 2003, grain production has stabilized slightly over 1800 millions of tons. In 2000, 2001, 2002 and

2003, grain stocks have been dropping, resulting in a global grain harvest that was short of consumption by 93 millions of tons in 2003. The Earth's average temperature has been rising since the late 1970s, with nine of the 10 warmest years on record occurring since 1995. In 2002, India and the United States suffered sharp harvest reductions because of record temperatures and drought. In 2003, Europe suffered very low rainfall throughout spring and summer, and a record level of heat damaged most crops from the United Kingdom and France in the Western Europe through Ukraine in the East. Bread prices have been rising in several countries in the region. Figure 3 finds it truly.

Figure 3 : Shortage in Grain Production

III.III. Crop Development Models

Models for climate behavior are frequently inconclusive. In order to further study effects of global warming on agriculture, other types of models, such as crop development models, yield prediction, quantities of water or fertilizer consumed, can be used. Such models condense the knowledge accumulated of the climate, soil, and effects observed of the results of various agricultural practices. They thus could make it possible to test strategies of adaptation to modifications of the environment. Because these models are necessarily simplifying natural conditions, it is not clear whether the results they give will have an in-field reality. However, some results are partly validated with an increasing number of experimental results. Other models, such as insect and disease development models based on climate projections are also used.

Scenarios are used in order to estimate climate changes effects on crop development and yield. Each scenario is defined as a set of meteorological variables, based on generally accepted projections. For example, many models are running simulations based on doubled carbon dioxide projections, temperatures raise ranging from 1°C up to 5°C, and with rainfall levels an increase or decrease of 20%. Other parameters may include humidity, wind, and solar activity. Scenarios of crop models are testing farm-level adaptation, such as sowing date shift, climate adapted species, irrigation and fertilizer adaptation, resistance to disease. Most developed models are about wheat, maize, rice and soybean.

- *Temperature potential effect on growing period*: Duration of crop growth cycles is above all, related to temperature. An increase in temperature will speed up development. In the case of an annual crop, the duration between sowing and harvesting will shorten (for example, the duration in order to harvest corn could shorten between one and four weeks). The shortening of such a cycle could have an adverse effect on productivity because senescence would occur sooner.
- *Effect of elevated carbon dioxide on crops*: Carbon dioxide is essential to plant growth. Rising CO_2 concentration in the atmosphere can have both positive and negative consequences. Increased CO_2 is expected to have positive physiological effects by increasing the rate of photosynthesis. Currently, the amount of carbon dioxide in the atmosphere is 380 parts per million. In comparison, the amount of oxygen is 210,000 ppm. This means that often plants may be starved of carbon dioxide, due to the enzyme that fixes CO_2, rubisco also fixes oxygen in the process of photorespiration. The effects of an increase in carbon dioxide would be higher on C3 crops (such as wheat) than on C4 crops (such as maize), because the former is more susceptible to carbon dioxide shortage. Studies have shown that increased CO_2 leads to fewer stomata developing on a plant which leads to reduced water usage. Under optimum conditions of temperature and humidity, the yield increase could reach 36%, if the levels of carbon dioxide are doubled. Further, few studies have looked at the impact of elevated carbon dioxide concentrations on whole farming systems. Most models study the relationship between CO_2 and productivity in isolation from other factors associated with climate change, such as an increased frequency of extreme weather events,

seasonal shifts, and so on. In 2005, the Royal Society in London concluded that the purported benefits of elevated carbon dioxide concentrations are likely to be far lower than previously estimated when factors such as increasing ground-level ozone are taken into account.

- *Effect on quality*: According to the IPCC, the importance of climate change impacts on grain and forage quality emerges from new research. For rice, the amylose content of the grain—a major determinant of cooking quality—is increased under elevated CO_2. Cooked rice grain from plants grown in high-CO_2 environments would be firmer than that from today's plants. However, concentrations of iron and zinc, which are important for human nutrition, would be lower. Moreover, the protein content of the grain decreases under combined increases of temperature and CO_2. Studies using free-air concentration enrichment (FACE) have shown that increases in CO_2 lead to decreased concentrations of micronutrients in crop plants. This may have knock-on effects on other parts of ecosystems as herbivores will need to eat more food to gain the same amount of protein.

Studies have shown that higher CO_2 levels lead to reduced plant uptake of nitrogen (and a smaller number showing the same for trace elements such as zinc) resulting in crops with lower nutritional value. This would primarily impact on populations in poorer countries less able to compensate by eating more food, more varied diets, or possibly taking supplements. Reduced nitrogen content in grazing plants has also been shown to reduce animal productivity in sheep, which depend on microbes in their gut to digest plants, which in turn depend on nitrogen intake.

IV. AGRICULTURAL SURFACES AND CLIMATE CHANGES IN OTHER COUNTRIES

Climate change may increase the amount of arable land in high-latitude region by reduction of the amount of frozen lands. A 2005 study reports that temperature in Siberia has increased three degree Celsius in average since 1960 (much more than the rest of the world). However, reports about the impact of global warming on Russian agriculture indicate conflicting probable effects: while they expect a northward extension of farmable lands, they also warn of possible productivity losses and increased risk of drought.

Sea levels are expected to get up to one meter higher by 2100, though this projection is disputed. A rise in the sea level would result in an agricultural land loss, in particular in areas such as South East Asia. Erosion, submergence of shorelines, salinity of the water table due to the increased sea levels, could mainly affect agriculture through inundation of low-lying lands. Low lying areas such as Bangladesh, India and Vietnam will experience major loss of rice crop if sea levels are expected to rise by the end of the century. Vietnam for example relies heavily on its southern tip, where the Mekong Delta lies, for rice planting. Any rise in sea level of no more than a meter will drown several km^2 of rice paddies, rendering Vietnam incapable of producing its main staple and export of rice.

- *Erosion and fertility*: The warmer atmospheric temperatures observed over the past decades are expected to lead to a more vigorous hydrological cycle, including more extreme rainfall events. Erosion and soil degradation is more likely to occur. Soil fertility would also be affected by global warming. However, because the ratio of carbon to nitrogen is a constant, a doubling of carbon is likely to imply a higher storage of nitrogen in soils as nitrates, thus providing higher fertilizing elements for plants, providing better yields. The average needs for nitrogen could decrease, and give the opportunity of changing often costly fertilization strategies. Due to the extremes of climate that would result, the increase in precipitations would probably result in greater risks of erosion, whilst at the same time providing soil with better hydration, according to the intensity of the rain. The possible evolution of the organic matter in the soil is a highly contested issue: while the increase in the temperature would induce a greater rate in the production of minerals, lessening the soil organic matter content, the atmospheric CO_2 concentration would tend to increase it.
- *Potential effects of global climate change on pests, diseases and weeds*: A very important point to consider is that weeds would undergo the same acceleration of cycle as cultivated crops, and would also benefit from carbonaceous fertilization. Since most weeds are C3 plants, they are likely to compete even more than now against C4 crops such as corn. However, on the other hand, some results make it possible to think that weed killers could gain in effectiveness with the temperature increase.
- Global warming would cause an increase in rainfall in some areas, which would lead to an increase of atmospheric humidity and the duration of the wet seasons. Combined

with higher temperatures, these could favor the development of fungal diseases. Similarly, because of higher temperatures and humidity, there could be an increased pressure from insects and disease vectors.

- *Glacier retreat and disappearance*: The continued retreat of glaciers will have a number of different quantitative impacts. In areas that are heavily dependent on water runoff from glaciers that melt during the warmer summer months, a continuation of the current retreat will eventually deplete the glacial ice and substantially reduce or eliminate runoff. A reduction in runoff will affect the ability to irrigate crops and will reduce summer stream flows necessary to keep dams and reservoirs replenished.

Approximately 2.4 billion people live in the drainage basin of the Himalayan Rivers. India, China, Pakistan, Afghanistan, Bangladesh, Nepal and Myanmar could experience floods followed by severe droughts in coming decades. In India alone, the Ganges provides water for drinking and farming for more than 500 million people. The west coast of North America, which gets much of its water from glaciers in mountain ranges such as the Rocky Mountains and Sierra Nevada, also would be affected.

Ozone and UV-B

Some scientists think agriculture could be affected by any decrease in stratospheric ozone, which could increase biologically dangerous ultraviolet radiation B. Excess ultraviolet radiation B can directly effect plant physiology and cause massive amounts of mutations, and indirectly through changed pollinator behavior, though such changes are simple to quantify. However, it has not yet been ascertained whether an increase in greenhouse gases would decrease stratospheric ozone levels. In addition, a possible effect of rising temperatures is significantly higher levels of ground-level ozone, which would substantially lower yields.

ENSO effects on agriculture: ENSO (El Niño Southern Oscillation) will affect monsoon patterns more intensely in the future as climate change warms up the ocean's water. Crops that lie on the equatorial belt or under the tropical Walker circulation, such as rice, will be affected by varying monsoon patterns and more unpredictable weather. Scheduled planting and harvesting based on weather patterns will become less effective.

From Table 1 it is found that agricultural GDP will increase in near future if all the bad impacts of climate change be overcome by most of the nations. Areas such as Indonesia where the main crop consists of rice will be more vulnerable to the increased intensity of ENSO effects in the future of climate change. It is found that the effects of future ENSO patterns on the Indonesian rice agriculture using [IPCC]'s 2007 annual report and 20 different logistical models mapping out climate factors such as wind pressure, sea-level, and humidity, and found that rice harvest will experience a decrease in yield. Bali and Java, which holds 55% of the rice yields in Indonesia, will be likely to experience 9-10% probably of delayed monsoon patterns, which prolongs the hungry season. Normal planting of rice crop begins in October and harvest by January. However, as climate change affects ENSO and consequently delays planting, harvesting will be late and in drier conditions, resulting in less potential yields.

TABLE 1

Current and Future Agricultural GDP

('000 million 1990 US$)

Continent	*2000*	*2050*	*2100*
Africa	84	202	416
Asia	492	1 125	2 259
Latin America	89	213	441
W. Europe	191	337	542
E. Europe	231	421	694
N. America	127	224	360
Oceania	16	28	47
Total	1 230	2 551	4 759

* Eastern Europe includes the former Soviet Union.

V. MAJOR FINDINGS

From the above analysis it is found that climate change has some huge impacts on agriculture. The major findings are:

(i) The agricultural sector is a driving force in the gas emissions and land use effects thought to cause climate change. In addition to being a significant user of land and consumer of fossil fuel, agriculture contributes directly to greenhouse gas emissions through practices such as rice production and the raising of livestock; according to the Inter-governmental Panel on Climate Change, the three main causes of the increase in greenhouse gases observed over the past 250 years have been fossil fuels, land use, and agriculture.

(ii) The planet's major changes to land cover since 1750 have resulted from deforestation in temperate regions: when forests and woodlands are cleared to make room for fields and pastures, the albedo of the affected area increases, which can result in either warming or cooling effects, depending on local conditions. Deforestation also affects regional carbon reuptake, which can result in increased concentrations of CO_2, the dominant greenhouse gas. Land-clearing methods such as slash and burn compound these effects by burning biomatter, which directly releases greenhouse gases and particulate matter such as soot into the air.

(iii) Adoption concerns how quickly farmers take advantage of new technologies. Adaptation is a direct response to global warming whereas adoption is an ongoing modernization process largely independent of global warming. In order to understand how climate will affect agriculture in developing countries, it is critical to address adoption. The bulk of climate change is not expected for many decades. The agriculture sector that will be most affected is therefore one that exists far into the future. Adoption is critical because it will determine whether this future sector is a modern capital-intensive industry or similar to the production processes currently in place in most developing countries.

(iv) Most climate models agree that the temperature increases will be larger in the higher latitudes and that they will be greater at night than during the day. That is, global warming will increase average temperatures but it will also decrease the range of temperatures both through the day (diurnal cycle) and across latitudes. Most climate models, however, disagree about the remaining seasonal and geographic patterns of climate change. That is, the models do not provide a consistent pattern of seasonal changes nor do they predict consistent local patterns of change. Local temperatures and precipitation can vary across a much wider range than average annual global temperatures. What will happen to local farmers over time will depends on what happens to local climate outcomes? Thus, outcomes for individual farmers are highly uncertain.

(v) The literature to date suggests that global warming will not damage aggregate global food supplies over the next century. Existing models predict that there will be sufficient food to feed future populations, even with global warming. Global warming is not expected to affect aggregate production in most developing countries. However, it is likely that global warming will increase production in most temperate and polar countries leading to small increases in overall supply. The resulting small reductions in the price of food may hurt developing country producers although it will help consumers. Despite this overall positive assessment, it is expected that productivity will fall in selected locations across the planet because of reductions in rainfall and increased temperatures. To the extent that there are subsistence farmers in these areas, they will be vulnerable to these adverse conditions. Global warming may act like long-term climate variability. Some areas will do worse with warming. The major difference is that these areas may be permanently harmed.

References

Adams, R., Glyer, D. and McCarl, B., 1989, "The Economic Effects of Climate Change in US Agriculture: A Preliminary Assessment" in D. Tirpak and J. Smith (eds.) : *The Potential Effects of Global Climate Change on the United States: Report to Congress*, Washington, DC: US Environmental Protection Agency, EPA-230-05-89-050.

Allen, Jr. L. *et al.*, 1996, "The CO_2 Fertilization Effect: Higher Carbohydrate Production and Retention as Biomass and Seed Yield", in F. Bazzaz and W. Sombroek (eds.): *Global Climate Change and Agricultural Production*, Wiley and FAO, Rome.

Antle, J., 1995, "Climate Change and Agriculture in Developing Countries", *American Journal of Agricultural Economics*, Vol. 77, pp. 741-46.

Bennholdt-Thomsen, V., 1982, "Subsistence Production and Extended Reproduction: A Contribution to the Discussion about Models of Production", *Journal of Peasant Studies*, Vol. 9, pp. 241-54.

Brown, Paul, 2005, "Frozen assets", *The Guardian*, 30th June.

Cline, W., 1996, "The Impact of Global Warming on Agriculture: Comment", *American Economic Review*, Vol. 86, pp. 1309-12.

Darwin, R., 1999, "The Impact of Global Warming on Agriculture: A Ricardian Analysis: Comment", *American Economic Review*, Vol. 89, pp. 1049-52.

Darwin, R. *et al.*, 1999, "Climate Change, World Agriculture, and Land Use", in F. Frisvold and B. Kuhn (eds.): *Global Environmental Change and Agriculture*, Edward Elgar Publication, UK.

Dinar, A. and R. Mendelsohn, 1999, "Climate Change, Agriculture and Developing Countries: Does Adaptation Matter?" *The World Bank Research Observer*, Vol. 14, No. 2, pp. 277-93.

Dinar, A., and D. Zilberman, 1991, "The Economics of Resource-Conservation, Pollution-Reduction Technology Selection, The Case of Irrigation Water", *Resources and Energy* , Vol. 13, pp. 323-48.

Dinar, A., R. Mendelsohn, R. Evenson, J. Parikh, A. Sanghi, K. Kumar, J. McKinsey, S. Lonergan (eds.), 1998, *Measuring the Impact of Climate Change on Indian Agriculture*, World Bank Technical Paper No. 402, Washington, D.C.

Dowlatabadi, H. and M. Morgan, 1993, A Model Framework for Integrated Assessment of the Climate Problem, *Energy Policy*, Vol. 21, pp. 209-21.

Fankhauser, S., 1995, *Valuing Climate Change: The Economics of the Greenhouse*, Earthscan, London.

Food and Agriculture Organization of the United Nations (FAO), 1992, *Agrostat*, Rome, Italy.

Food and Agriculture Organization of the United Nations (FAO), 1996, *Agro-ecological zoning: Guidelines*, (FAO Soils Bulletin 73), Rome, Italy.

Hennessy, K. *et al.*, 2007, "Australia and New Zealand", in *Climate Change 2007: Impacts, Adaptation and Vulnerability*, Contribution of Working Group II to the Fourth Assessment Report of the IPCC [M.L. Parry *et al.* (eds.)], Cambridge University Press, Cambridge, U.K., and New York, N.Y., U.S.A., pp. 507-40.

Hope, C., J. Anderson, P. Wenman, 1993, "Policy Analysis of the Greenhouse Effect: An Application of the PAGE Model", *Energy Policy*, Vol. 21, pp. 327-38.

IPCC, 1996a, Houghton, J., L. Meira Filho, B. Callander, N. Harris, A. Kattenberg, and K. Maskell, (eds.): *Climate Change 1995: The State of the Science*, Cambridge University Press: Cambridge.

IPCC, 1996b, Watson, R., M. Zinyowera, R. Moss, and D. Dokken (eds.): *Climate Change 1995: Impacts, Adaptations, and Mitigation of Climate Change: Scientific-Technical Analyses*, Cambridge University Press: Cambridge.

IPCC, 2007, *Special Report on Emissions Scenarios.*

Kaiser, H, 1999, "Assessing Research on the Impacts of Climate Change on Agriculture", in F. Frisvold and B. Kuhn (eds.): *Global Environmental Change and Agriculture*, Edward Elgar Publication, UK.

Kaiser, H., Riha, S., Wilkes, D., and Sampath, R., 1993, "Adaptation to Global Climate Change at the Farm Level", in H. Kaiser and T. Drennen (eds.): *Agricultural Dimensions of Global Climate Change*, St. Lucie Press: Delray Beach, FL.

Kumar, K. and Parikh, J., 1998, "Climate Change Impacts on Indian Agriculture: The Ricardian Approach", in A. Dinar, R. Mendelsohn, R. Evenson, J. Parikh, A. Sanghi, K. Kumar, J. McKinsey, S. Lonergan (eds.): *Measuring the Impact of Climate Change on Indian Agriculture*, World Bank Technical Paper No. 402, Washington, D.C.

McKinsey, J. and Evenson, R., 1998, "Technology-Climate Interactions: Was The Green Revolution in India Climate Friendly?" in A. Dinar, R. Mendelsohn, R. Evenson, J. Parikh, A. Sanghi, K. Kumar, J. McKinsey, S. Lonergan (eds): *Measuring the Impact of Climate Change on Indian Agriculture*, World Bank Technical Paper No. 402, Washington, D.C.

Mendelsohn, R. and A. Dinar, 1999, "Climate Change Impacts on Developing Country Agriculture", *World Bank Research Observer*, No. 14, pp. 277-93.

Mendelsohn, R., W. Nordhaus, and D. Shaw, 1994, "The Impact of Global Warming on Agriculture: A Ricardian Analysis", *American Economic Review,* Vol. 84, pp. 753-71.

Mimura, N. *et al.*, 2007, "Small Islands", in *Climate Change 2007: Impacts, Adaptation and Vulnerability,* Contribution of Working Group II to the Fourth Assessment Report of the IPCC [M.L. Parry *et al.* (eds.)], Cambridge University Press, Cambridge, U.K., and New York, N.Y., U.S.A., pp. 687-716.

Nordhaus, W., 1994, *Managing the Global Commons: The Economics of Climate Change* , MIT Press, Cambridge, MA.

Pan, S., 2002, "Environmental Rules & WTO— Preparatory Work for Labour Standards: A Note", *BEA Conference Volume, No. 22, February*, pp. 69-73, Kolkata.

———, 2006, "Emerging Agricultural Diversity", in P.K. Pal (ed): *Economic Growth and Development: Emerging Issues*, pp. 99-109, Deep and Deep Publications, New Delhi.

———, 2010a, "How FDI Affects the Environment in India: Consequences from Recent Performance", in R.K. Sen (ed): *Environment and Sustainable Development in India,* Deep & Deep Publications, New Delhi.

———, 2010b, "Problem of Water Scarcity and Integrated Water Resource Management in Indian Perspective: An Overview, *Artha Beekshan*, Vol. 19, No. 3, Kolkata.

Rao, C.H. Hanumantha, 2006, *Agriculture, Food Security, Poverty, and Environment: Essays on Post-Reform India*, Oxford University Press, New Delhi.

Reilly, J., 1995, "Climate Change and Global Agriculture: Recent Findings and Issues", *American Journal of Agricultural Economics*, Vol. 77, pp. 727-33.

Reilly, J., N. Hohmann, and S. Kane, 1994, "Climate Change and Agricultural Trade: Who Benefits, Who Loses?" *Global Environmental Change,* Vol. 4, No. 1, pp. 24-36.

Sanghi, A. and R. Mendelsohn, 2000, *The Climate Sensitivity of Indian and Brazilian Agriculture,* Yale FES, New Haven, CT, USA.

Schneider, S.H. *et al.*, 2007, "Assessing Key Vulnerabilities and the Risk from Climate Change", in *Climate Change 2007: Impacts, Adaptation and Vulnerability,* Contribution of Working Group II to the Fourth Assessment Report of the IPCC [M.L. Parry *et al.* (eds.)], Cambridge University Press, Cambridge, U.K., and New York, N.Y., U.S.A., pp. 779-810.

Van de Guijin, S. *et al.*, 1996, "The Effects of Elevated CO_2 and Temperature Changes on Transpiration and Crop Water Use", in F. Bazzaz and W. Sombroek (eds.): *Global Climate Change and Agricultural Production*, Wiley and FAO, Rome.

Winters, P. *et al.*, 1999, "Climate Change and Agriculture: Effects on Developing Countries", in F. Frisvold and B. Kuhn (eds.): *Global Environmental Change and Agriculture*, Edward Elgar Publication, UK.

Chapter 28

Greenhouse Gas Emissions and Sinks at River Basin

A Case Study of Tungabhadra, in Southern India

M.S. Umesh Babu, E.T. Puttaiah

ABSTRACTS

Kyoto Protocol of 1997 has clearly laid the "Common but Differentiated Responsibilities to reduce the climate change". Central to this reduction in Green House Gases (GHGs) emission from all the member nations. Accordingly, guidelines were developed to measure GHGs emissions at national levels by various agencies and IPCC. On the other hand, in the event of negative implication due to climate change, it would be a particular river basin which will bore it, and not the entire nation. Thus, GHGs reduction should be at macro-level and adoption measures at micro-level. However, the major issues are uncertainty at micro level and also emissions of GHGs which are contributing to global warming.

To study these issues in detail we have selected a Tungabhadra river basin in Southern India, which is primarily constituted with agrarian, industrialization and urbanization activities. We have estimated GHGs emissions from four sectors, viz. Energy, Industry, Agriculture and Waste Management and sink from Forestry and also, estimated potential for carbon sequestration. End calculations have showed that carbon sequestration due to natural forest is higher than the emission rates.

In this paper, we argue how potential gains from various sequestration measures could be routed for Millennium Development Goals.

Keywords: Climate Change, IPCC Guidelines, Greenhouse Gas Emissions, Inventory, Sequestration potential

INTRODUCTION

Energy from the sun reaches the earth and portion of it is reflected back (albido). Some of the atmospheric gases such as water vapor, carbon dioxide trap some of this outgoing energy, in other words, they act like glass panels of a greenhouse. Without this natural "greenhouse effect," temperatures in the atmosphere would be much lower than the current temperatures. However, problems may arise when the atmospheric concentration of greenhouse gases increases. The rising concentrations of greenhouse gases (GHGs) such as carbon dioxide (CO_2), methane (CH_4) and nitrous oxide (N_2O) from various anthropogenic activities in the atmosphere is reported by several studies. As apprehended, since the beginning of the industrial revolution, atmospheric concentrations of carbon dioxide have increased nearly 30 per cent, methane concentrations have more than doubled and nitrous oxide concentrations have risen by about 15 per cent. Subsequently, the global mean surface temperature have risen by 0.4–0.8°C (Subhod Sharma, *et al.*, 2006). As a result of this warming, 20th century's 10 warmest years have occurred in the last 15 years. Of these, 1998 was the warmest year on the record. The snow cover in the Northern Hemisphere has decreased. Globally, sea level rise of 4-8 inches was observed over the past century. The frequency of extreme rainfall events has increased throughout the world.

This paper attempts to estimate the Greenhouse Gases at micro-level, i.e. at a level of river basin. Its main focus is on the inventory of GHGs emissions and sinks and the potential for carbon sequestration.

GREENHOUSE GAS INVENTORIES

A greenhouse gas inventory is an accounting the amount of greenhouse gases emitted or removed from atmosphere over a specific period of time. It also provides information on the activities that result in emissions and removals. It is a very good tool for the policy-makers to track emission trends, develop strategies and policies and evaluate progress to reduce GHG emmisions, on the other hand, scientists can use greenhouse gas inventories as inputs to atmospheric and economic models.

INDIAN SCENARIO

India occupies 2.4 per cent of the world's geographical area but supports nearly 17 per cent of its population and emits less than 5 per cent of GHG emissions. GHG emissions per capita in India are very low (fifth position in the world average), on the other side, India, being the world's second most populous country with a burgeoning middle income population with increasingly energy-intensive lifestyles. It is vulnerable to climate change on several aspects, directly, the impacts on coastal districts, which are very densely populated (above 500 persons/ km^2) with over a 100 million people inhabiting them. Indirectly also, India is highly vulnerable to climate change as its economy is heavily reliant on climate sensitive sectors like agriculture. Though agriculture contributed only 22 per cent to India's GDP in 2001-02, about 68 per cent of the country's workforce is employed in this sector. As per the regional model (HadRM2, IS92a scenario), the projections of climate variables for the 2050s, under the IS92a scenario of GHG emissions are summarized below:

- An all-round increase in temperature and a general increase in precipitation in monsoon season.
- A large spatial variation in the relative increase in monsoon precipitation
- An overall decrease in the number of rainy days over a major part of the country
- An overall increase in the rainy days intensity by 1-4 mm/ day
- An increase in the temperature of the order of 2-4°C over the southern region which may exceed 4°C over the northern region

With these projections, India seriously concerned with the possible impacts of climate change, such as:

- Water stress and reduction in the availability of fresh water due to potential decline in rainfall.
- Threats to agriculture and food security, since agriculture is monsoon dependent and rainfed agriculture dominates in many states.
- Shifts in area and boundary of different forest types and threats to biodiversity with adverse implications for forest-dependent communities.
- Adverse impact on natural ecosystems, such as wetlands, mangroves and coral reefs, grasslands and mountain ecosystems.
- Adverse impact of sea-level rise on coastal agriculture and settlements.
- Impact on human health due to increase in vector and water-borne diseases, such as malaria.

- Increased energy requirements and impact on climate-sensitive industry and infrastructure.

As a proactive member, India has acceded to the Kyoto Protocol in August 2002 and has fulfilling its obligations as per UNFCCC, in addition to active participation in the Clean Development Mechanism (CDM) measures.

STUDY AREA

The Tungabhadra river basin is one of the main sub-basin of the Krishna river basin in peninsular India and it stretches over an area of about 47,827 km^2 (1.45% of the Indian total geographical area) in both the riparian states of Karnataka (81.1% of the basin) and Andhra Pradesh (18.89% of the basin). The major tributaries of the river are Bhadra, Tunga, Hagari and Varada. The river Tungabhadra joins river Krishna in Andhra Pradesh and ends up in the Bay of Bengal. The climate in the basin is characterized by sub-tropical conditions with considerable rainfall in the mountains of the Western Ghats and arid conditions in the basin (Bouwer *et al.*, 2006). The average normal rainfall is 1024 mm and actual average rainfall is 1028 mm (in the year 2005) while the annual average temperature reaches 26.7°C (Bouwer *et al.*, 2006).

TABLE 1
Features of Study Area

Characteristics	*Status*
Population (2001 census)	7.9 millions (15.25 % decadal growth)
Density per sq. km.	262 numbers
Literacy in % (average)	53.86
Origing of River	Western Ghat
Major Crops	Paddy, Jowar, Maize, Ragi, Bajra, Sunflower, Groundnut etc.

Objective

- Estimation of level of GHG contribution to global warming in the river basin aspects.
- Estimate the potential for Carbon sequestration or carbon trading.

Methodology and Data Sources

The methodology prescribed by the IPCC (Revised, 1996) for National Greenhouse Gas Inventories is used the below said equations are adopted for estimation.

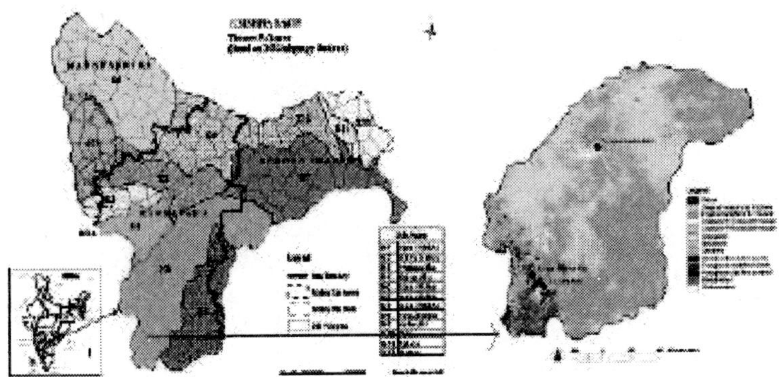

Tungabhadra Basin Map

Figure 1 : Study Area

Equation 1: Rice Cultivation

Methane emissions from rice fields is calculated by using the below said equation

$$F_c = EF \times A \times 10^{-12}$$

where,

F_c = estimated annual emission of methane in Gg/yr;

EF = methane emission factor integrated over integrated cropping season, in g/m^2; and

A = annual harvested area in m^2/yr.

Equation 2: Domestic Livestock

Methane emissions from enteric fermentation and manure management are calculated by applying an emissions factor to the number of animals of each livestock type in the basin to produce a total. Emission factors are considered from the developing country (India).

Nitrous oxide emissions from Animal Waste Management Systems are calculated by the following equations :

$$1.\ Nex_{(AWMS)} = \sum_{(T)} [N_{(T)} \times Nex_{(T)} \times AWMS_{(T)}]$$

Where,

$Nex_{(AWMS)}$ = N excretion per Animal Waste Management System (kg/yr)

$N_{(T)}$ = number of animals of type T

$Nex_{(T)}$ = N excretion of animals of type T in the country (kgN/animal /yr)

$AWMS_{(T)}$ = fraction of $Nex_{(T)}$ that is managed in one of the different distinguished animal waste management systems for animals of type T, and

T = type of animal category.

2. $N_2O_{(AWMS)} = \sum[Nex_{(AWMS)} \times EF_{3(AWMS)}]$

Where,

$N_2O_{(AWMS)}$ = N_2O emissions from all Animal Waste Management Systems in the country (kgN/yr)

$Nex_{(AWMS)}$ = N excretion per Animal Waste Management System (kg/yr); and

$EF_{3(AWMS)}$ = N_2O emissions factor for an AWMS (kg N_2O -N/kg of $Nex_{(AWMS)}$).

Equation 3: Agricultural Soils

Estimation of Nitrous Oxide emissions from synthetic fertilizers applied for cultivation is calculated by below equation

$$T\,N_2O = (A \times B) \times 10^{-6}$$

Where,

$T\,N_2O$ = Total Nitrous Oxide emissions from the synthetic fertilizers used in Gg

A = Total synthetic fertilizers applied (in Kg N/yr), and

B = Emissions factors for direct emissions (in Kg N_2O-N/Kg N).

Equation 4: Wood

Wood harvested for fuelwood, commercial timber and other uses is also estimated as significant quantities may be gathered informally for traditional fuelwood consumption. Carbon Dioxide emissions from the fuel wood is calculated by the below shown equation

$$CO_2\,E = FW_{HH} \times FW_{PC} \times FW_{EF}$$

Where,

$CO_2\,E$ = Carbon Dioxide Emissions in Gg

FW_{HH} = Number of households use fuelwood

FW_{PC} = Per capita consumption of Fuelwood in tons/year

FW_{EF} = Carbon Dioxide emissions factor for fuelwood burnt in kg/tons

Equation 5: Electricity

Carbon Dioxide emissions from the electricity is calculated by the below shown equation.

$$CO_2E = E_{Con} \times E_{EF}$$

Where,

CO_2E = Carbon Dioxide Emissions in Gg

E_{Con} = Electricity consumption from different sectors in KW

E_{EF} = Carbon Dioxide Emissions Factor 0.80 kg/KW

Equation 6: Fossil Fuels

The IPCC methodology breaks the calculation of carbon dioxide emissions from fossil fuel combustion by the following equations in different 6 steps such as:

Step 1: Estimate Apparent Fuel Consumption in Original Units
Step 2: Convert to a Common Energy Unit
Step 3: Multiply by Emission Factors to Compute the Carbon Content
Step 4: Compute Carbon Stored
Step 5: Correct for Carbon Unoxidised
Step 6: Convert Carbon Oxidized to CO_2 Emissions in Gg

$$CO_2E = F_{Con} \times F_{EF} \times F_{CS} \times F_{CU}$$

Where,

CO_2E = Carbon Dioxide Emissions

F_{Con} = Fuel Consumption in tones

F_{EF} = Fuel Emission Factor

F_{CS} = Fuel Carbon Stored

F_{CU} = Fuel Carbon Unoxidized

Equation 7: Industries

The general methodology employed to estimate emissions associated with each industrial process involves the product of activity level data, e.g., amount of material produced or consumed, and an associated emission factor per unit of consumption/production according to,

$$Total_{ij} = A_j \times EF_{ij}$$

Where,

$Total_{ij}$ = the process emission (tons) of gas i from industrial sector j

A_j = the amount of activity or production of process material in industrial sector j (tons/yr.)

EF_{ij} = the emission factor associated with gas i per unit of activity in industrial sector j (ton/ton)

Equation 8: Solid Waste Management

Methane emissions calculated from the solid waste disposals into the sites :

$$CH_4 \text{ (Gg/yr)} = (MSWT \times MSWF \times MCF \times DOC \times DOCF \times F \times 16/12 - R) \times (1 - OX)$$

Where,

MSWT = total MSW generated (G/yr)
MSWF = fraction of MSW disposed to solid waste disposal sites
MCF = methane correction factor (fraction)
DOC = degradable organic carbon (fraction)
DOCF = fraction DOC simulated
F = fraction of CH_4 in landfill gas (default is 0.5)
R = recovered CH_4 (Gg/yr)
OX = oxidation factor (default is 0)

Equation 9: Land Use Change and Forestry

To calculate the net uptake of CO_2, the annual increment of biomass in plantations, forests which are logged or otherwise harvested, the growth of trees in villages, farms and urban areas and any other significant stocks of woody biomass are estimated.

TABLE 2

Data Sources from Different Departments

Department	*Data Collected*
Indian Oil Corporation (IOC) Bangalore/Dealers	Consumption of petrol, Diesel and LPG
Karnataka Food and Civil Supplies	Kerosene consumption and LPG connections
Karnataka State Pollution Control Board (KSPCB)	GHGs emitted Industries details and annual production on their industries and Effluent details of the Industries
Director of Agriculture	Agricultural Crops and production details and Nitrogen consumption details
Directorate of Economics and Statistics	Population details
Forest Department/Social Forestry/ Farm Forestry/Agro forestry	Forest cover details, plantation, reforestation, etc.
Directorate of Municipal Administration	Municipal Solid Waste Details and their disposal Methods
Horticulture	Details of plantations
Transport	Vehicles details
Director of Animal husbandry	Domestic livestock details by the census

RESULTS AND DISCUSSION

GHG Emissions

GHG emissions from the basin are calculated as per the IPCC methodology and same are discussed in the following sections.

Energy

The main energy resources in study area can be grouped into three broad categories, viz. commercial (coal, oil, natural gas, electricity); non-commercial (fuel wood, dried cow dung cakes, vegetable wastes); renewable (solar and wind power). In India, over 65 per cent of the total energy consumption is met by commercial energy sources, and the remaining 35 per cent comes from non-commercial and renewable sources. The rate of coal consumption in production of electricity, overall, for India is of the order of 0.77–0.85 kg/kWh and on average CO_2 emissions are about 0.80 kg/KWh[1]. At household level, nature and extent of energy use is directly related to the income sources, with marginal sections depending on kerosene and fuel wood. Growth in household incomes and in urbanization has been accompanied by a change in fuels to LPG to electricity from fuel wood and kerosene. Urban home largely depends on cooking gas for its energy requirement, 85-90 per cent of the energy demand of a rural home is dependent on fuel wood. Solar energy is found to be limited to the water heating purposes and on pilot basis for lighting purposes.

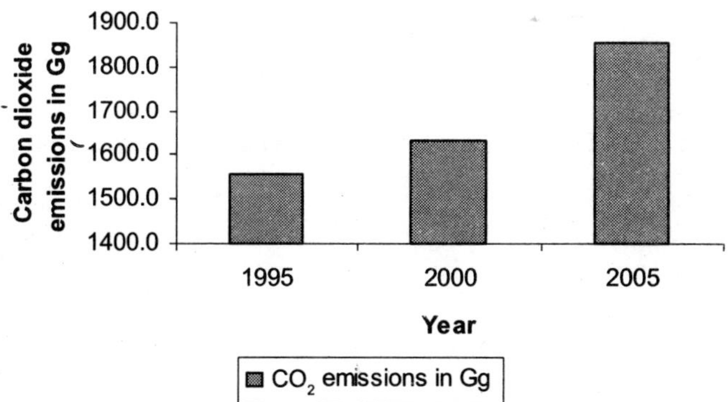

Figure 2 : Carbon Dioxide Emissions from fossil fuel burning in Gg

Electricity Consumption and Carbon Dioxide Emissions

Fuel wood

The result in the year 1995 and 2000 shows same emissions due to the household assumptions are made according to 2001 census.

TABLE 3

Electricity Consumption in the Year 2005

(in Lakh Units)

Sectors	*Household consumption*	*Industrial Commercial*	*IP sets*	*Others*	*Total*	CO_2	*Emissions @ 0.80kg/ KWh in Gg[1]*
Electricity consumption in lakh units	1822.31	708.15	356.05	461.38	464.61	3812.5	305.27

Source : District at a Glance.

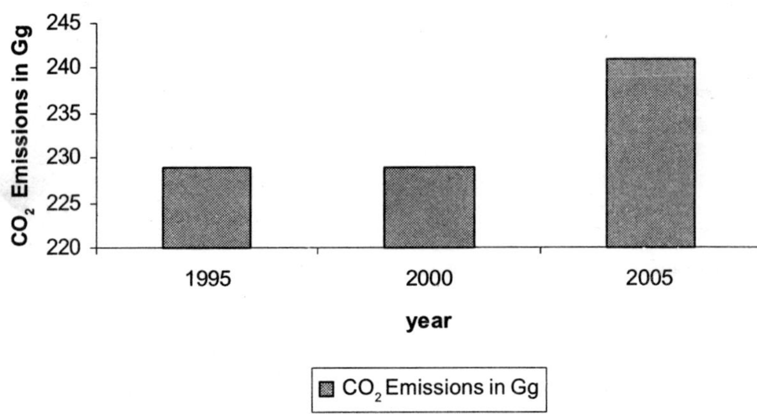

Figure 3 : GHG Emissions from Fuel Wood Consumption in Gg

Fossil fuel consumption emits 77.2 per cent of Carbon Dioxide in energy sector due to increase of automobiles and stationary sources. Electricity consumption from domestic, industrial, commercial, street lights, IP sets and others contributes 12.7 per cent and burning of fuel wood for domestic purpose contributes 10 per cent.

Industrial Source

Industrial process is one among the GHG sources, which chemically or physically transform materials and the following are the details of same:

Agriculture

Emissions from agriculture are accounted from all the four sectors, viz., flood irrigation of rice, application of nitrogen fertilizers, field

TABLE 4

Carbon Dioxide Contribution from Energy Sector

Year	*Energy*	*Carbon Dioxide Emissions in Gg*	*Carbon Dioxide Emissions in percentage*
1995	Fossil Fuel	1556.2	87.18
	Wood	228.94	12.82
	Total	1785.14	100
2000	Fossil Fuel	1632.2	87.72
	Wood	228.4	12.28
	Total	1860.6	100
2005	Fossil Fuel	1853	77.23
	Electricity	305.27	12.72
	Wood	241.07	10.05
	Total	2399.34	100

TABLE 5

CO_2 Emitted from the Iron and Steel Industries in Gg

Year	*CO_2 Emitted in Gg*
1995	3423
2000	3423
2005	3423

Note : We have taken the average estimated production of the industries so all the years emissions are same.

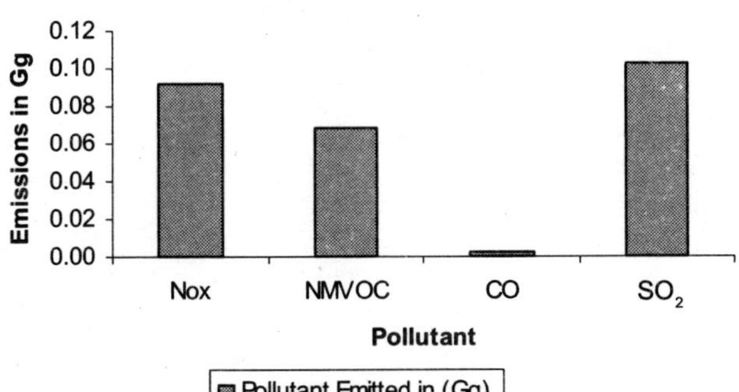

Figure 4 : NMVOC, Nox , CO and SO_2 Emissions from Iron and Steel Industries

Note : We have taken the average estimated production of the industries so emission is same in each year.

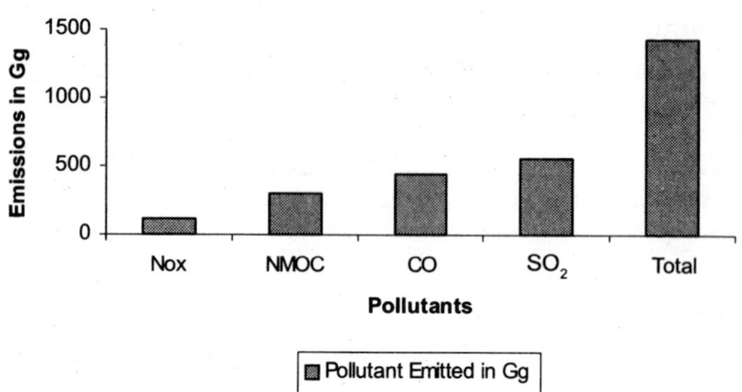

Figure 5 : NMVOC, Nox , CO and SO_2 Emissions from Paper and Pulp Industries

burning of agricultural residues and animal husbandry. Tungabhadra river basin has a huge command area and most of the area is dominated by the flooded rice cultivation and other water intensive crops. Methane emissions from flooded rice cultivation is calculated and shown in below Figure 6.

Methane Emissions from Rice Cultivation

As can be seen from the Figure 6, the emission of methane from rice field is directly proportional to the area sown under wet system. The fluctuating rate of emission is reflection on the area cultivated and the reasons cited were bad monsoon, lack of surface water, etc.

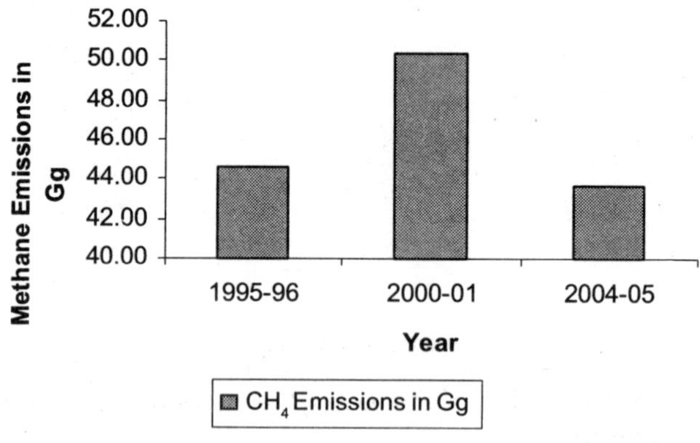

Figure 6 : Methane Emissions from Rice Cultivation

Field Burning of Agricultural Residues

Large quantities of agricultural residues are produced from farming systems world-wide. Burning of crop residues in the field is a common agricultural practice in developing countries.

Nitrous Oxide emitted during the burning of agricultural waste such as sugarcane and maize leaves. Field burning of agricultural residues contributes less emission than other emissions.

TABLE 6

Nitrous Oxide Emissions from Field Burning in Gg

Variables/Year	*1995-96*	*2000-01*	*2004-05*
CO	24.75	30.44	24.84
N_2O	0.04	0.05	0.04
Nox	1.41	1.73	1.41

Agricultural Soil (Applications of Nitrogen Fertilizers)

Nitrogenous fertilizers play an import role in increasing crop yields. Use of nitrogenous fertilizer, increases nitrous oxide emissions from soil and water through nitrification and denitrification processes.

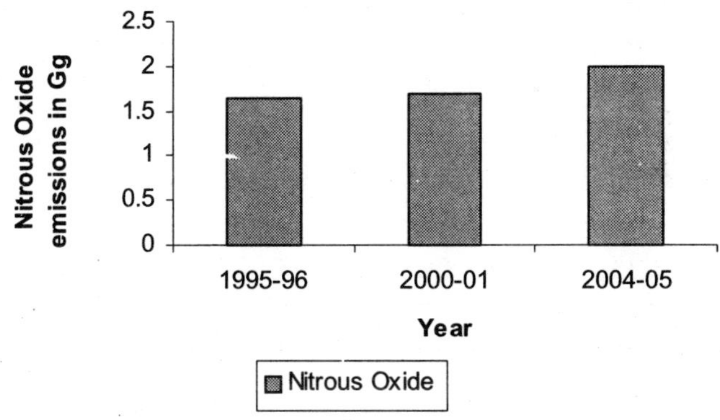

Figure 7 : Nitrous Oxide Emissions from Agricultural Soil

Nitrogen fertilizers usage is increasing year to year due to decline in soil fertility and production. Therefore, the emissions rate also increases.

Animal Husbandry

Methane from enteric fermentation is produced in herbivores as a by-product of the digestive process by which carbohydrates are broken down by micro-organisms into simple molecules for absorption into the blood-stream. Both ruminant animals (e.g., cattle, sheep) and some non-ruminant animals (e.g., pigs, horses), which produce methane, though ruminants are the largest source. The amount of CH_4 that is released depends upon the type, age and weight of the animal and the quantity and quality of the feed consumed.

Enteric Fermentation

Methane from enteric fermentation is produced in herbivores as a by-product of the digestive process by which carbohydrates are broken down by micro-organisms into simple molecules for absorption into the blood-stream. Both ruminant animals (e.g., cattle, sheep) and non-ruminant animals (e.g., pigs, horses) produce methane, although ruminants are the largest source. The amount of CH_4 released depends upon the type, age and weight of the animal and the quantity and quality of the feed consumed. In addition to enteric fermentation, methane is emitted from manure management due to anaerobic decomposition as well.

Manure Management

Methane will emitted during anaerobic decomposition of manure. The amount of methane produced depends on the number of animals, the amount of feed consumed. Environmental conditions (temperature and moisture) play important role. Nitrous Oxide emitted during manure storage, handling and application through nitrification or denitrification. Indirect sources of nitrous oxide include volatilization and subsequent atmospheric deposition of ammonia (NH_3) and Nitrogen Oxides (NO_X).

TABLE 7

CH_4 Emissions from both EF and MM in Gg

Year	*1995-96*	*2000-01*	*2004-05*
Non Dairy Catle	40.8	46.6	47.0
Buffalo	49.7	46.9	55.5
Sheep	4.8	4.9	6.1
Goats	2.6	3.0	3.3
Swine	0.2	0.4	0.2
Poultry	0.2	0.1	0.2
Total	98.2	101.9	112.4

A Nitrous Oxide emission in the year 2000 was high because number of cattle and poultry was high compared to 1995 and 2005. The number of livestock is declining because of decrease in grazing land,

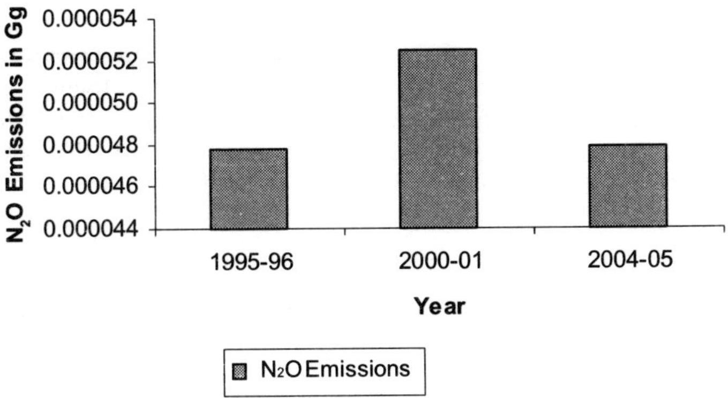

Figure 8 : N_2O Emissions from Domestic Livestock in Gg

migration of people and increased mechanization in the agricultural sector.

Carbon Dioxide (Equivalent) Contribution from Agricultural Sector

TABLE 8

Carbon Dioxide (Equivalent) Emissions from Agricultural Sector

Year/Variables	*1995*		*2000*		*2005*	
	In Gg	*In %*	*In Gg*	*In %*	*In Gg*	*In %*
Rice Cultivation	937.2	26.70	1059.19	28.09	918.13	23.32
Filed Burning of Agricultural Residues	36.81	1.05	45.27	1.20	36.93	0.94
Agricultural Soils	509.82	14.52	526.79	13.97	621.71	15.79
EF and MM	2026.56	57.73	2139.27	56.74	2359.73	59.94
Total	3510.39		3770.52		3936.5	

Animal husbandry contributes 60 per cent of total agricultural emissions because many households are rearing livestock for agriculture and dairy purposes. Flooded rice cultivation contributes 23 per cent due to over-cultivation of rice. Agricultural residues emit 16 per cent due to excess use of synthetic fertilizers. Field burning contributes less because some part of the basin has sugarcane cultivation.

Waste Management—Land Disposal of Solid Waste

Anaerobic decomposition of organic matter by methanogenic bacteria in solid waste disposal sites results in the production of CH_4 to the atmosphere. Solid waste for the purpose of GHG emissions is defined as: household waste; yard/garden waste; and commercial/ market waste. Globally, it is estimated to account for about 5 to 20 per cent of global anthropogenic CH_4 emissions (US EPA, 1994; IPCC, 1992). According to IPCC guidelines urban waste is managed under Solid Waste Disposal Site (SWDS) and rural population is not accounted because dispose of waste in such a way that Methane emissions are extremely low.

TABLE 9

CH_4 and CO_2 (equivalent) Emissions from Municipal Waste

Year	*CH_4 Emissions in Gg*	*CO_2 (Equivalent) Emissions*
1995	2.71	57
2004	3	63

Note : Population for the year 1995 and 2005 is taken from the 1991 and 2001 Census respectively because these are the recent data available for those years.

GHGs Contribution

India has experienced a drastic growth in fossil fuel CO_2 emissions, and is rated as the 6th largest contributor of CO_2 emissions and China the 2nd. However, per capita emissions of CO_2 is 0.93 tons per annum only, which is well below the world average of 3.87 tons per annum.

As the study findings indicate that agriculture is the largest contributor of Carbon Dioxide emissions in our study area contributing as much as 40.3 per cent of the total emissions. Industrial sector constitutes the next major contributor, accounting for nearly 35.1 percent. Emissions from Industrial sector mainly come from Iron and Steel and Paper and Pulp. The energy sector contributes at the third position approximately 24.6 per cent, consisting of emissions from road transport, burning of traditional bio-mass fuels, electricity and fugitive emissions from oil and natural gas. Solid Waste contributes in fourth position approximately 0.6 per cent, emissions include from municipal solid waste disposal sites for urban population. A comparasion is given in Table 9.

Sinks—Land Use (Forestry)

To calculate the net uptake of CO_2, the annual increment of biomass in plantations, forests which are logged or otherwise harvested, the growth of trees in villages, farms and urban areas and any other significant stocks of wood biomass had been estimated. Wood harvested for fuel, commercial timber and other uses is also estimated as significant

TABLE 10

Total Carbon Dioxide Emitted by Different Sectors in Gg

Sectors	*Years*								
	1995	*2000*	*2005*	*1995*	*2000*	*2005*	*1995*	*2000*	*2005*
	Emissions in Gg	*Emissions in %*	*Per Capita Emissions in Tons*	*Emissions in Gg*	*Emissions in %*	*Per Capita Emissions in Tons*	*Emissions in Gg*	*Emissions in %*	*Per Capita Emissions in Tons*
(1)	*(2)*	*(3)*	*(4)*	*(5)*	*(6)*	*(7)*	*(8)*	*(9)*	*(10)*
Agriculture	3510.4	40.0	0.526	3770.5	41.4	0.56	3936.5	40.3	0.50
Energy	1785.1	20.3	0.267	1860.6	20.4	0.28	2399.3	24.6	0.30
Industries	3423.0	39.0	0.513	3423.0	37.6	0.51	3423.0	35.1	0.43
Waste	57.0	0.6	0.009	57.0	0.6	0.01	63.0	0.6	0.01
Total	8775.5		1.314	9111.1		1.36	9758.8		1.24

Note : Per capita emissions rate decreased in the year 2005 due to increase in population (15 per cent).

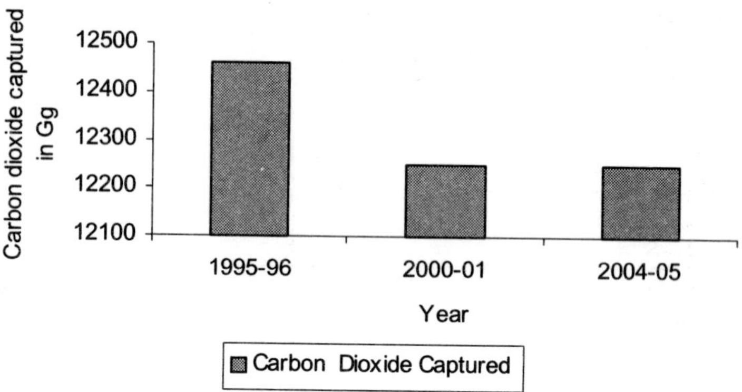

Figure 9 : Carbon Capture from Land-use Change and Forestry

Note : Carbon Dioxide Capture is more in the year 1995-96 because, forest cover was high compared other years.

quantities may be gathered informally for traditional fuel wood consumption. In this case the commercial statistics should be supplemented by FAO fuel wood consumption data.

Sinks through Water

Solubility of Carbon dioxide is higher at low temperature so colder oceans can absorb more CO_2 than warmer ones. In the study area, the reservoirs called Bhadra and Tungabhadra are the major and it can store up to 207 TMC of water at given time. These reservoirs fill up in the rainy season and dried up in summer. Due to lack of data pertaining to solubility, this aspect was not considered.

Potential for Carbon Sequestration

Tungabhadra basin has 2,848 kms (including distributary length also) of canal length, which can grow plantations/vegetations all along the sides of canal (each side 2 line, totally 4 line along the canal) it captures 6.2 lakh tons carbon per year and it may earn Rs. 2492.5 lakhs per year and the total estimated cost for canal side plantation is Rs. 17.8 lakh (one time cost). One ton of captured carbon yields almost 10 US $, thus a total of 5.33 lakh ha. of waste land available in the basin if converted into plantation (one ha. 100 plants or one acre 40 plants) or green cover can captures 14.5 million tons per year and may earn 1,459.4 lakh US $ per year (Rs. 58,379.9 lakhs). The total estimated cost of the plantation would be Rs. 1,664.9 lakh (one time investment). Total road (national, state, district and rural) length is 70,149.75 km in the basin, thus each side of the road (one line each side) can be converted into plantation, which may captures 7.67 million tons of carbon per year and can earn 767.43 lakh US$ per year (Rs. 30,697.5 lakhs). The total estimated costs of road side plantation could be approximately Rs. 438.43 lakh (one time investment).

TABLE 11

GHG Inventory for the years 1995, 2000 and 2005 in Gg

GHGs Sources and Sinks/Year	*1995*	*2000*	*2005*
I. Emissions			
1. Energy	1785.1	1860.6	2399.3
2. Industry	3423.0	3423.0	3423.0
3. Agriculture	3510.4	3770.5	3936.5
4. Waste	57	57	63
Total	8775.5	9111.1	9821.8
II. Sinks			
Forest 12460.16	12248.54	12248.54	
III. Net Carbon Dioxide Removals (+)/Absorption (-)			
Net Carbon Dioxide Removal (+)/Absoption(-)	3684.7	3137.4	2426.7

Note : Positive (+) sign indicates Removals and Negative (-) Absorption.

CONCLUSION

Greenhouse gases emissions from the agriculture sector is contributing on an average of 40.3 per cent in the basin due to over cultivation of paddy and burning of sugarcane residues. Industrial sector is contributing on an average of 35.1 (in 2005) per cent because it has major Iron and Steel and Paper and Pulp industries in the basin. Energy sector is emitting on an average of 24.6 per cent from burning of fossil fuels, electricity and fuel wood. Waste (Municipal Solid Waste) sector is emitting less (0.6 per cent) compare to all other sectors because Tungabhadra basin has 28 per cent of the urban population and rest is village population. This study identified that amount of greenhouse gas emissions is lesser than the sinks, because part of this basin belongs to Western Ghat where vegetation cover is high. Hence, overall GHGs contribution is nil. The above Table 11 shows the inventory of GHGs for the year 1995, 2000 and 2005.

Environment-friendly and green energies such as solar and hydro will be adopted at micro-level rather than fossil fuels for domestic and commercial activities. Better livestock and agricultural managements such as Systematic Rice Intensification (SRI), decomposition of residues instead of burning on field, use organic manure rather than applying synthetic fertilizers, adopt a biogas plants instead of dumping cow dung as a dry lots etc., will reduce the methane emissions. Introducing an advanced technology in the engines of automobiles will reduce the carbon dioxide emissions. The basin has almost 5.33 lakh hectares of wasteland in and around it can be used for plantations and will capture almost three thousands Gg of Carbon and it will benefit the local economy directly and indirectly. Clean Development Mechanism (CDM) is adopted by the UNFCCC it would provide a large opportunity to obtain external financial support and technology (for high yields) to promote the commercial forestry strategy aimed at meeting the domestic biomass demand as well as to reduce CO_2 emissions and remove CO_2 from atmosphere.

Note

1. Shiv Pratap Raghuvanshi *, Avinash Chandra, Ashok Kumar Raghav (2005), Carbon dioxide emissions from coal based power generation in India, Elsevier Ltd.

References

Association of American Geographers (2003). Global Change and Local Places Research Team, Global Change and Local Places, Estimating Understanding and Reducing GHG. Cambridge University Press.

Bhargava, G. (Unpublished). Environment and Its Global Implications, Theory and Practice, Kalpaz Publications, Delhi.

Bouwer, L.M., Aerts, J.C.J.H., Droogers, P. and Dolman, A.J. (2006). Detecting the Long-term Impacts from Climate Variability and Increasing Water Consumption on Runoff in the Krishna River basin (India). Hydrology and Earth systems Sciences 10: 703-13.

Burroughs, W. (2001). Climate Change Multidisciplinary Approach. The Press Syndicate of the University of Cambridge.

Burroughs, W. (2003). Climate In to the 21st Century. United Kingdom at the University Press, Cambridge.

Environment Protection Agency (EPA) (1998). Inventory of U.S. Greenhouse Gas Emissions and Sinks 1990-96. United States of Environmental Agency.

EPA (Unpublished). Solid Waste Management and Greenhouse Gases A life cycle Assessment of Emission and Sinks, Environment Protection Agency (EPA).

EPA (2002). U S Greenhouse Gas Inventory, EPA F 430-F-02-008

Grifin, J.M. (2003). Global Climate Change the Science, Economics and Politics, Edward Eglar Publishing Ltd.

Giambellula, T.W. and Henderson-Sellers, A. (1996). Climate Change Developing Southern hemisphere Perspectives, John Wiley and Sons Publishers.

Halmam, M.M. (2000). Greenhouse Gas, Carbon Dioxide Mitigation Science and Technology, Lewis Publishers.

IPCC (1996). Revised Guidelines for National Greenhouse Gas Inventory.

Proceedings (2000). The National Workshop on Environment Statistics. Central Statistical Organizations Ministry of Central and Programme Implementation, Government of India, Hyderabad.

Sathaye, J.A. and Ravindranath, N.H. (Unpublished). Climate Change Mitigation in the Energy and Forestry Sectors of Developing Countries, Centre for Ecological Sciences, IISc. Bangalore.

Sharma, S., Bhattacharya, S. and Garg, A. (2006). A Greenhouse Gas Emissions from India: A Perspective, *Current Science,* 90.

Shukla, P.R. (2006). India's GHG Scenarios: Aligning, Development and Stabilization Paths, Current Science 90.

Warrick, B. and Jager, D. (2001). The Greenhouse Effect Climate Change and Ecosystems, *Bioenergy News,* 5.

White, J.C. (1993). Global Energy Strategies Living with Restricted Greenhouse Gas Emissions, Plenum Press.

Chapter 29

Environmental Management

Is it the Key to Success for Future Sustainability of the Earth?

ASIM K. KARMAKAR

ABSTRACT

We all know that through the actions of the IMF, WTO and its precursor GATT, the World economy entered a period aptly described as the post-war economic boom. With the advent of this economic boom also came the explosion of productions; on its heels flourished the wave of consumerism, with little regard for the environmental consequences. To meet the demand of increasing production as well as consumerism, a clear picture of harms they have breathed is also noticeable. Noticeable is the concentration of Greenhouse Gas Emissions and other environmental hazards that have increased over the years owing to human activities. Scientific evidence suggests that a continued increase of that gas concentration and other types of pollutions are likely to have significant effects on the climate. The age of environmental activism was thus born. Environmental and social costs of the progress of consumerism that is accelerating on a global scale with the advent of sophisticated products and services, economic growth being the paramount driver in this context, are being questioned and challenged. It is unquestionable that modern economic growth has raised levels of atmospheric CO_2—leading to a rise in the Earth's surface temperature and the threat of climate change inviting disparities in energy use, CO_2 emissions, and living standards between rich and poor. Thus, the central dichotomy that every country is facing today is how to maintain economic growth and progress while protecting and enhancing the environment. The growing demands for conservation of nature and its sustainability have driven country after country to deliver solutions using rules, regulations and legislations.

The paper, however, is an attempt to provide fresh perspective on global warming and climate change. It essentially forwards some points. First, that especially due to momentum in economic and demographic processes, it is sure that there will be a major rise in atmospheric CO_2 during the 21st century. Second, that the coming rise in global temperatures will be faster than anything that human populations have experienced ever. Finally, the agricultural, political, economic, demographic, social and other consequences of future climate change could be very considerable. In a more and more populous world of eight to ten billion people, adverse development could well occur on several fronts simultaneously, and to cumulative adverse effect. Notice that in 2000 the world's population of 6.09 billion was releasing about23.2 billion tons of CO_2 through the combustion of fossil fuels—implying an average annual per capita emissions figure of about 3.8 metric tons. The two very poor regions like the sub-Saharan Africa and that of South-central Asia (which includes India, Pakistan, and Bangladesh) taken together, during the period 2000-50, are projected in the Table to account for about two-thirds of the growth in world population over this time period. And for the developed regions too demographic growth produces a 16% rise in total emissions (i.e. from 12.8 to 14.8 billion tons.) In this context it is worth-noting that projections made by the International Energy Agency (2005) suggest that between 2005 and 2030 energy-related CO_2 emissions may rise by 52%—implying an annual growth rate of 1.7%.

In the above backdrop, the present paper discusses in Section II the impacts of anthropogeneric actions on the environment. Aim of this section is to introduce different types of pollution effecting the environment and living organisms. Issues such as global warming, ozone depletion, acid rain and air, water and soil pollution are reviewed. If the eco-systems are damaged or polluted, we ourselves as human beings still being a single species living within a series of these eco-systems linked at a global level are affected lock, stock and barrel. This section also has focused on the mechanism of pollution and the key issues related to them. Section III is a review of different protocols that have been proposed by national regulators in a world-wide effort to preserve and protect the planet for present and future generations. The protocols like the Montreal Protocol 1987 and the Kyoto Protocol 1997, and the convention like the Vienna Convention, the Rio declaration on environment and development clearly show that governments have become more and more environmentally conscious and decided for an improvement in environment which is a must. The focuses for these two protocols are ozone depletion and global warming and reduction of specific pollutants. This section also gives some viewpoints of the opponents who rejected these treaties as a 'deeply flawed agreement'. Finally conclusions offer brief thoughts on the future of the environment and its sustainability.

"I will be the gladdest thing
Under the Sun!
I will touch a hundred flowers
And not pick one."
— From the poem 'Afternoon on a Hill' by Edna St. Vincent Millay (1892-1950).

I. INTRODUCTION

We all know that through the actions of the IMF, WTO and its precursor GATT, the World economy entered a period aptly described as the post-war economic boom. With the advent of this economic boom also came the explosion of productions; on its heels flourished the wave of consumerism, with little regard for the environmental consequences. To meet the demand of increasing production as well as consumerism, a clear picture of harms they have breathed is also noticeable. Noticeable is the concentration of Greenhouse Gas Emissions and other environmental hazards that have increased over the years owing to human activities. Scientific evidence suggests that a continued increase of that gas concentration and other types of pollutions are likely to have significant effects on the climate. The age of environmental activism was thus born. Environmental and social costs of the progress of consumerism that is accelerating on a global scale with the advent of sophisticated products and services, economic growth being the paramount driver in this context, are being questioned and challenged. It is unquestionable that modern economic growth has raised levels of atmospheric CO_2—leading to a rise in the Earth's surface temperature and the threat of climate change inviting disparities in energy use, CO_2 emissions, and living standards between rich and poor. Thus, the central dichotomy that every country is facing today is how to maintain economic growth and progress while protecting and enhancing the environment. The growing demands for conservation of nature and its sustainability have driven country after country to deliver solutions using rules, regulations and legislations.

The paper, however, is an attempt to provide fresh perspective on global warming and climate change. It essentially forwards some points. First, that especially due to momentum in economic and demographic processes, it is sure that there will be a major rise in atmospheric CO_2 during the 21st century. Second, that the coming rise in global temperatures will be faster than anything that human populations have experienced ever. Finally, the agricultural, political, economic, demographic, social and other consequences of future climate change could be very considerable. In a more and more populous world of eight to ten billion people, adverse development could well occur on several fronts simultaneously, and to cumulative adverse effect. Notice that in 2000 the world's population of 6.09 billion was releasing about 23.2 billion tons of CO_2 through the combustion of fossil fuels—implying an average annual per capita emissions figure of about 3.8 metric tons. The two very poor regions like the sub-Saharan Africa and that of South-Central Asia (which includes India, Pakistan, and Bangladesh) taken together, during the period 2000-50, are projected in the Table to account for about two-thirds of the growth in world population over this time period. And for the developed regions too demographic growth produces a 16% rise in total emissions (i.e. from 12.8 to 14.8 billion tons.) In this context it is worth noting that projections made by the International Energy Agency (2005) suggest that between 2005 and 2030 energy-related CO_2 emissions may rise by 52%—implying an annual growth rate of 1.7%.

In the above backdrop, the present paper discusses in Section II the impacts of anthropogeneric actions on the environment. Aim of this section is to introduce different types of pollution effecting the environment and living organisms. Issues such as global warming, ozone depletion, acid rain and air, water and soil pollution are reviewed. If the eco-systems are damaged or polluted, we ourselves as human beings still being a single species living within a series of these eco-systems linked at a global level are affected lock, stock and barrel. This section also has focused on the mechanism of pollution and the key issues related to them. Section III is a review of different protocols that have been proposed by national regulators in a world-wide effort to preserve and protect the planet for present and future generations. The protocols like the Montreal Protocol, 1987 and the Kyoto Protocol, 1997, and the convention like the Vienna Convention, the Rio declaration on environment and development clearly show that governments have become more and more environmentally conscious and decided for an improvement in environment which is a must. The focuses for these two protocols are ozone depletion and global warming and reduction of specific pollutants. This section also gives some viewpoints of the opponents who rejected these treaties as a 'deeply flawed agreement'. Finally, conclusions offer brief thoughts on the future of the environment and its sustainability.

II. THE MAIN CONCERN OF CLIMATE CHANGE

The main concern of climate change centres on the global

warming effect. The predicted warming is called greenhouse effect which is directly associated with human industrial activities, particularly the burning of fossil fuels such as oil, coal and natural gas. The earth's global average temperature during the last two centuries has risen by between 0.4-0.8°C. The primary cause is the release of large quantities of greenhouse gases in the atmosphere.

The main gas of concern is carbon dioxide (CO_2), a product of the burning of fossil fuels and non-renewable biomass. Carbon dioxide plays an important role within the atmosphere providing a means of transferring carbon within ecosystems, allowing photosynthesis to take and allowing human to breathe.

CO_2 is not alone in creating a warming effect; methane (CH_4), Nitrous Oxide (N_2O), Ozone and CFC gases all create, to a larger or smaller extent, the same result. Most of these gases exist naturally in the atmosphere and are responsible for making the planet habitable. Without greenhouse gases, the Earth would be 33°C colder and the diversity of life would not have developed. At the other end, large amount of these gases trap heat and increase the Earth's temperature.

The Greenhouse Effect

The greenhouse effect is a relatively simple and well understood natural phenomenon. It was originally described in 1827 by the physicist-mathematician Jean Baptiste Fourier. The effect takes its name from its resemblance to a garden greenhouse, where glass panels let in visible radiation and hinder the exit of long wave thermal radiation. This increases the indoor temperature. The Earth receives a relatively constant amount of energy in the form of 'incoming' radiation. Light consists of visible spectrum, ultra violet and infrared. Some of this energy is reflected directly back out to space by the atmosphere and the surface, but approximately 70 per cent is absorbed. This absorption raises the temperature of the Earth's surface and consequently that of the atmosphere. The same amount of energy is emitted back into space as thermal or 'outgoing' radiation. This balance between incoming and outgoing energy is necessary to maintain average temperature levels of approximately 15°C.

Concern about the greenhouse effect generally focuses on the increase in the concentration of greenhouse gases from anthropogenic process. When greenhouse gas concentration increases, more heat is captured causing an increase in Earth's average temperature and that will raise sea levels through melting of the polar ice and thermal expansion of the oceans thus having disastrous consequences for people living in low-lying countries or coastal regions.

Were the rate of temperature increase to be rapid, organisms would not be able to adapt quickly enough, jeopardizing the survival of numerous spices.

Approximately 80 per cent of the extra CO_2 originates from fuel combustion such oil, coal and gas. The remaining 20 per cent is emitted from 'slash and burn' deforestation methods, natural forest fires, volcanic activity and other changes in the tropics. About 55 per cent of the CO_2 is absorbed by the oceans, by northern hemisphere forest and more generally by plant growth. Overall, the concentration of CO_2 has increased since pre-industrial revolution times by 33 per cent (*Lomborg*, 2001).

Since 1751 approximately 305 billion tons of carbon have been released into the atmosphere from the consumption of fossil fuels and cement production. Half of these emissions have occurred since the mid 1970s. The 2003 global fossil-fuel CO_2 emission estimate, 7303 million metric tons of carbon, represents an all-time high and a 4.5 per cent increase from 2002.

Greenhouse gases affect global temperatures both directly and indirectly. Direct effects happen because the gas itself is a greenhouse gas and indirectly because the climate acts as a natural chemistry set. The atmospheric transformation processes that break the gases into less harmful compounds can produce other greenhouse gases.

Although different scenarios have been published regarding global warming and temperature rises, it is difficult to make accurate predictions. Some scenarios are more dramatic; others are more optimistic. These perspectives stem from differences with respect to predicted CO_2 levels. The Inter-governmental Panel on Climate Change (IPCC) prediction of an increase of 1.5-4.5°C in global temperature has remained constant since 1970. No predictions over the last 25 years have suggested that conditions will improve, rather they confirm an evermore pessimistic view of future temperature change.

The Ozone Hole

The catalytic destruction of ozone from depleting compounds (ODCs) in the stratosphere originally was theorized and proposed in 1974 by Mario Molina and Sherwood Rowland. The ozone hole above Antarctica was first confirmed in 1982 by the British Antarctic Survey Team and was brought to light in the British Science Journal '*Nature*' in 1985. Since then, research on the principles of ozone depletion have made significant advances and helped to create a global environmental awareness and consensus for action. Within two years of its publication in '*Nature*', the Montreal Protocol in 1987 took steps to ban ODCs such as CFCs.

Significance of Ozone

Ozone, derived from the Greek 'azein' ('to smell'), is a natural gas in the earth's atmosphere. It is considered to be a trace gas, naturally existing in extremely small amounts. Ozone is a 'Janus faced' molecule. In the lower atmosphere it is a toxic oxidizing agent that causes lung irritation. In the upper atmosphere it is transformed into a guardian of life on the earth. Approximately 90 per cent of ozone lies within the stratosphere. Maximum ozone concentration occurs in the middle of the stratosphere in the so-called ozone layer.

Sunlight consists of radiation of various wavelengths that can be harmful to humans, such as ultraviolet. One of the impacts of the increased Ultraviolet-B radiation is an increased susceptibility of the immune system, reducing natural defences against infectious and fungal diseases (*Martens*, 1998). The World Health Organisation claims a 10% decrease in stratospheric ozone would cause an additional 300,000 non-melanoma and 4,500 melanoma skin cancers and between 1.6 and 1.75 million more cases of cataract each year. This is nothing but the ozone hole effect on health.

Homogeneous Ozone Depletion

Human activities are causing an imbalance between natural production and destruction of stratospheric ozone. Nitrous oxide from increased fertilizer usage results in an estimated 0.28 per cent increase in NO_2 concentrations every year. The production of foam and refrigerant coolant relied heavily on compounds that containing the ozone depleting chemicals chlorine (Cl_2) and bromine (Br_2). When emitted into the atmosphere, these compounds migrate from the troposphere through the tropopause and accumulate in the stratosphere. Here they are broken down by ultraviolet radiation.

It takes many years for the ODCs to reach the stratosphere and undergo destruction. Despite the abandonment of these materials by the older industrialized nations, this long life cycle allows the release of chlorine to continue for many decades.

Another mechanism for ozone depletion involves hydrogen chloride (HCl). HCl is formed by reactions with methane, a gas created from varied sources such as natural gas emissions, cattle, coal mining and oil extraction and refining. Besides there are Heterogeneous Ozone Depletions phenomena likely to occur in the polar regions and over middle latitudes. The theories suggest that the total concentration of all chlorine containing compounds can deplete ozone considerably more under certain conditions like those that prevail in the Arctic during winter and summer. These conditions can make HCl and $ClONO_2$

become very active Cl and ClO as they react to aerosol surfaces and therefore increase ozone depletion. It is to be noted that the scale of the ozone hole and its proximity to Australia, South Africa, Chile and Argentina. Weather systems drive the boundary gaseous areas out and mix low ozone and high ozone concentrations. This makes it very important for humans to protect against high ultraviolet (UV) radiation.

Effect of Ozone Depletion

Depletion of the ozone allows greater amounts of radiation to reach the Earth's surface. But not all radiation has the same effect on living organisms. Partial exposures to UV radiation is necessary for the formation of Vitamin D in humans. There are three classes of UV radiation: UV-A (wavelength above 320 nanometres), UV-B (290-320 nm) and UV-C (40-290 nm). The latter two can be extremely harmful and cause severe biological injury.

UV-B radiation

UV-B radiation is responsible for DNA damage. The main effects on human are skin cancer, cataracts and immuno-suppression. Cataracts and cataract-related blindness is related to long-term and cumulative exposure to UV-B. Acute exposure can cause 'snow-blindness.

Air Pollution

Of all the different types of pollution, air contaminants are the most significant. However, since the Industrial Revolution, air pollution has increased dramatically.

The principal air pollutants are:

- Particles (smoke and soot from industrial plants, diesel engines and volcanic ash);
- Sulphur dioxide (SO_2) that creates acid rain;
- Ozone (O_3);
- Lead;
- Nitrogen oxides (NO and NO_2, NO_3);
- Carbon monoxide (CO)

The main sources of these pollutants are transportation (cars, trucks and motorcycles), aviation and industry. This type of pollution can be seen in all major modern cities and has been cited as a key factor in the growing incidence of pulmonary disorders, particularly amongst children.

Climatologists are stressing that the roots of both global dimming causing pollutants and global warming causing greenhouse gases have to be dealt with together and soon.

Acid Rain

Acid rain occurs when sulphur dioxide (SO_2) and nitrogen oxides (NO_x) react with water, oxygen and other chemicals in the atmosphere to form acidic compounds such as sulphuric or nitric acid. The main source of SO_2 and NO_x is fossil fuel combustion.The phenomenon of acid rain becomes noticeable in different parts of the globe.

River Pollution

Apart from soil pollution (created by the metallurgical industries in the 20th Century, happened in Europe, Eastern North America, Ex-Soviet Union States, Japan and Love Canal areas located in Niagara Falls region of New York State, the last being the most widely reported case of the chemical pollution), water pollution, oil pollution in the Seas and Oceans, pollution in seas and oceans itself, pollution in rivers is of major concern because they form the major source of water for drinking, agriculture irrigation, and everyday use for human activities. In order for water to be suitable for use there are some certain criteria. Water must not contain E-coli bacteria or any other water-borne enzymes and viruses. There is some correlation between the wealth of a country and how polluted its rivers are. This is due to the fact that poorer countries have not invested in the technologies required to treat waste and hence dump them untreated directly into rivers.

Chemicals are another issue concerning river pollution. Fertilizers used in agriculture illustrate these concerns. A principle constituent of these products is Nitrogen which if allowed to accumulate in drinking water leads to health problems such as stomach cancer.

III. GLOBAL ENVIRONMENTAL INITIATIVES

The last two decades, in fact, have been marked by the endeavours of the United Nations and European Union to produce legislation that would help the conservation of the environment at a global scale. Although the first step towards a sustainable environment and future were done back in the 1970s, the efforts were only isolated in countries holding International meetings and mainly discussing environmental stewardship. The decade of 1980s was marked with a serious approach to look into environmental issues. Events such as the discovery of the ozone zone hole over Antarctica, made people realize that if certain measures were not introduced, the Earth could become a much more hostile home. The United Nations Environment Programme (UNEP) and the United Nations Framework Convention on Climate Change (UNFCC) produced a series of protocols, such as the Montreal and Kyoto Protocols, which are considered to be the most significant. These can be seen as the most systematic efforts of world governments to create a sustainable future. Parallel to this endeavour the EU has created its own set of directives which attempt to minimize the anthropogenic impact on the environment and promote sustainability.

Viennna Convention

In 1985, the UNEP Governing Council set up a working group with the objective to create a Protocol for the Protection of Ozone Layer. The Protocol was signed in Vienna by 20 nations.

The Vienna Convention provided the foundations for the Montreal Protocol in 1987. For the first time nations agreed in principle to tackle a global environmental problem before its effects were felt or even scientifically proven (UNFCC, 2002). Later on, in the journal '*Nature*". Dr. Joe Farman, a British scientist, wrote about severe ozone depletion in the Antarctic and his findings would be confirmed by American satellite observations. This resulted in governments taking dramatic, transnational, measures in the form of the Montreal Protocol—a Protocol for reductions of ozone depleting substances agreed in Montreal in 1987, subsequently revised in the London, 1990, Copenhagen, 1992, Montreal, 1997 and Beijing, 1999 amendments, adding more chemicals to the list of controlled substances. Following the amendments, the developing countries were allowed to produce up to 10 per cent of their calculated level of emissions for domestic and essential use.

Montreal Protocol

Ozone depletion was no longer a theory but a real phenomenon that could not be dismissed and which posed a significant threat to the environment. Within a year of the ozone hole's discovery, the Vienna Convention took place, taking the first steps towards relating ozone depletion to human health.

Within the next two years, governments from all over the globe were called to take drastic measures to control and minimize the effect of ozone depletion layer resulting from human activities. On 1 January 1989 the protocol came into force. All the parties agreed to freeze production and consumption of CFCs and halons.

Rio Declaration on Environment and Development

The United Nations Conference on Environment and Development also known as the 'Earth Summit' took place in Rio de Janeiro between 3 and 14 June 1992. This was the largest-ever meeting of world's leaders who came from 179 countries to discuss and take action on the environment. The Summit essentially reaffirmed the UNCED

TABLE 1

Estimates of Regional and Globalization and Global Emissions of CO_2 Produced by the Combustion of Fossil Fuels for Around the year 2000 with Projected Calculation for 2050

Region	*Population (millions) 2000*	*Per capita CO_2 emissions (metric tons) 2000*	*Total CO_2 emissions (million metric tons) 2000*	*Projected Population (millions) 2050*	*Totals CO_2 emissions (million metric tons) 2050*
(1)	*(2)*	*(3)*	*(4)*	*(5)*	*(6)*
Developing Regions					
Sub-Saharan Africa	670	0.9	613.8	1692	1,550.1
North Africa/West Asia	335	4.3	1,430.8	628	2,682.2
Eastern Asia	1479	3.4	5,044.6	1587	5,412.9
South-central Asia	1485	0.9	1,368.2	2495	2,298.8
South-eastern Asia	519	1.3	696.1	752	1,008.6
Central America and Caribbean	174	2.8	481.2	256	707.9
South America	349	2.2	771.9	527	1,165.6
Sub-Total	5011	2.1	10,406.6	7937	14,826.2
Developing Regions					
Europe	729	8.4	6,106.2	653	5,469.6
North America	315	20.0	6,294.5	438	8,752.3
Oceania	31	11.8	365.0	48	5,65.1
Sub-Total	1075	11.9	12,765.7	1139	14,787.0
World	6086	3.8	23,172.2	9076	29,613.2

Sources : World Resources Institute (2003: 258-59); United Nations (2005).

Declaration adopted at Stockholm on 16 June 1972 and sought to build on it (UNEP, http://www.unep.org/). Concern had been growing over the previous two decades about the huge discrepancy between the North and the South. This inequality was seen as symptomatic of a degrading global relationship between environmental stakeholders which could not be sustained for the long-term without radical realignment. The challenge in this concept is how to make more people more aware of environmental issues and at the same time move towards sustainable forms of development and lifestyles.

The goal of the Rio Declaration was to 'establish a new equitable global partnership through the creation of new levels of cooperation among States, key sectors of societies and people', in order to 'work towards international agreements which respect the interests of all and protect the integrity of the global environmental and developmental system, recognizing the integral and interdependent nature of the Earth our home' (UNEP, 1992).

The Declaration consisted several principles of which one important one was that all countries would have the right to exploit their resources without causing damage to the environment outside their borders and they should create international laws that should compensate those countries whose environment had been damaged because of their negligence.

Another was that, in order to achieve sustainable development, environmental protection should be an integral part of the development process. This development process should be aiming for the eradication of poverty. Therefore, public awareness of the environment was to be encouraged by government and people should put pressure on societies to pursue sustainable development. Hence, nations should create national laws that would protect the environment from pollution and the polluter should bear the cost of pollution. Cooperation between nations should be part and parcell to create an open international economic system to support economic growth and sustainable development in all countries. These environmental policies should not create barriers to international trade and countries should exchange technological knowledge to achieve sustainability. Finally, respecting international laws by all nations, development and environmental protection should be independent.

Kyoto Protocol

On 11 December 1997, with an important meeting of the Climate Convention looming in Kyoto, Japan, with targets and timetables for reducing emissions of greenhouse gases, most notably CO_2 from fossil-fuel combustion, on its agenda, stalling move were being set in motion by a resolution in the US Congress. In short, building on the UN Framework of the Convention Climate Change the Kyoto Protocol 'broke new ground with its legally-binding constraints on greenhouse gas emissions and its innovative "mechanisms" aimed at cutting the cost of curbing emissions.' The Protocol has set out a list of policies.

These policies are:

- enhancing energy efficiency;
- protecting and enhancing greenhouse gas sinks;
- promote sustainable agriculture;
- promoting renewable energy, carbon sequestration and other environmentally-friendly technologies;
- removing subsidies and other market imperfections for environmentally-damaging activities;
- encouraging reforms in relevant sectors to promote emission reductions;
- tackling transport sector emissions;
- controlling methane emissions through recovery and use in waste management (UNFCC, 2002).

However, the Treaty is limited in both in its effect on global warming and the coverage it gives to carbon emission. The Treaty includes not the international transport. Besides, countries like China and India claim that climate change was just a scare, a plot of the wealthy countries to keep them poor. On the other hand they made it clear that they themselves would accept no cuts unless similar cuts were imposed on developing nations as well.

When the US and Canada insisted upon Majority World countries pledging reductions, Zhong Shukong, the Chinese delegation leader—replied: "In the developed world only two people ride in a car and you want us to give up riding the bus!". He was right: at that time Los Angeles alone had more cars than the whole of China.

In addition to the above criticisms, antagonists of the Kyoto Protocol conclude that the Kyoto Protocol is an impractical policy focused on achieving an unrealistic and inappropriate goal. An incomplete list of key arguments against the Kyoto structure includes the following:

(i) The Kyoto Protocol is defective on both efficiency criteria because it omits a substantial fraction of emissions and has no plans beyond the first period;

(ii) The most fundamental defect of the Kyoto Protocol is that the policy lacks any connection to ultimate economic or environmental policy objectives;

(iii) International permit runs the risk of being highly inefficient, given uncertainties in the marginal cost of abating greenhouse gas emissions ... it would probably generate large transfers of wealth between countries; and

(iv) The Kyoto Protocol can only work if it includes an elaborate and expensive international mechanism for monitoring and enforcement.

Despite its ambition, the Treaty can only be seen as a first step in the confrontation of global warming. The Protocol does not consider emissions of oxides of nitrogen and water vapour which, at high altitude, can cause more damage to the atmosphere than carbon dioxide (*Royal Commission on Environmental Pollution*, 2002). Despite these limitations the Kyoto protocol finally came into operation in 16 February 2005.

IV. CONCLUSION

In what follows is that the environmental future of the Earth is bleak if we do not properly manage it. As time progresses, the negative impact of environment will come to dominate all regions, both rich and the poor. The irony is that the earliest and the heaviest burdens will fall on the poorer countries who are, in no way, held responsible for the problems cited above. The challenge of climate change, if not abated and mitigated, will magnify the distress of humanity with deprivation, drought, flood, sickness and other health hazards. Only by coming onto a common pedestal of cosmopolitan understanding that our beloved Earth is in danger through rapid environmental changes and also by understanding the network of invisible forces and directing them can we move forward towards true environmental management, where markets flooded with the wave of consumerism and the nations upon which the nature is 'red with tooth and claws', can be balanced.

REFERENCES

Chatterjee, Biswajit (2011) : *Climate Change, Trade & Natural Disasters,* Deep & Deep Publications, New Delhi

Chatterjee, Biswajit and Asim Karmakar (2003) : "Environment and Development in Open Economy" in G.S. Monga (ed.), *Environment and Development,* Deep & Deep Publications, New Delhi

Godrej, Dynar (2006) : *The No-Nonsence Guide to Climate Change,* New Internationalist Publications Ltd., Oxford.

International Energy Agency (2005) : *World Energy Outlook, 2005,* Paris: International Energy Agency.

Lomborg, B. (2001) : *The Sceptical Environmentalist,* Cambridge, Cambridge University Press.

Martens, Pim (1998) : *Health and Climate Change*, Earthscan.

Newlands, David J. and Mark J. Hooper (2009) : *The Global Business Handbook: The eight Dimensions of International Management*, Gower Publishing Company, USA.

Royal Commission on Environmental Pollution (2002) : *The Environmental Effects of Civil Aircraft in Flight*, London.

Chapter 30

India's Vulnerability to Climate Change

Dhulasi Birundha Varadarajan and K. Sadasivam

INTRODUCTION

Recognition of climate change as a significant global environmental challenge has its origin in the adoption of the United Nations Framework convention on Climate Change (UNFCCC) in 1992. This paper highlights the extent of India's vulnerability to climate change. Government of India, 2003, the growth in commercial energy-use is closely related to the growth of industry, transport and urban areas. India accounted for 3.5 per cent of world commercial energy demand in 2001 reaching 314.7 million tons of oil equivalent. A large part of India's population does not have access to commercial energy. India's per capita energy consumption of 482 kgoe is 29 percent of the world average of 1,671 kgoe and compare poorly with some other developing countries. India's energy use efficiency in 1999 (WDI, 2002), which is lower than some industrialized nations but more than double of energy efficiency has been achieved since 1980. The energy elasticity (energy growth rate/economic growth rate) was more than unity for the period 1953-2001. However, elasticity less than unity for the period 1991-200 due to improvement in the efficiency of energy-use and structural change in the economy (GOI, 2003).

TRENDS AND PROJECTIONS FOR GLOBAL CLIMATE CHANGE

The earth warmed significantly over the last 100 years. The global average temperature has risen about 0.7 or about 1.3. Global temperature since 2000 have been particularly warm-sic of the seven warmest years on record have occurred since 2000. There is also evidence that the rate of warming currently about 0.13C per decade is increasing. The IPCC estimates the during 21st century global average temperature will rise between 1.1°(2.°F) and 6.4°(11.°F) with the range more likely to be between 1.8°(3.°F) 4.°C (7.°F).

The magnitude of actual warming and other effects will depend upon the level at which atmospheric concentrations of CO_2 and other green house gases and ultimately stabilized. The current atmospheric CO_2 concentration is around 380 ppm. When we consider the contribution of other green house gases, the over all effect is equivalent to a concentration of 430 ppm of CO_2 refereed to as CO_2e.

Stabilizing greenhouse gas concentrations at 450 ppm CO_2e would be 90 percent likely to eventually result in a temperature increase between 1°0 and 3.8°C with current greenhouse gas concentrations in the atmosphere at 430 ppm CO_2e referred to stabilizing greenhouse gas concentration at 450 ppm CO_2e would be 90 percent likely to eventually result in a temperature increase between 1° and 3.8° with current concentration in atmosphere at 450 ppm would be extremely challenging (Johan than M. Harris and Brain Roach).

INDIA'S VULNERABILITY OF CLIMATE CHANGE

India has reasons to be concerned about climate change. A vast population depends on climate sensitive sectors like agriculture, forestry and fishery for livelihood in the country. The adverse impact of climate change in the form of declaiming rainfall and rising temperature and thus the increases severity and livelihood in the economy. To add to this, poor infrastructure facilities, weak institutional mechanisms, lack of financial resources and vast sectoral and regional variability adversely affect the adaptive capacity of the country to climate change. Climate change would represent additional stress on the ecological and social-economic systems that are already facing tremendous pressure due to rapid industrialization, urbanization and economic development.

REASONS WHY INDIA SHOULD BE CONCERNED ABOUT CLIMATE CHANGE

India is a large developing country with nearly 700 millions of rural population directly depending on climate sensitive sectors and natural resources such as water, biodiversity, mangroves coastal zones, grasslands for their subsistence and livelihood. Further, the adaptive capacity of dry land farmers, forest, dwellers, and nomadic shepherds is very low (Ravindranath, 2000). Climate change is likely to impact all the natural eco-systems as well as socio-economic systems as shown by the national communication report of India to the UNFCCC (MOIB, 2004). The latest high resolution climate change scenarios and projections for India, based on regional climate modeling (RCM) system, known as PRECIS developed by Hadley center and applied for India using IPCC scenarios A_2 and B_2 shows the following:

- an annual mean surface temperature rise by the end of century, ranging from 3 to 5°C under A_2 scenario and 2.5 to 4.5°C under B_2 scenario with warming more pronounced in the northern parts of India.
- A 20% rise in all India summer monsoon rainfall and further rise in rainfall is projected over all states except Punjab, Rajasthan and Tamil Nadu, which show a light decrease.
- Extremes in maximum temperature are also expected to increase and similarly extreme precipitation also shows substantial increase particularly over the west coast of India and west central India.

IMPACTS

Temperature, Precipitation and Occurrence of Extreme Events

Several severe examples of the consequences of climate change are available from around the globe. Nine of the hottest years recorded in more than a century have occurred since 1988. Worldwide, July 1988 was the hottest month ever. In 1998, India experienced its worst hot spell in 50 years, which took a toll of over 3,000 lives. The tropical cyclone or Orissa in 1999 took a toll of about 10,000 lives. Himalayas and glaciers are retreating at the rate of 18 m per year in Gangotri.

Under future scenarios of increased GHG concentrations, a marked increase in rainfall and temperature is projected into the 21st century. India's climate could become warmer by 2.33 to 4.78°C under condition of doubling of CO_2 concentration (Logern, 1998). An increase in annual temperature of 0.7 to 1.0°C by 2040 is predicted with respect to the 1980s (Lal *et al.*, 1995). There is an overall decline in the number of rainy days over a major part of the country. This decline is more in the western and central parts (by more than 15 days while near for foothills of Himalayas) and in north-east India, the number of rainy days may increase by 5 to 10 days (Kumar *et al.*, 2003).

The impact of climate variability and change, climate policy responses, and associated socio-economic development will affect the opportunities for and success of climate policies. In particular, the socio-economic and technological characteristics of different development paths will strongly affect missions, the rate and magnitude of climate change, climate change impacts, the capability of adapt and the capacity to mitigate.

WATER RESOURCE

India's rich water resources are unevenly distributed and result in spatial and temperature portal shortage the demand for water has increased tremendously over the year due to an increasing population, expending agriculture, and rapid industrialisation which are responsible for considerable imbalance in the quantity and quality of water resources. According to the Ministry of water resources the amount of water available per person in India is decreasing steadily from 3450 cm in 1951 to 1,250 cm in 1999 and is expected to decline future to 760 per person in 2050.

TABLE 1

Future GHG Emission Projections under BAU Scenario, CO_2 Equivalent Terms

CO_2 equivalent Emission	*2000*	*2010*	*2020*	*2030*	*CAGR**
Carbon Dioxide (CO_2)	956	1507	2080	2572	3.35
Methane (CH_4)	391	422	462	529	1.01
Nitrous Oxide (N_2O)	95	156	214	250	3.26
Perflurocarbns (HFCs)	7	10	15	24	4.13
Sulfur Hexafluoride (SF_6)	4	15	56	110	11.96
Total CO_2 equivalent	0.3	4.4	12	21	15.12
	14.54	2115	2839	3507	2.98

Note : *Compounded Annual Growth Rate During 2000-30.
Source : Shukla *et al.*, 2003, p. 153.

Lower rainfall and more evaporation would have the dire consequence of less runoff, substantially changing the availability of freshwater in the watersheds, decline of soil moisture and increasing aridity-level of hydrological zones. By the year 2050, the average annual runoff in the river Brahmapurtra will decline by 14 percent (TERI, 2004). The hydrological cycle is likely to be altered and the severity of droughts and intensity of floods in various parts of India is likely to increase. Further, a general reduction in the quantity of available runoff is predicted.

Agriculture

High climate sensitive Indian agriculture 65 percent of which is in rain-fed areas, contributes nearly 25 percent of GDP, employs 65 percent of the total work force and accounts for 13.3 percent of total exports together with allied activities (GOI, 2002). Several studies predict that despite substantial increase in national foodgrain production the productivity of some important crops such as rice and wheat could decline considerable with climate change (Achanta, 1993). Due to a 2 to 3.5°C rice in temperature accompanied by a 7 percent to 25 percent change in precipitation, farmers may be losing net revenue between 9 percent and 25 percent which may adversely affect GDP by 1.8 percent to 3.4 percent (Kumar and Parikh, 1998; Sanghi *et al.* 1997). Brazil, India and many sub-Saharan African countries could lose out to climate change: winners could include Russia, China, Canada and Argentina: there will be serious consequences for food security in the south India stands to lose a massive 125 mt equivalent to some 18 percent of it rain-fed cereal production (Fisher *et al.*, 2001). The major impact will be on rain-fed crops.

In India, the estimated total requirement for foodgrains would be more then 250 mt by 2010. The gross arable area is expected to increase from 191 to 215 million ha. by 2010, which would require an increase of cropping intensity to approximately 150 percent (Sinha *et al.* 1998). Because land is fixed resources agriculture the need for more food in India can be met only through higher yield per unit of land, water, energy and time such as through precision farming. A 1.5 degree centigrade rise and tow mm increase precipitation could result in a decline in rice yields by three to 15 percent.

Another study (Kumar and Parikh, 1998) indicates that the economic impacts would be significant even after accounting for farm level adaptation. The net loss in net revenue at the farm level is estimated to range between nine percent and 25 percent for a temperature rise of two to 3.5°C. Furthermore a 2°C rise in mean temperature and a seven percent increase in mean precipitation would reduce net revenues by 12.3 percent for the country as a whole.

Crop productivity will fall especially in non-crepitated land as temperatures rise for all of South Asia by as much as 1.2 degrees C on average by 2040 and even greater crop loss of over 25 percent as temperatures rise up to 5.4 degree C by the end of the century. This means an even lower caloric intake of India's vast rural population, already pushed to the limit with the possibility of starvation in many rural areas of dependent on rainfall for their crops.

Forestry

Forest in India plays a crucial socio-economic role contributing 2.37 percent to GDP (GOI, 2003). Nearly 55 million people in India depend upon non-timber forest products a critical component for their sentence changes in forestry could have profound implication for traditional livelihoods, industry biodiversity, soil and water resources and hence, agricultural productivity. Moreover, climate change induced effects will aggravate the existing stress due to non-climate factors, such as land-use change is expected to affect the boundaries of forest types and areas, primary productivity, species populations and migration the occurrences of pests and diseases and forests regeneration. The increase in GHGs also affects species composition and the structure of ecosystems that can thrive and the amount of plant tissues that can be sustained (Melillo *et al.*, 1996).

In a case of Kerala, the precipitation effectiveness index was linked to the net primary productivity of teak plantation results indicate that soil moisture is likely to decline and in turn reduce teak productivity from 5.40 cubic meters (m^3)/ha. to 5.07 m^3 ha.

Also, the productivity of moist deciduous forests could decline from 1.8 m^3/ha. (Achanta and Kanitkar, 1996s). According to the IPCC reports, even with a modest global warming of 1-2° most forest ecosystems will be impacted through changes in forest species composition, bio-diversity and plant productivity.

In India, nearly 2,00,000 villages are located inside or on the fringes of forest. According to some estimates, nearly 2000 million people depend on forests for their livelihood; with more than 30 million people depend on forest for their livelihood, with more than 30 million people directly involved in gathering and trading non-timber forest products such as fruits seeds, flowers, leaves, honey and gum. Thus, any impact on forest vegetation and bio-diversity will have adverse implications for the livelihoods of forest dependent communities currently the forest area, vegetation and bio-diversity are being subjected to degradation and loss due to human and livestock pressers.

RENEWABLE SOURCES OF ENERGY

India is endowed with abundant natural and renewable resources of energy and is more of the major countries harnessing renewable resources. India is second only to China in installing biogas plants and improved wood stoves, the use of which results in the saving of over 16 million tons of fuel wood every year. Rural distribution of LPG for cooking has been extended to 7 million customers. The world's largest solar steam cooking system for 15,000 people has been installed in Tirumala, Andhra Pradesh and is working well. Small hydro projects of 538 MW capacities are under consideration in 15 potential states (GOI, 2004). India has been able to achieve significant capacity addition of 1.367 MW through wind farms and ranks fifth in the world after

Germany, US, Spain, and Denmark. The country also has significant potential of ocean thermal, sea wave and tidal power.

Sea Level

Coastal belts are more prone to the devasting impacts of global warming. Assessments based on limited evidence show that one metre sea-level rise can lead to welfare loss of $ 1259 million in India equipment to 0.36 percent of GNP. Short-term data analysis show that, by 2050, cumulative mean sea level elevation in the Bay of Bengal near the Indian sundarbans may be closer to one metre with an anticipated loss of 15 percent land area by 2020. Over one decade, mangrove area in Sundarbans has declined from 420 hectares (1987) to 212 hectares (1997).

Himalayas Glaciers

Anil Kularkani of the space application center in Ahamedabad along with other scientists examined the data for 466 glaciers in the Himalayas and found that their surface has shrunk by about 21 percent since 1961. As the glaciers retreated they become more fragmented. The smaller glaciers were more sensitive to global warming, said the scientist.

Policy responses to Climate include

Two types of measures can be used to address climate change: preventive measures tend to lower or mitigate the green home effect, and adaptive measures deal with the consequence of the greenhouse effect and trying to minimize their impact.

Preventive measures include

- Reducing emissions of greenhouse gases, either by reducing the level of emissions-related economic activities or by shifting to more energy efficient technologies that would allow the same level of economic activity at lower level of co-emissions.
- Enhancing carbon sinks forests recycle CO_2 into oxygen, preserving forested areas and expanding reforestation have a significant effect on net CO_2 emissions.

Adaptive measures include

- Construction of dikes and sea walls to protection against rising sea-level and extreme weather events such as floods and hurricanes.
- Shifting cultivation patterns in agriculture to adapt to changed weather conditions in different areas and relocating people away from low lying coastal areas.
- Creating institutions that can mobilize the needed human,

material and financial resources to respond to climate-related disasters.

POLICY TOOLS

Carbon Taxes

The release of greenhouse gases in the atmosphere is a clear example of native externality that imposes significant costs on a global scale. A standard economic remedy for internalizing external costs is a per unit tax on the pollutant. In this case what is called for is a carbon taxes, levied exclusively on carbon-based fossil fuels in proportion to the amount of carbon associated with their production and use.

Tradable Permits

An alternative to a carbon tax is a system of tradable carbon permits. A carbon trading scheme could be national in scope, or include several countries. Under an international permit system each nation world be allocated a certain permissible level of carbon emissions. Permits are allocated to individual carbon emitting sources in each nation. Firms are able to trade permits freely among themselves and nations and firms could also receive credit for financing carbon reduction efforts in other countries. From an economic point of view, the advantage of a tradable permit system is that it world encourage the least cost carbon reduction options to be implemented.

There are other policy tools

- Shifting subsidies from carbon-based to non-carbon-based fuels
- The use of efficiency standards to require utilities and major manufactures to increase efficiency and renewable content in power sources.
- Research and development promoting the commercialization of alternative technology.
- Technology transfer to developing nations

CONCLUSION

- Strong political will commitment and research and development alone can generate energy that is free from emission and safe to the environment. At production-level appropriate and environment-friendly sources of energy have to be found and tapped. It is disturbing to not that at distribution-level the transmission loss is estimated at as high as of 40 percent in India. Replacing the old transmission instruments, repairing the appropriate ones and maintaining the distribution system also need to be looked into for saving and utilizing the energy on sustainable manner. Besides the civil society corporate sector and the state have a vibrant role to identify environment-friendly sources of energy to meet the increasing demand. The community as a larger role to play in judiciously using the energy thereby reducing the wastage in both domestic and commercial consumption.

- Sustainable development needs careful planning and implementation of energy production and utilization. It has already created some social tension when larger projects resulted in evaluation of social cost and benefit and suitable assessment of environmental implication on the local communities in the project areas. So far there is no clear evaluation method that has been evolved due to the involvement of complicated and sensitive issues. This needs thorough understanding and holistic approach. Notwithstanding the fact, there is a growing demand for energy, the community's interest should not be compromised for a long-term sustainable development.
- India as an emerging economic force should forcefully argue its case in the multilateral conventions. It has a genuine right to force the developed world to pay for the damage already done to the society. Possible extent it can switch over to carbon free technologies to generate energy for its growing requirements. Since a sizable Indian population lives on agriculture sector, any adverse climate conditions will affect the very survival of the people. In a nutshell it is desired to minimize the environmentally hazardous methods for generating energy, reduce the economic cost (both for production and distribution) develop feasible technology and increase the efficiency level in the production process of the economic system. As it has been discussed already the researchers, planners, administrators and the civil society have to consciously work to generate energy from carbon free technology. India has an advantage with the availability of environment-friendly renewable resources. Proper planning and implantation alone are required to do this.
- Global climate change has emerged as a threat to sustainability. In the absence of adaptation and mitigation strategies climate change can seriously damage agriculture, water resources, forests, coastal areas and health, etc., in the Indian economy. The impact of vulnerability is not only

decided by the extent of climate change but also by the robustness of the developmental process in the economy. The quality of development would provide an insurance against the impact of climate change and increase adaptive commitment in the next commitment period of the Kyoto Protocol and it has therefore initiated domestic efforts for enhancing adaptive capacity and mitigation HGH emissions. However, there is a long way to go to achieve the desired results.

India should be alert, active and assertive in its global participation and ask for risk minimization in the south, over the much-hyped cost minimization of GHG mitigation in the north. Responsibility should not be shirked in the name of affecting the economy or the provision of equal per capita rights.

References

Achanta, A.N. (1993): 'An Assessment of Potential Impact of Global Warming on Indian Rice Production' in A.N. Achanta (ed.). *The Climate Change Agenda: an Indian Perspective,* Tata Energy Research Institute, New Delhi.

Achanta, A. and R. Kanetkar (1996): *Impact of Climate Change on Forest Productivity: A Case Study of Kerala*, India, Paper presented at the Asian and Pacific workshop *On Climate Change, Vulnerability and Adaptation Assessment,* Manila, Philippines, 15-19 January.

Cline, William R. (1992): The Economics of Global Warming, Washington DC, Institute for International Economics.

Fisher, Gunther, Mahendra Sha, Harrij van Velthuizen and Freddy O Nachtergaele (2001): Global Agro-Ecological Assessment for Agriculture in the 21st century, International Institute of Applied Systems Analysis: Austria, pp. 27-31, Cited in Down to Earth, August 15, 2001, pp. 42-43.

Frant Hauser, Samuel (1995): Valuing Climate Change: The Economic of the Greenhouse London: Earthscans Publications.

Gupta, Vitaya (2005): Climate Change and Domestic Mitigation Effects, *Economics and Political Weekly*, Vol. XL, No. 10, pp. 981-87.

GoI (2002): Economic Survey 2001-02, Government of India, New Delhi.

———, (2003): Tenth Five Year Plan 2002-07, Planning Commissions, Government of India, India, New Delhi.

———, (2004): India, 2004 *A Reference Manual,* Publication Division, Ministry of Information and Broadcasting, Government of India, New Delhi.

India's Initial National Communications to the United Nations Framework Convention on Climate Change, Ministry of Environment and Forest, New Delhi (2004).

Inter-governmental Panel on Climate Change (2001): IPCC, Climate Change, 2001, The Scientific Basis, Cambridge, UK Cambridge Press, 2001.

Inter-governmental Panel on Climate Change (2007): IPCC, Climate Change, 2007, The Physical Science Basis Cambridge.

Janarthanan, M. Harris and Brain Roach, (2007): The Economics of Climate Change, Global Development and Environment Institute, Tuffs University MA 02155.

Jayant, Sathaye P.R. Shukla and N.H. Ravindranath (2006): "Climate Change Sustainable Development and India: Global and National Concerns". *Current Science,* Vol. 90, No. 3.

Kumar, K.S. and J. Parikh (1998) 'Climate Change impacts in India Agriculture: The Ricardian Approach' in a Dinar, *et al.* (Eds.), Measuring the Impact of Climate Change in Indian agriculture, World Bank, Washington DC.

Lal, M.U., Cubasch, R. Voss and J. Waszkewitz (1995): 'Effects on Transient Increase in Greenhouse Gases and Sulphate Aerosols on Monsoon Climate; Current Science, 69, pp. 752-63.

Logern, S. (1998): Climate warming and India in Dinar *et al.* (Eds.) Measuring the impact of climate change on Indian Agriculture, World Bank Washington.

Melillo, J.M.I.C., Printice, G.D. Farquhar, E.D. Schuze, and O.E. Sala (1996): Terrestrial Biotic responses to environmental change and feedbacks to climate in J.T. Houghton, L.G. Meria Fiho, B.A. Calendar, N. Harris, A. Kattenberg and K. Maskell (eds.), Climate Change 1995: The Science of Climate Change, Contribution of Working Group to Second Assessment Report of Intergovernmental Panel on Climate Change, Cambridge University Press, Cambridge, U.K. and New York, US, pp. 444-516.

Ravindranath, N.H. and Sathaye, J. (2002): Climate Change and Developing Countries, Kulwer Academic Publishers, Dordrecht, Netherlands.

Sukla, P.R. Amit Garg and Manmohan Kapshe (2003): Greenhouse Gas Emission Scenario in Shukla *et al., op. cit.,* pp. 128-58.

Sinha, S.K.M. Rai and G.B. Singh (1998): Decline in Productivity in Punjab and Haryana: A myth or reality? Indian Council of Agricultural Research (ICAR) Publication, New Delhi, India, p. 89.

TERI (1996): The Economic Impact of a one meter sea level rise on the India Coastline: Method and Case Studies, Report Submitted to the Ford Foundation, Tata Energy Research Institute, New Delhi.

WDI (2002): World Development Indicators, The World Bank, Washington, DC.

Chapter 31

Trade and Climate Change Nexus

A Way Forward

DEBESH BHOWMIK

ABSTRACT

Trade and climate change nexus has gained increasing attention after Kyoto Protocol. Trade intersect with climate change in a multitude ways. Literature indicate that more open trade is likely to increase CO_2 emissions as a result of increased the scale effect, the technique effects and the composition effect. The researches of Frankel and Rose (2005), Cole and Elliot (2003a), McCarney and Adamowicz (2005), Managi (2004, 2008), Roberts and Grimes (1997), Moomaw and Unruh (1997), Peters and Hertwich (2008) and Verburg (2008) verified the nexus statistically taking data for a long period and number of countries. Even Huang *et al*. (2008) test for the existence of an environmental Kuznets Curve for GHG emissions in transition countries and (Kyoto Protocol) Annex II countries and conclude that the evidence for most of these countries does not support the environmental Kuznets Curve Hypothesis. WTO with its Articles XX, XXIV, XX(b) and XX(g) and MEA can maintain and control the nexus with good co-operation of UNFCCC. To realize UNEP's CDM , the trade policies and climate policies should have good coherence towards long-term objectives of all the conventions of WTO. But, from Doha to Copenhagen, no concrete decisions were taken by the developed and developing countries to fulfil Kyoto Protocol. Besides, the International Chamber of Commerce, International Maritime Organisation, and the trading blocs must have specific role in formulating climate-trade nexus policies because trade and climate policies may also intersect with each other as a matter of law. WTO members should have good co-ordination in framing climate-friendly commodity trade rule, coding system, harmonized commodity description, and tariff rules including carbon tax for post-2012 world minimizing unequal exchange and development.

INTRODUCTION

The Stern Review[1] calculated that the impact of climate change would be a loss of at least 5% of global GDP each year and could be reach as much as 20% of global GDP at 2.5°C and 75-250 million African people would experience water stress in 2020 and some African countries would suffer from a 50% decline in agricultural yields. The required change in consumption and production patterns will neither be easy to achieve nor occur without seriously altering global trade patterns. Both climate change and measures taken to combat it will thus have an impact on international trade.

Trade intersects with climate change in a multitude of ways. In part, this is due to the innumerable implications that climate change may have in terms of its potential impacts and the profound regulatory and economic changes that will be required to mitigate and adapt to these impacts. Climate change is expected to have an impact on trade infrastructure and trade transportation routes. Literature indicates that more open trade is likely to increase CO_2 emissions as a result of increased economic activity (the scale effect). On the other hand, trade opening could facilitate the adoption of technologies that reduce the emission-intensity of goods and the production process (the technique effect) and lead to a change in the mix of production from energy-intensive to less energy-intensive sectors if it is where it has a comparative advantage (the composition effect). Although most studies to date have found that the scale effect tends to outweigh the technique and composition effects in terms of CO_2 emissions, it remains difficult to determine in advance the magnitude of each of these three effects, and therefore estimating the overall impact of trade on greenhouse gas emissions can be challenging.

Trade and climate change are linked in multiple ways in the domestic and international rules and institutions because climate change is already affecting the productive base of international trade which may help or hinder climate efforts by transferring climate-friendly technologies or increasing transport-related emissions. In the realm of climate policy such as new regulations or standards may affect trade and competitiveness. On the other hand, trade policies may influence economic activities and associated GHG emissions. Trade and climate policies may also intersect with each other as a matter of law. Domestic climate measures and climate negotiations are likely to be scrutinized in the WTO rules. The linkage is essential in preamble of the WTO in the interest of developing countries and achieving sustainable development.The existing WTO agreements such as trade in goods, trade in services and protection of intellectual property transverse the territory covered by climate issues and institutions. Doha Work Programme in relation to agriculture, industrial products and environmental goods and services may also affect efforts to mitigate and adapt to climate change. WTO's dispute settlement body may also come into play in the event that climate-related trade disputes cannot be addressed through diplomatic or other channels.

The association between trade and climate change measures in the climate regime is governed by, among others, Art. 3.5 of the UNFCCC which states that "measures taken to combat climate change, including unilateral ones, should not constitute a means of arbitrary or unjustifiable discrimination or a disguised restriction on international trade." This reflects Art. XX of the General Agreement on Tariffs and Trade (GATT), which allows WTO Members to adopt measures that may be inconsistent with their WTO obligations if such measures are, *inter alia*, "necessary to protect human, animal or plant life or health" or are related "to the conservation of exhaustible natural resources if such measures are made effective in conjunction with restrictions on domestic production or consumption", provided that these measures "are not applied in a manner which would constitute a means of arbitrary or unjustifiable discrimination between countries where the same conditions prevail, or a disguised restriction on international trade."

The climate-trade nexus has gained increasing attention after Kyoto Protocol in relation to unilateral action and multilateral efforts in GATT Article XX[2] and in regional agreements in Article XXIV, for the interpretation of the causal link required to justify environmental measures under Article XX(b) and, by implication, XX(g). However, the relevance of WTO rules to climate change mitigation policies, as well as the implications for trade and the environmental effectiveness of these measures, will very much depend on how these policies are designed and the specific conditions for implementing them.

The discussion on trade-and-climate-change is also expanding the notion of what constitutes "unfair" trade. For decades, international trade law, as reflected in both the WTO system and the domestic law of most trading nations, has recognized that pricing imports below certain levels (whether due to "dumping" by foreign exporters or subsidies provided by foreign governments) is a form of "unfair" trade that should be redressed where it harms domestic industries. This notion of unfair trade is based purely on how an imported product is priced. Climate change concerns are now expanding the notion of unfair trade to take into account how imported products are made-specifically, the volume and nature of the greenhouse gases associated with their manufacture.

TRADE AND CLIMATE CHANGE : THE LINK

Opening up of trade can affect the amount of emission in three principal ways , e.g., (1) the scale effect, (2) the composition effect , and (3) the technique effect. *The scale effect* refers to the expansion of economic activity arising from trade opening, and its effect on GHG emissions.

This increased level of economic activity will require greater energy-use and will therefore lead to higher levels of GHG emission.

The *composition effect* describes the way that trade opening changes the structure of a country's production in response to changes in relative prices, and the consequences of this on emission levels. Changes in the structure of a liberalizing country's production will depend on where the country's "comparative advantage" lies. The effect on a country's GHG emissions will depend on whether a country has a comparative advantage in emission-intensive sectors and whether these sectors are expanding or contracting. The composition of production in an economy that is opening its markets to trade may also be a response to differences in environmental regulations between countries (resulting in the "pollution haven hypothesis", which suggests that high-emission industries may relocate to countries with less stringent emission regulation policies).

Finally, the *technique effect* refers to improvements in the methods by which goods and services are produced, so that the emission intensity of output is reduced. This is the principal way in which trade opening can help mitigate climate change. A decrease in GHG emission intensity can come about in two ways, namely, (a) More open trade can increase the availability and lower the cost of, climate-friendly goods and services which may highlight the importance of negotiations with WTO and (b) as income levels rise because of trade opening and populations may demand lower GHG emissions. Rise in income leads to improvement in environment, thereby government formulates tax and regulatory measures to meet public demands. Only if such measures are put in place will firms adopt cleaner production technologies, so that a given level of output can be produced with fewer greenhouse gas emissions. It has been pointed out, however, that the positive link between per capita income and environmental quality may not necessarily apply to climate change. Since the scale and technique effects tend to work in opposite directions, and the composition effect depends on the comparative advantage of countries and on differences in regulations between countries, the overall impact of trade on greenhouse gas emissions cannot be determined *a priori*. The net impact of greenhouse gas emissions will depend on the magnitude or strength of each of the three effects, and ascertaining this requires detailed empirical analyses.

We have information on the impact of trade on GHG emission where the environmental assessments of trade agreements that have been undertaken by countries engaged in negotiating multilateral and bilateral free trade. The WTO has been notified of 243 regional trade agreements (RTAs) which are currently in force. Before examining the results of these assessments, it may be helpful to consider their limitations. First, nearly all of these assessments involve an analysis of the likely economic, environmental, and frequently also social impacts of the trade agreements. As such, the conclusions reached by the studies, if any, are about the anticipated rather than actual impacts of the trade agreements. Only a few of the environmental assessments analyze climate change impacts; most deal with local or domestic environmental impacts. This is because a study of climate change impacts was not included in the scope of the assessment, or because the impact the trade agreement would have on GHG emissions was considered to be negligible. Furthermore, although some of the assessments apply the scale, composition and technique effect framework, few are able to provide quantitative estimates of the impact, each of these effects would have as a result of the trade agreement.

THE ESTIMATION OF THE NEXUS

Torras and Boyce (1998), who examined the available evidence on international variations in seven indicators of air and water quality, find that the degree of inequality in incomes, as well as inequalities in levels of literacy, political rights and civil liberties, had a substantial impact on the quality of environmental protection in low-income countries. Countries which had a more equitable distribution of income and which achieved greater equality in literacy, political rights and civil liberties tended to have better environmental quality. This evidence suggests that increases in income from trade opening may not translate into environmental improvements if the economic benefits are not shared more equitably among the population. While trade openness does not appear to be a major factor in income inequality in developed countries, the evidence is more mixed for developing countries.

Frankel and Rose (2005) examine evidence from several countries on the relationship between seven indicators of environmental quality (including CO_2 emissions) and trade openness, for a given level of per capita income. In the case of CO_2 emissions, their data allows them to study nearly 150 countries. Their paper also takes account of the possible "endogeneity" of trade and per capita income through the use of instrumental variables techniques. The endogeneity of trade openness and per capita income arises because there may be a two-way rather than a one-way relationship between trade openness and per capita income. In other words, while more openness may increase per capita income, the latter may in turn lead to increased trade. When this endogeneity is not taken into account, the statistical results of their study indicate that trade openness would lead to increased CO_2 emissions. On the other hand, by including the possible endogeneity of trade in their calculations, the detrimental effect of trade openness on CO_2 emissions becomes statistically insignificant. Nevertheless, Frankel and Rose conclude that the main instance where trade and growth might have a detrimental effect is on the amount of CO_2 emissions. They recognize that this is due to the global nature of the externality (emissions are released into the global commons and the costs of the pollution are partly borne by foreigners) and that, as a result, CO_2 emissions are unlikely to be addressed by national environmental regulations.

Cole and Elliott (2003a) consider the effect of trade openness on four environmental indicators, including carbon dioxide (CO_2) emissions. The data on CO_2 which they used covered 32 developed and developing countries during the period 1975-95. They find that, overall, more trade openness would be likely to increase CO_2 emissions, due to a large scale effect and only a small technique effect (in other words, increased trade openness would lead to increased production and therefore increased emissions, without a large enough increase in the use of emission-reduction technologies to counter such growth).

McCarney and Adamowicz (2005) use "panel data" for 143 countries, spanning the period 1976 to 2000, to examine the link between trade openness and CO_2 emissions. Their results indicate that more open trade significantly increases emissions of CO_2, although they are not able to give a breakdown of the overall outcome into the individual contributions of the scale, composition and technique effects. Managi (2004) uses data for 63 developed and developing countries over the period 1960 to 1999 to examine the link between trade openness and levels of CO_2 emissions as in the Frankel and Rose, the possibility of endogeneity between trade openness and income was taken into account in the estimation. The results of the study suggest that further trade opening would result in increased emissions with an estimated elasticity (a measure of the responsiveness of CO_2 emissions to trade openness) of

0.579, with a greater contribution from the scale effect than from the technique effect. A later study by Managi *et al.* (2008) suggests that the impact of trade openness on CO_2 emissions may differ between developed countries (OECD members) and developing countries. They estimate the overall impact of trade openness on emission levels of carbon dioxide and sulphur dioxide, and on levels of biochemical oxygen demand (BOD)—a measure of the amount of oxygen used by micro-organisms while breaking down organic matter in water, which is used as an indicator of pollution levels. They use panel data on CO_2 and SO_2 emissions of 88 countries from 1973 to 2000 and the BOD levels of 83 countries from 1980 to 2000. The econometric analysis they employed allows them to correct for the endogeneity of income and trade and enabled them to distinguish between the short-term and long-term relationships between trade and CO_2 emissions. They find that trade openness reduces CO_2 emissions in OECD countries because the technique effect dominates the scale and composition effects, but that it has a detrimental effect on carbon dioxide emissions in non-OECD countries, where the scale and composition effects prevail over the technique effect. They also find that the long-term impact of trade on CO_2 emission levels is large, although it is small in the short-term. Roberts and Grimes (1997) looked at a larger sample of countries (147 countries) and a longer time period (1962-91). They find that the relationship between CO_2 emissions per unit of GDP and level of economic development has changed from essentially linear in 1962 (i.e. each change in GDP and in level of economic development gave rise to a constant amount of change in CO_2 emission levels), to strongly curvilinear in 1991 (in other words, the amount of change in emission levels became larger and larger in response to each change in GDP and economic development level).They also find that, during a brief period in the early 1970s, and increasingly since 1982, the environmental Kuznets curve reached statistical significance. They conclude, however, that this is not the result of groups of countries passing through stages of development, but is due to improvements in production efficiency in a small number of developed countries, combined with reduced efficiency in low and middle-income countries. The study by Shafik (1994) shows that although some environmental indicators (such as water and sanitation) improve with rising incomes, others (such as particulates and sulphur dioxide) initially deteriorate before eventually improving, while others become steadily worse (carbon dioxide emissions, dissolved oxygen in rivers and municipal solid wastes, for example). Moomaw and Unruh (1997) identify 16 countries (a subset of OECD members) that demonstrate sustained income growth with stable or decreasing levels of CO_2 emissions per capita. Using these 16 countries, they then compare two models—an environmental Kuznets curve and what they term a structural transition model of per capita CO_2 emissions and per capita GDP—to see which model best matches the experience of these 16 countries. The structural transition model attempts to find a sudden change in the pattern of the data and relate it to some precipitating event. They find that improvements in levels of CO_2 emissions per capita are not correlated with income levels, but are closely linked with historic events related to the oil-price shocks of the 1970s and to the policies that followed. They conclude that an environmental Kuznets curve for CO_2 emissions and income does not provide a reliable indication of future behaviour. Using more recent data from the period 1990 to 2003, Huang *et al.* (2008) test for the existence of an environmental Kuznets curve[3] for greenhouse gas emissions in transition economies and (Kyoto Protocol) Annex II countries. They conclude that the evidence for most of these countries does not support the environmental Kuznets curve hypothesis.

Peters and Hertwich (2008a, 2008b), for example, have suggested that 21.5 per cent of global CO_2 emissions are as a result of international trade. Similar studies have been undertaken by the Stockholm Environment Institute and the University of Sydney to estimate the CO_2 emissions associated with UK trade (*Wiedmann et al.*, 2007). What these studies have in common is that they calculate emissions associated with consumption rather than those resulting from production. Since consumption, by definition, involves trade (consumption = production + imports - exports), it is necessary to estimate the CO_2 emissions "embodied" in trade.

The IMO study also compares the "carbon emission efficiency" index (which the IMO defined as CO_2 efficiency = CO_2/ton-kilometre) of ships with that of other modes of transport. The lower this index, the more carbon-emission efficient a mode of transport is. It found that shipping has the lowest value of the index among the different modes of transport. The study also compares the share of each mode of transport in the sector's total CO_2 emissions and arrives at figures similar to those given by the IPCC report (*Kahn Ribeiro et al.*, 2007): road transport accounted for the biggest share of emissions, with 72.6 per cent, followed by international shipping with 11.8 per cent, then aviation with 11.2 per cent and finally rail transport, which accounted for 2 per cent.

A recent paper by Verburg *et al.* (2008) employs the Global Trade Analysis Project (GTAP) computable general equilibrium model, together with the Integrated Model to Assess the Global Environment (IMAGE), to simulate the long-term consequences on global emission levels of removing all trade barriers in agriculture, particularly in the milk-livestock sector. The study first establishes a baseline: the "business as usual" scenario. According to this baseline, by 2050, carbon dioxide (CO_2), methane and nitrous oxide (N_2O) emissions will be 63 per cent, 33 per cent and 20 per cent higher, respectively than their levels in 2000.The authors simulate various trade opening scenarios. Full agricultural trade opening would add an extra 50 per cent of CO_2 emissions to the baseline scenario by 2015. This initial increase in emissions would, however, decrease over time, so that CO_2 emissions in 2050 would be 30 per cent lower compared to the baseline. Full liberalization would lead to an additional 1.7 per cent of N_2O emissions by 2015, but this would later fall, so that by 2050 N_2O emissions would be equal to the baseline. Methane is the only emission which would have increased, by an extra 5 per cent over the baseline level, in 2050. Despite these complicated evolutions over time, the authors suggest that the full liberalization scenario would lead to a small overall increase in greenhouse gases by 2050.

THE ROLE OF WTO AND THE NEXUS

In June 2009, the WTO and the United Nations Environment Programme (UNEP) released a joint study, *Trade and Climate Change,* the first comprehensive study done by the WTO Secretariat that examines the nexus between trade and climate change At the release event for the WTO/UNEP report, Pascal Lamy, Director-General of the WTO, said that an international agreement on climate change should come first before the WTO would begin work on determining the WTO compatibility of trade measures related to climate change. He emphasized that the relationship between trade and climate change would be best defined by an international accord on climate change that embraces all polluters. Lamy asserted that WTO members want trade addressed as part of an overall post-Kyoto treaty, and they do not want a separate Geneva-based WTO negotiation on permissible trade-related climate measures. The climate regime itself could act multilaterally to create norms on trade and climate. In fact, recently there has been some movement to do so. For example, the post-Kyoto regime may establish non-binding principles for the use of trade measures for climate change and those principles could be considered by a WTO panel when a dispute arises. The WTO as an institution might prefer that a trade-related dispute pertaining to a MEA[4] be resolved within the relevant MEA before landing on the WTO's doorstep. However, it is WTO members, not the WTO as an institution, that decide the forum for bringing disputes. It seems likely that many WTO members will want to use the tried and true machinery of WTO panels and the Appellate Body when they bring disputes. (*Briefing Paper 11, German Institute of Development*, 2010).

WTO bodies provide an important forum to debate policy measures: the Committee on Trade and Environment, for example, could discuss, among others, trade-related measures that could help support climate change mitigation and adaptation or to what extent trade is affected by requirements for emissions reduction and energy efficiency. Price-based mechanisms, such as a carbon tax on fossil fuels or a tax on energy, have been employed in several countries over the past two decades as a means of internalizing the environmental cost of greenhouse gas emissions. More recently, attention has focused on emission trading schemes. These involve fixing a capon total emission levels, translating this into allowances to cover emissions, and creating a market to trade these allowances at a price determined by the market. More empirical work on the economic implications and environmental effectiveness of emission trading schemes would be useful. There is considerable debate on the extent to which certain industrial sectors may be economically affected by carbon-constraining domestic policies, and in particular, by emission trading schemes. Policies aimed at preventing carbon leakage (i.e. the risk that energy-intensive industries will simply relocate to countries with less rigid emission regulations) and at protecting competitiveness in these sectors are also under discussion. Government policies range from exemptions from participation in emission trading schemes to the use of border trade measures. The debate on the potential use of border measures has highlighted the formidable difficulties involved in applying such measures. These include the challenge of precisely assessing the quantity of CO_2 emitted during a product's production, which may depend on the company and the country, and the difficulty of measuring the economic impact of an emission trading scheme on a particular industry. Further research on methodologies to address these difficulties could be useful to policy-makers. There are a vast number of views expressed by academics, policy-makers, and various stakeholders on how trade is affected by measures to mitigate climate change, and on the extent to which these measures are consistent with WTO rules. A number of GATT and WTO rules deal specifically with many of the economic and regulatory instruments used in a number of countries. However, the relevance of WTO rules to climate change mitigation policies, as well as the implications for trade and the environmental effectiveness of these measures, will very much depend on how these policies are designed and the specific conditions for implementing them. (*WTO, UNEP*, 2009)

The WTO's "trade and environment-*acquis* " has evolved over the past 15 years in a number of landmark trade and environment disputes, but it has by no means stopped evolving and there are still important questions to be answered. Trade and climate issues may put those questions on the forefront. That goes for the relationship between unilateral action and multilateral efforts, for the relationship between the general exceptions in GATT Article XX and the exception for regional agreements in Article XXIV, for the interpretation of the causal link required to justify environmental measures under Article XX(b) and, by implication, XX(g) Compared to the relationship between the WTO and the Cartagena Protocol on Bio-safety, the relationship between the WTO and the Kyoto Protocol (and the broader climate regime) appears to be relatively harmonious and there has so far been no direct collision at the level of rules, and no instances of "forum shopping" by Parties with axes to grind. Yet this is not necessarily an indication that the two regimes are perfectly harmonized. In the following summary, I show that the potential conflict has been managed through a particular transnational discourse of economy environment integration that constructs an open and expanding international economy as compatible with the goal of curbing global aggregate emissions of greenhouse gases A comparison of the principles of the two regimes reveals four significant points of overlap.[1] The endorsement of sustainable development—mentioned in the preamble of the Marrakesh Agreement[5] Establishing the WTO as well as in several articles of the United Nations Framework Convention on Climate Change (UNFCCC);[2] Despite certain common features and shared views on the importance of sustainable development, a fundamental difference exists between the UNFCCC and the WTO regimes.[3] Support for an open economic system including avoidance of any arbitrary or unjustifiable discrimination or disguised restrictions of trade—a principle inherent in the provisions of the trade regime, and explicitly mentioned in Article 3.5 of the UNFCCC; and [4] The maintenance of economic growth—this is mentioned throughout the UNFCCC and basic to, though implicit, in the liberalization agenda of the WTO. Charnovitz (2003) has pointed out that climate change presents an extreme case of market failure—namely the failure to incorporate the damage done by GHG emissions into the prices of goods and services—and that a classic role for governments is to correct market failures. However, governments normally want great flexibility in the choice of national instruments to correct market failures, because they need to balance the economic characteristics of alternative measures against their political acceptability. By contrast, the trade rules embodied in the GATT and the WTO presuppose a world of market economies and attempt to discipline government failures that lead to economic distortions with the flavor of mercantilism and protectionism. This fundamental difference entails potential conflicts between the two regimes. In the absence of clearer guidelines than now exist, it is difficult to predict whether various policy options would be compatible with WTO rules. For example, it remains uncertain whether border adjustments are allowable for carbon taxes or permits that are based on energy consumed or carbon emitted, either in making a product or inputs to the product. It is also unclear whether the Agreement on Technical Barriers to Trade (TBT) would allow standards and labeling requirements based on production and processing methods (PPMs) that do not affect the physical characteristics of the product. The prior record of panel and Appellate Body decisions on these and other climate-related questions is sparse. If the rule book is filled out through case by-case litigation, it could be years before an overall framework is established. Moreover, case outcomes may depend heavily on how disputed measures are designed and implemented, making for a pretty complicated rule book (*Hufbauer and Kim*, 2009).The issue of climate change, in itself, is not part of the WTO's ongoing work programme and negotiation agenda. However, the WTO's rules and institutions are relevant because climate change measures and policies intersect with international trade in a number of different ways. In the context of the Doha Round, ministers have called for the liberalization of environmental goods and services. The mandate of negotiations stipulates "the reduction, or as appropriate, elimination of tariff and non-tariff barriers to environmental goods and services". These negotiations could result in fewer and lower barriers to trade in environmental goods and services, and therefore improve global market access to more efficient, diverse, and less expensive goods and services, including goods that can contribute to climate change mitigation and adaptation. Another question addressed in the Doha Round[6] is the relationship between the WTO and multilateral environmental agreements (MEAs), such as the UNFCCC. In this area of the negotiations, WTO members have focused on the means for further strengthening cooperation between the WTO and MEA secretariats, as well as promoting coherence and mutual supportiveness between the trade and climate regimes. With regard to the liberalization of environmental goods and services, as well as to the WTO-MEA relationship, the Doha mandate provides an unprecedented opportunity for the multilateral trading system to contribute to furthering mutual supportiveness of trade and environment. Significant work has been carried out in the Special Session of the Committee on Trade and Environment (CTE in Special Session), which is the negotiating group responsible for overseeing discussions relating to the trade and environment mandate. However, many issues have yet to find a resolution and the outcome of the negotiations remains elusive.

In this context, the WTO environmental goods and services negotiations have a role to play in improving access to climate-friendly goods and technologies. In the CTE in Special Session a number of countries have identified a broad range of goods serving various environmental purposes, including mitigation of climate change. For instance, the following categories of goods have been discussed: water and waste-water management; air pollution control; management of solid and hazardous waste; renewable energy production; heat and energy management; cleaner or more resource-efficient technologies and products; and environmental monitoring, analysis assessment. Climate-friendly goods and technologies are contained in several of these categories, particularly in the category of renewable energy. Environmental services are covered as part of the services negotiations under Article XIX of the General Agreement on Trade in Services (GATS), which form an integral part of the negotiating framework under the Doha Development Agenda. In the negotiations on environmental services, WTO members are seeking specific commitments on activities which may be directly relevant to policies aimed at mitigating climate change.

Trade-exposed and energy-intensive activities in the carbon-constrained economies might face competitiveness distortions coming from the lack of similar efforts elsewhere. The key to successfully designing a post-2012 climate change regime, therefore, would be devising an equitable burden sharing mechanism through financing and technology transfer, while addressing competitiveness and carbon leakage concerns. One rapidly emerging approach is an international sector-based arrangement, which might open an avenue to engage major developing country economies early on and address leakage concerns if well-designed. For instance, if emissions leakage is only likely to occur in a few large energy-intensive sectors in relatively few countries, a government-led sectoral agreement could improve the effectiveness of reducing greenhouse gases (GHGs), complementing an international agreement including national emissions targets. What implications for climate change policy in a post-2012 world can be drawn from these examples of environmental cooperation mechanisms? First, if there is mutual desire among the Parties concerned, environmental cooperation mechanisms can be used to address specific issues such as climate change.

As seen above, the scope of cooperation can vary ranging from the implementation of Parties' international environmental commitments, the adoption of common policies for the protection of the environment, the conservation of natural resources, joint communications on objects of common interest and exchanges of information about national positions in international forums to cooperation on the transfer of specific technologies. The language identifying technologies can be as broad as "environmental" or "clean" technologies or it can be more specific. For instance, the environmental cooperation mechanisms can be targeted at the transfer of certain climate-friendly technologies in which relevant Parties have a common interest, and which might be useful to reduce emissions in energy-intensive sectors. (*Zhang*, 1998)

The WTO may affect technology transfer in three main areas:

The environmental goods and services negotiations offer potential to support transfer of low carbon and energy efficient technologies. The WTO intellectual property (TRIPS) agreement may also affect technology transfer.

The WTO Working Group on Trade and Transfer of Technology is examining trade and technology linkages, but has so far achieved little in terms of concrete recommendations on steps that might be taken within the mandate of the WTO to increase flows of technology to developing countries. Reforming subsidies, Protecting positive subsidies, Addressing perverse subsidies, Enhancing WTO rules, Enhancing policies and measures, Managing competitiveness concerns.

- *Industry-level approaches* facilitate an integrated discussion of climate change and trade in the context of specific "carbon-intensive" industries
- *Enabling measures* such as finance and technology transfer provided on a measurable, reportable and verifiable basis should be established before consideration of unilateral measure.
- *Compliance measures* in any future regime may also prevent large states from free-riding on the emissions reductions of other states.

There is a two-fold rationale for reducing tariffs and other trade barriers regarding climate-friendly goods and technologies. First, reducing or eliminating import tariffs and non-tariffs barriers for these types of products should reduce their price and therefore facilitate their deployment at the lowest possible cost. Access to lower-cost and more energy-efficient technologies may be particularly important for industries which must comply with climate change mitigation policies that place the burden of emission reductions on the emitters. Moreover, trade liberalization of climate-friendly goods, in particular in developing countries, could help increase local capabilities for innovation and adaptation of domestic technology rather than foster dependence on transfer of foreign technology. Trade opening could then facilitate the integration of small and medium-sized enterprises into related global supply chains, thereby increasing employment and reducing poverty. Moreover, trade liberalization of climate-friendly goods, in particular in developing countries, could help increase local capabilities for innovation and adaptation of domestic technology rather than foster dependence on transfer of foreign technology. Trade opening could then facilitate the integration of small and medium-sized enterprises into related global supply chains, thereby increasing employment and reducing poverty. A review of several developing-country case studies has noted a significant shift in the structure of these countries' environmental goods and services industries, from traditional "end-of-pipe" activities to the use of cleaner technologies that reduce pollutants at source. Several other studies have further noted that many developing countries, such as China, Republic of Korea, Malaysia, India and Indonesia, have emerged as leading producers in clean energy sectors, such as wind and solar energy or efficient lighting. A number of developing countries have a significant export interest in certain product lines which are included in the category of renewable energy. For example, in 2007 the following developing economies were among the top five exporters for at least one HS 6-digit sub-heading in the category of renewable energy products: Brazil; China; Hong Kong, China; India; Republic of Korea; Malaysia; Mexico; Turkey; Singapore; South Africa; and Thailand. In addition, the following countries were among the top ten exporters: Argentina, Jordan, the Philippines, Saudi Arabia, and the United Republic of Tanzania. Five developing economies are among the top ten exporters of the entire renewable energy category of goods: China; Hong Kong, Mexico; Singapore; and Thailand. Moreover, several developing countries are the top exporter of one or more product lines in the renewable energy category. For instance, Mexico is the top exporter of the product line which covers solar water heaters (HS 841919), while China is the top exporter of lines which include wind turbine towers (HS 730820), static converters that change solar energy into electricity (HS 850440), solar batteries for energy storage in off -grid photovoltaic systems (HS 850720), and concentrator systems used to intensify solar power in solar energy systems (HS 900290).

For certain countries, exports in the renewable energy category represent a substantial part of their overall exports. For instance, in

2007, about 2 per cent of China's exported goods figured in renewable energy product lines, while both Mexico and Thailand's exports of these goods amounted to 2.2 per cent. It should also be noted that in 2007, world exports of goods contained in the 30 product lines (HS 6) of the renewable energy category amounted to US$ 189 billion (i.e. they accounted for 1.5 per cent of world exports). Developing countries' exports in the same category amounted to US$ 59 billion and their imports amounted to US$ 69 billion. Finally, the trade of climate-friendly goods has seen a considerable increase in the past few years. For instance, between 1997 and 2007 exports of goods contained in the product lines listed in the renewable energy category grew by 598 per cent in developing countries and by 179 per cent in developed countries, representing 62 per cent and 29 per cent of annual average growth respectively. It should of course be noted that the price of climate-friendly goods is not the only factor that affects the diffusion of these technologies. A number of authors have pointed to other important factors, such as a country's gross domestic product, its level of foreign direct investment and the regulatory framework for climate change action. (*WTO-UNEP Report*, 2009).

POLICIES RELATING TO NEXUS

Issues that link trade and climate change policy reflect in many ways the policy considerations that underlie how developing countries view these two policy regimes. Negotiations are taking place among countries which are Parties to the United Nations Framework Convention on Climate Change (UNFCCC) under the Bali Action Plan (BAP)[7] adopted by the UNFCCC Conference of the Parties (COP) in Bali, Indonesia, in December 2007, for the purpose of arriving at an agreed outcome that would serve as the basis for long-term global cooperative action in enhancing the full, effective and sustained implementation of the UNFCCC. At the World Trade Organization (WTO), countries that are Members of the WTO have been engaged in trade negotiations that commenced in December 2001 under the WTO Doha Ministerial Declaration and which places the needs and interests of developing countries at the heart of the negotiations.

Climate change policy (including those shaped pursuant to the UNFCCC) with respect to climate adaptation and mitigation becomes an important element in a developing country's development policy tool box. The underlying treaty regime and negotiating mandates for both the current trade and climate change negotiations provide ample basis for such an approach by developing countries. In fact, sustainable development is the foundation for effective societal responses to trade and climate change challenges. an UNDP study on transfer of low carbon technologies to developing countries points out that it is questionable whether technology transfer under stringent IPR regimes in developing countries can have long term benefits for the recipient developing country because recipient firms in these countries may be less likely to gain access to the underlying knowledge that is necessary to develop technological capacity within the recipient country, and thus it can retard the recipient country's long-term ability to absorb and innovate on the basis of new low carbon technologies, which is critical for their sustainable development. Reinaud points out that "the vast majority of allowances under existing ETS (emissions trading systems) are currently distributed free to trade-exposed sectors" (such as cement, iron and steel, aluminium, chemicals), on the basis of the application of both eligibility criteria and distribution formulae. The definition of both the eligibility criteria to be able to receive emission rights and the formulae for the distribution of such emission rights often involves a political and policy-driven process. That is, both the criteria and the distribution formulae depend on governments' political and policy assessment of, *inter alia:* the level of the national emissions cap that underlies the ETS (and thus the amount of emission rights that can be allocated and, ultimately, traded under that cap); and the industrial sectors and the emitters therein that would benefit from emission allowances (or that should be kept "competitive" with their non-carbon constrained competitors by effectively subsidizing the emitters' cost of compliance with emission limits through the free allocation of emission rights). Finally, emissions trading cannot be divorced from the GHG emission reduction commitments that developed countries have to agree to and comply with under the Kyoto Protocol. Much of the emission allowances that can be traded will come from the Certified Emission Reductions (CERs) that can be generated from projects implemented under the Kyoto Protocol's Clean Development Mechanism (CDM).[8]

What implications for climate change policy in a post-2012 world can be drawn from these examples of environmental cooperation mechanisms? If there is mutual desire among the Parties concerned, environmental cooperation mechanisms can be used to address specific issues such as climate change.

The environmental cooperation mechanisms can be targeted at the transfer of certain climate-friendly technologies in which relevant Parties have a common interest, and which might be useful to reduce emissions in energy-intensive sectors.

There are two ways in which co-operation makes a difference and can help clarify the legal position of countries when developing and implementing climate change policies that might also restrict or distort international trade: (1) International co-operation can influence the applicability of international trade law: Co-operation between States can make specific provisions of the WTO-regime or even whole agreements under the WTO non-applicable altogether. That is to say agreements adopted outside the WTO could over-ride the WTO Agreements.

(2) International co-operation can influence the interpretation of international trade law: International co-operation efforts on climate change policies may not formally override WTO-law but still ensure that environmental considerations established outside of the WTO will be taken into account in the interpretation of WTO-rules. This way, international co-operation may effectively shield climate change policy measures from an invalidation through international trade law.

In short, the reality of international co-operation as enshrined in the many international environmental agreements outside of the WTO can serve as justification for environmental measures affecting international trade. *(*Articles 1, 3, 4.1, 4.2, 7.2 UN Charter, and Principle 7 of Rio).[9]

Trade restrictive environmental measures—including PPM-based measures—can be justified under the provisions of the GATT if such measures have been agreed and negotiated on a multilateral basis and are therefore not an expression of unilateralism. Furthermore, unilateral measures might be acceptable if they are adopted after serious and substantial efforts to reach an international agreement with states whose rights under the WTO will be affected by an environmental policy measure and if further co-operative efforts are pursued—WTO-issues raised when states engage in emissions trading. By co-operating in the WTO, the UNFCCC, in the context of the Kyoto Protocol, or in other international, regional, and bilateral a government can directly influence the application and interpretation of relevant WTO-provisions and thereby raise the likelihood that climate change policies will be regarded as WTO-compatible in case of a challenge under the WTO's Dispute Settlement Mechanism.

The International Chamber of Commerce (*ICC*, 2009) is among those who strongly support the implementation of an open trade policy for dealing with climate change. Open trade policies are believed to provide conducive framework for global efforts in securing a sustainable energy supply important for economic growth and, at the same time, increase access to innovation/technology and the related investment which are needed for addressing environmental impacts from economic activities (*ICC*, 2009; *Low*, 2009). This argument is supported by

Lichenko and O'Brien (2009) who viewed the interlink between climate change and trade liberalization in the context of meeting growing energy needs and achieving sustainable development and economic growth through the use of climate-friendly technologies. The impacts of open trade and climate change policies require the government, private sector and civil society to understand their capacity in adapting to the policies *(Pratiwi*, 2010).

The IMO base scenario predicts that emissions from international shipping will increase from current levels by between 125.7 per cent and 218 per cent by the year 2050. The base scenario also predicts that increases over the next decade (to 2020) will range from a low of 9.7 per cent to a high of 25.5 per cent above current emission levels. The major part of international trade is transported by sea (90 per cent when calculated by weight and 70 per cent when calculated by value). This proportion appears to have remained steady, at least since 2000, despite the rapid growth in the use of air transport. Among the different modes of transport, shipping is also the most efficient in terms of carbon dioxide emissions. It is important to take this into account when assessing the contribution of trade to transport related emissions. However, it should not be taken as a cause for complacency since there are warning signs that, without significant policy or regulatory changes, CO_2 emissions from international shipping will rise by significant amounts in the next four decades. There appear to be two likely effects of climate change on international trade.

First, climate change may alter countries' comparative advantages and lead to shifts in the pattern of international trade. This effect will be stronger on those countries whose comparative advantage stems from climatic or geophysical reasons. Countries or regions that are more reliant on agriculture may experience a reduction in exports if future warming and more frequent extreme weather events result in a reduction in crop yields. Warming need not always produce negative impacts on exports, since it may succeed in increasing agricultural yields in other regions.

In relation to climate change, such regulations and standards intend generally to: (i) improve the energy efficiency of products and processes; and (ii) reduce their energy consumption and/or the quantity of greenhouse gases emitted during the production of a product, or emitted while it is being used. Moreover, some regulations and standards are being developed to facilitate the adaptation to the consequences of climate change. The literature indicates that more open trade is likely to increase CO_2 emissions as a result of increased economic activity (the scale effect). On the other hand, trade opening could facilitate the adoption of technologies that reduce the emission-intensity of goods and their production process (the technique effect) and lead to a change in the mix of production from energy-intensive to less energy-intensive sectors if it is where it has a comparative advantage (the composition effect). Although most studies to date have found that the scale effect tends to outweigh the technique and composition effects in terms of CO_2 emissions, it remains difficult to determine in advance the magnitude of each of these three effects, and therefore estimating the overall impact of trade on greenhouse gas emissions can be challenging.

WTO bodies provide an important forum to debate policy measures: the Committee on Trade and Environment, for example, could discuss, among others, trade-related measures that could help support climate change mitigation and adaptation or to what extent trade is affected by requirements for emissions reduction and energy efficiency. Price-based mechanisms, such as a carbon tax on fossil fuels or a tax on energy, have been employed in several countries over the past two decades as a means of internalizing the environmental cost of green house gas emissions. More recently, attention has focused on emission trading schemes. These involve fixing a cap on total emission levels, translating this into allowances to cover emissions, and creating a market to trade these allowances at a price determined by the market.

Lest, it would encourage the countries, both the developed and the developing ones, to embrace the new paradigm of Preferential Trade Agreements (PTAs). The European Union (EU), the North Atlantic Free Trade Agreement (NAFTA), the Association of South-East Asian Nations (ASEAN), etc. have had created large trading blocks, spurring others on the same path. There are relative merits and demerits of a country entering into any Preferential Trade Agreement. However, it is apprehended that the alliance between PTA countries could create stronger protectionist lobbies which may be more difficult to dismantle; these protectionist groups may become politically more formidable. By locking a country into an inefficient production pattern, it handicaps its ability to adjust to a structure that would be more competitive under global free trade. Climate Change and Business and Industry Environment An interface between climate change and present day business is another important dimension. To effectively meet the challenges of climate change, the business and the industries, especially in the manufacturing sector, need to reduce GHGs emissions. They need to undertake carbon trading by investing in the environment-friendly technologies, besides putting in place clean development mechanisms. In order to retain competitiveness of products in a rapidly changing global economy, they need to eschew energy intensive products and processes and adopt environment friendly changes in their practices, promoting energy efficient technologies, newer greener manufacturing methods, etc.

Issues that link trade and climate change policy reflect in many ways the policy considerations that underlie how developing countries view these two policy regimes. Climate change policy (including those shaped pursuant to the UNFCCC) with respect to climate adaptation and mitigation becomes an important element in a developing country's development policy toolbox The underlying treaty regime and negotiating mandates for both the current trade and climate change negotiations provide ample basis for such an approach by developing countries. In fact, sustainable development is the foundation for effective societal responses to trade and climate change challenges. Art. 2 on the objective of the UNFCCC requires that global climate actions to stabilize atmospheric concentrations of GHGs (such as the mitigation actions of developed countries under Art. 4.2(a) and (b) and the Kyoto Protocol 7) must be done within such timeframes as would allow ecosystems to adapt, secure food supplies, and allow for sustainable development to take place.

Governments are seriously committed to addressing the challenge of climate change are well advised to bring the Kyoto Protocol into force, since the adoption of this multilateral agreement will considerably reduce remaining uncertainties about the WTO-compatibility of domestic and international climate policy measures. Furthermore, if in force, the Kyoto Protocol would constitute an important international forum in which progressive governments can move forward to address specific issues and tensions that may arise between international trade law and the climate change regime. In contrast, failure to bring the Kyoto Protocol into force will raise the likelihood of trade-conflicts arising from climate change policies. This would not only put at risk the effectiveness of domestic climate change programmes. In some cases it would also require resorting to second best climate policy-options, which would substantially raise the aggregate economic costs for "internalising" climate change externalities. Furthermore, a range of international trade disputes about climate change measures would pose a very serious challenge to achieving a mutually supportive relationship between trade and climate change policies. An open and transparent international economic system, with a long-term commitment to sustainable development is, however, key for a rapid diffusion of new climate change-related knowledge and technologies and for lower aggregate costs of stabilising greenhouse gas concentrations in the atmosphere at a level that may suffice to prevent dangerous anthropogenic interference with the climate system.

Efforts to agree on a post-2012 agreement should be further strengthened: a legally binding global climate deal with full coverage and participation would reduce the political pressure in favour of border adjustments.In this context,the purpose of the Post-2012 Climate and Trade Policies workshop can be classified into three-fold:

(1) To examine potential key elements of a future climate regime that have implications for the international trading system;

(2) To discuss various approaches through which the world trade regime can address or otherwise affect climate change in a post-2012 world; and

(3) To explore opportunities for synergistic interplay between the negotiating agendas of both regimes, i.e. the Doha Agenda and the Bali Road Map.

In formulating the tariff policy, the following are needed for implementation.

(i) The Harmonized Commodity Description and Coding System (Harmonized System)

(ii) Consolidation and Bound Tariffs

(iii) Deconsolidation and Compensation

(iv) Clean technology trade would greatly benefit from a systematic alignment of harmonization standards

(v) Members may extend a Harmonized System number to eight or ten digits

(vi) HS and Montreal Protocol: Creation of subheadings to cover various mixtures containing halogenated derivatives of methane, ethane or propane

Therefore, the story of Doha and Copenhagen both suggest the need for a paradigm shift in geo-political relationships—we are at a crossroads where the world's large, dynamic economies will either share the political burden of meaningful multilateral trade and climate change concessions or the two processes will fail. If that turns out to be the case, the costs for the mice could hardly be higher. In a Post-2012 World, a climate policymaker might ask: "How can the trade regime contribute to climate change mitigation and adaptation?" And in cases where it cannot: "How do we ensure that trade law does not stand in the way of climate change action?" In contrast, for a trade policy-maker, the question might be: "How can we ensure the climate is protected without creating trade distortions or unfair trade practices?" Finally, others would stress that this is not only about climate and trade, but ask: "What does this all mean for development?"

CONCLUSION

It is very much doubtful and debatable whether climate friendly goods and services are traded under free trade regime and may produce more welfare gain. Protection and tariff measures in environment-friendly goods and services under WTO serve the interest of developing countries. But, the agreement on the trade and environment from Doha to Copenhagen went against the developing countries irrespective of CDM and low emitters of GHG and persisting USA's constant neglect towards Kyoto Protocol. The paradigm shifts of production, consumption and trade as a result of climate change or vice versa is not only depend on the governance of WTO alone but also is an issue of the international co-operation and integration towards green world for equity, justice and a safer place to live in narrowing unequal exchange and development. It is hoped that the future research will overcome the essence of the dominance of developed countries over the developing nations on carbon trade and climate friendly goods and services including its international finance to fulfill UNEP completely.

Notes and References

1. *Stern Review: The Economics of Climate Change*, Executive Summary, available at http://www.hm-treasury.gov.uk/independent_reviews/stern_review_economics_climate_change/sternreview_index.cfm+ The paper has been written to show my gratitude, respect and honour to Prof. Raj Kumar Sen who is my friend, Philosopher and guide.
2. For a detailed explanation of Article XX, see WTO (2002).
3. Ssee Aldy, J.E. (2005), "An Environmental Kuznets Curve Analysis of U.S. State-Level Carbon Dioxide Emissions", *Journal of Environment and Development*, 14:1, pp. 48-72.
4. For further reading see Murase, S. (2003). WTO/GATT and MEAs: Kyoto Protocol and Beyond: GETS/FTC/GISPRI Project Pershing, J. and Mackenzie. 2004. Removing Subsidies.Leveling the Playing Field for Renewable Energy Technologies. Thematic Background Paper. International Conference for Renewable Energies, Bonn, 2004.
5. In Morocco, the agreement was signed in 1994 but came into effect on 1.1.1995 by WTO.
6. Ministerial discussions have taken place in Cancun (2003), Geneva (2004), HongKong (2005), Geneva (2006, 2008), for details see www.wto.org.
7. It was adopted by decision 1/CP.13 of the COP-13, for further study See for details in unfcc.int//cp_bali_action.pdf.
8. Defined in Art. 12 of the Kyoto Protocol,see in details in cdm.unfcc.int, also see Yamin, F. (1998): Issues and Options for Implementation of the Clean Development Mechanism,London, UK.
9. See World Development Report, 1992.

References

Bhowmik, Debesh (2008): Global Warming, Sustainable Development and WTO. *91st Conference Volume of Indian Economic Association*, Mohanlal Sukhadia University, Udaypur, 2008, December, pp. 1041-53.

———, (2009): A Note on Climate Change Policy-Paper presented in International Seminar in RBU on 20-21, Feb.

———, (2010): Issues on Forest Management and Climate Change—Paper presented in International Seminar at RBU on 27-28, Feb.

———, (2010): Climate Change Debates and India's Response to International Climate Policy—Paper presented in conference of IEA at Nagarjuna University, Dec. 27-29, 2010.

———, (2010): An Introduction to Food Security and Climate Change—Paper Presented in a Seminar of BEA on 19/9/2010 at Netajinagar College, Kolkata.

Brewer Thomas L. (2008): US Climate Change Policy and International Trade Policy Intersections: Issues Needing Innovations for a Rapidly Expanding Agenda, February 12, Georgetown University.

Buck Matthias and Roda Verheyen (2001): International Trade Law and Climate Change—A Positive Way Forward, July.

Charnovitz, S. (2003): "Trade and Climate: Potential Conflicts and Synergies", in Aldy, J.E., Ashton, J., Baron, R., Bodansky, R., Charnovitz, S., Diringer, E., Heller, T.C., Pershing, J., Shukla, P.R., Tubiana, L., Tudela, F., and Wang, X., *Beyond Kyoto. Advancing the International effort against climate change*, Pew Centre of Global Climate Change, Arlington, VA, pp. 141-70.

Cole, M.A. and R.J.R. Elliott (2003): 'Do Environmental Regulations Influence Trade Patterns? Environment Institute at the University of York and Centre for Integrated Sustainability Analysis at the University.

Cosbey, A. (2008): Border Carbon Adjustment. Background paper to the Trade and Climate Change Seminar, June 18-20, 2008, Copenhagen. German Marshall Fund/Ministry of Foreign Affairs, Denmark.

Cosbey, A. (2009): Achieving Consensus: Multilateral Trade Measures in Post-2012, Scenarios. In UNEP/ADAM (Eds.) *Climate and Trade Policies in a Post-2012 World* (pp. 19-26). Geneva: United Nations Environment Programme/ADAM.

Cosbey, A. and Tarasofsky, R. (2007): *Climate Change, Competitiveness and Trade*, A Chatham House Report, London: Chatham House.

Frankel, J. and Rose, A. (2005): "Is Trade Good or Bad for the Environment? Sorting Out the Casuality", *Review of Economics and Statistics*, 87:1, pp. 85-91, Its Infrastructure", in Metz, B., Davidson, O.R., Bosch, P.R., Dave, R. and Meyer, L.A. (eds.), *Climate Change 2007: Mitigation. Contribution of Working Group III to the Fourth Assessment Report of the Inter-governmental Panel on Climate Change*, Cambridge University Press, Cambridge and New York, pp. 323-85.

German Institute of Development (2010): International Trade and Climate Change: Border Adjustment Measures and Developing Countries, *Briefing Paper-11*.

Grand Copthorne Waterfront (2009): Dialogue on Trade Policy, Climate Change and Sustainable Development for the Asia Pacific Region, 6-7, November, Singapore.

Hufbauer Gary Clyde and Jisun Kim (2009): The WTO and Climate Change: Challenges and Options, Peterson Institute of International Economics, September.

Huang, W.M., Lee, G.W.M. and Wu, C.C. (2008): "GHG Emissions, GDP Growth and the Kyoto Protocol: A Revisit of Environmental Kuznets Curve Hypothesis", *Energy Policy*, 36, pp. 239-47.

International Chamber of Commerce [ICC]. (2009): *Policy Statement: ICC Recommendations on Trade and Climate Change*. Document 103 / 291 rev4 final E. 23 February 2009 SB/CBS/amu/tl accessed December 27th, 2009 20:49 pm from www.iccwto.org

ICTSD (International Centre for Trade and Sustainable Development) (2008a): *Climate Change, Technology Transfer and Intellectual Property Rights*, Copenhagen Economics, 11 p.

ICTSD (2008b): *Liberalization of Trade in Environmental Goods for Climate Change Mitigation: The Sustainable Development Context*, Publication prepared for the Seminar on Trade and Climate Change, Copenhagen, 8 p.

IMO (International Maritime Organization) (2007): *International Shipping and World Trade: Facts and Figures*. Available at www. imo.org/includes/ blastDataOnly.asp/data_id% 3D23754/ International Shipping and World Trade-facts and figures.pdf.

IMO (2008): *Prevention of Air Pollution from Ships: Updated 2000 Study on Greenhouse Gas Emissions from Ships*, Phase 1 Report. MEPC 58/INF.6, International Maritime Organization, London.

IPCC (2007d): "Summary for Policymakers", in Parry, M.L., Canziani, O.F., Palutikof, J.P., van der Linden, P.J. and Hanson, C.E. (eds.), *Climate Change 2007: Impacts, Adaptation and Vulnerability, Contribution of Working Group II to the Fourth Assessment Report of the Intergovernmental Panel on Climate Change*, Cambridge University Press, Cambridge, 7-22.

IPCC (2007e): *Climate Change 2007: Mitigation. Contribution of Working Group III to the Fourth Assessment Report of the Intergovernmental Panel on Climate Change.*

Davidson, O.R., Bosch, P.R., Dave, R., Meyer, L.A. (eds.), Cambridge University Press, Cambridge and New York, NY.

IPCC (2007f): "Summary for Policymakers", in Metz, B., Davidson, O.R., Bosch, P.R., Dave, R., Meyer, L.A. (eds.), *Climate Change 2007: Mitigation. Contribution of Working Group III to the Fourth Assessment Report of the Intergovernmental Panel on Climate Change*, Cambridge University Press, Cambridge and New York, NY.

Kahn Ribeiro, S., Kobayashi, S., Beuthe, M., Gasca, J., Greene, D., Lee, D.S., Muromachi, Y., Newton, P.J., Plotkin, S., Sperling, D., Wit, R., and Zhou, P.J. (2007), "Transport and Its Infrastructure", in Metz, B., Davidson, O.R., Bosch, P.R., Dave, R. and Meyer, L.A. (eds.), *Climate Change 2007: Mitigation.*

Leichenko, R., and O'Brien, K. (2009): Mapping double exposure to climate change and trade liberalization as an awareness-raising tool. In A.G. Patt,

D. Schroter, R.J.T. Klein and A.C. de la Vega-Leinert (Eds.), *Assessing Vulnerability to Global Environmental Change* (pp. 173-93), London: Earthscan.

Low, P. (2009): Climate Change, Trade and WTO, Taillores, Power Point Presentation.

Managi, S. (2004): "Trade Liberalization and the Environment: Carbon Dioxide for 1960-99", *Economics Bulletin,* 17:1, pp. 1-5.

Managi, S., Hibiki, A. and Tsurumi, T. (2008): "Does Trade Liberalization Reduce Pollution Emissions, *Research Institute of Economy, Trade and Industry (RIETI), Discussion Paper Series* 08-E-013.

McCarney, G. and Adamowicz, V. (2005): The Effects of Trade Liberalization on the Environment: An Empirical Study, selected paper prepared for presentation at the Canadian Agricultural Economics Society Annual Meeting 6-8 July, San Francisco, California.

Moomaw, W.R. and Unruh, G.C. (1997): "Are Environmental Kuznets Curves Misleading Us? The Case of CO_2 Emissions", *Environment and Development Economics* 2:4, pp. 451-63.

Peters, G.P. and Hertwich, E.G. (2008a): "CO_2 Embodied in International Trade with Implications for Global Climate Policy", *Environmental Science & Technology,* 42:5, pp. 1401-07.

Peters, G.P. and Hertwich, E.G. (2008b): "Post-Kyoto Greenhouse Gas Inventories: Production Versus Consumption", *Climatic Change* 86, pp. 51-66.

Pratiwi, Sudhiari and Leonardo A.A.T. Sambodo (2010): The Challenges of Climate Change; Trade Liberalisation *v.* Opportunities for Micro and Small Enterprises in Developing Countries, Tech Monitor, Mar.-Apl.

Rajya Sabha Secretariat (2008): International Trade and Climate Change. *Policy Brief-16,* September.

Ranganathan Malini (2004): The Climate Change and Trade Regimes:Conflict or Compatibility?

EEP 131 Term Paper, December 9.

Reinaud, J. (2005), Industrial competitiveness under the European Union Emissions Trading Scheme, IEA information paper, 91 p.

Reinaud, J. (2008a), Climate Policy and Carbon Leakage, Impacts of the European Emissions Trading Scheme on Aluminium, IEA Information Paper, OECD/IEA, 43 p.

Reinaud, J. (2008b), *Issues behind Competitiveness and Carbon Leakages, Focus on Heavy Industry,* IEA Information Paper, OECD/IEA, 120 p.

Roberts, J.T. and Grimes, P.E. (1997), "Carbon Intensity and Economic Development 1962-91: A Brief Exploration of the Environmental Kuznets Curve", *World Development* 25:2, pp. 191-98.

Sampson, Gary P. (2000-04): Rules that Govern World Trade and Climate Change—The Importance of Coherence, *Discussion Paper.*

Shafik, N. (1994), "Economic Development and Environmental Quality: An Econometric Analysis", *Oxford Economic Papers,* 46, pp. 757-73.

Shukla, P.R., Tubiana, L., Tudela, F., and Wang, X., *Beyond Kyoto : Advancing the International Effort against Climate Change,* Pew Centre of Global Climate

Change, Arlington, VA, pp. 141-70, Testing Old and New Trade Theories', *The World Economy*, 26(8): 1163-86.

South Centre (2009): International Trade and Climate Change, *Policy Brief-16*, September.

South Centre (2009): Carbon-based Competitiveness,Trade and Climate Change Linkages: Perspectives of Developing Countries, Geneva.

Stern, Nicholas (2006): 'Stern Review on the Economics of Climate Change' (London: Her Majesty's Treasury).

Stern, Nicholas (2007): *The Economics of Climate Change: The Stern Review*, Cambridge: Cambridge University Press.

Stilwell, Matthew (20009): New Challenges of Global Governance Managing International Trade and Climate Change, Graduate Institute of International and Development Studies, Geneva.

Torras, M. and Boyce, J. (1998): "Income, Inequality and Pollution", *Ecological Economics*, 25:2, pp. 147-60.

UNEP (2009): Climate Change and Trade Policies in a Post-2012 World. The ADAM Project.

UNEP and WTO (2009): Trade and Climate Change. A Report the United Nations Environment Programme and the World Trade Organization. http://www.wto.org/ english/res_e/ booksp_e/ trade_climate_change_ e.pdf

UNFCCC (United Nations Framework Convention on Climate Change) (2009): *Second Synthesis Report on Technology Needs Identified by Parties Not Included in Annex I to the Convention.* Bonn.

Verburg, R., Woltjer, G., Tabeau, A., Eickhout, B. and Stehfest, E. (2008): *Agricultural Trade Liberalisation and Greenhouse Gas Emissions: A simulation Study Using the GTAP-IMAGE Modelling Framework*, LEI, The Hague.

Wiedmann, T., Wood, R., Lenzen, M., Minx, J., Guan, D. and Barrett, J. (2007): *Development of an Embedded Carbon Emissions Indicator—Producing a Time Series of Input-Output Tables and Embedded Carbon Dioxide Emissions for the UK by Using a MRIO Data Optimisation System*, Report to the UK Department for Environment, Food and Rural Aff airs by Stockholm.

World Bank (2010): *World Development Report*, 2010: Development and Climate Change.

World Bank (2007): International Trade and Climate Change-Economic Legal and Institutional Perspectives. http://www.mse.ac.in/trade/World%20Bank%20(2007).pdf

WTO (World Trade Organization) (2007): *Trade and Environment at the WTO.* Geneva.

Zhang, Zhong Xiang (1998): "Greenhouse Gas Emissions Trading and the World Trade System", *Journal of World Trade*, 32, 5: 219-39.

Chapter 32

Financing Environment in Orissa

Bimal K. Mohanty

The history of market economies during the nineteenth century has revealed three phases of economic transformation in succession, of course the phases are heterogeneously spaced. The sequence of events was the Industrial Revolution followed by revolution in technology invention and use, and lastly globalisation. Each phase of transformation though is fundamentally different from the other in terms its contents, they have a common chord of connection i.e., industrialisation. Industrialisation is believed to be one of the few means through which a nation can transform itself from a traditional agrarian society to a modern economy. The intention behind this is not only the self esteem but also the aspiration of the countries for merging with the mainstream of international economic trend and standard. Industrialisation is believed to be the most dominant parameter of economic growth. It not only acts as a vehicle for sustainable agriculture but also a powerful support for service-led growth. Industrialisation without environmental and ecological protection is expected to reverse the process of growth. In order to have a sustainable growth, it is essential that with progress of industrialisation, environment must be protected against degradation and depletion which requires financial commitments. Such commitments may come from quite a good number of sources of which state funding is the principal one. Having felt so, countries have provided for environmental financing in their annual budgets as a matter of routine. UNDP, in one of its studies (UNDP, 2008) has asserted that developed countries of the world spend between 3 and 5 per cents of their GNP on environmental protection and management. But most of the developing countries incur around 1 per cent of their GNP on environmental financing.

Environmental financing is undoubtedly different from other types of financing. It is a type of durable investment for the future. Normally what happens is financing a public project is treated as a means through which some end results are expected to be reaped. But in case of environmental financing, it is both end and means which has permanent consequence for the future. This inference has been strengthened and seems irrefutable with the acceptance of the global strategy of sustainable development in 1987. It was believed that with issuance of private property rights on environmental resources with individual freedom, private markets would lead to increased preservation of environmental resources. But this argument has lost its significance. In recent time, the alarming increase in global warming and catastrophic depletion in biodiversity has challenged the belief of the proponents of liberalism. It has been realised that state intervention is one of the unavoidable solutions to the environmental problems. Besides administration and management, public funding is essential for the preservation of environment and ecology which does not completely exclude the relevance of private financing and/or financing through public-private partnership (PPP) mode. But environmental interventions are very costly means. Public investment on environment is supposed to be lumpy and bulky and it requires relatively huge recurring expenses for constant monitoring. Since the resources available for the purpose are limited, it necessitates fixing the priorities. Since allocations in the budget for environment like other demands are made by voting in the Legislature, it is believed that such allocations flow into prioritised heads.

In the present paper an investigation has been made into the trend of state financing of environment in Orissa during 2000-01 – 2008-09. The period of study has been chosen on the ground of availability of accounts (confirmed) budget data on the one hand and the agenda of reforms that the Indian economy has completed in 1994-95.[1] Globalisation of the Indian economy has not only encouraged foreign direct investment (FDI) inflow but also the portfolio investment by foreign investors. Portfolio investments by foreign investors have tended towards those states which have both investment-friendly climate and the stock of relevant inputs including raw-materials to which they could have easy access. Orissa is enriched with vast stock of mineral and non-mineral resources. With a view to attracting investment on Mega Projects by both foreign and domestic investors, the Government of Orissa modified the Industrial Policy Resolution

2001 in 2004 which provides for special packages to such investors. As a consequence large number of foreign and domestic investors has not only signed Memorandum of Understandings (MoUs) with the Government of Orissa but some of them have already started their ventures. One of the concomitant outcomes is the environmental concern. In spite of the fact that the individual investors as a part of the agreements have put their own resources for the protection of environment that might have been endangered due to the functioning of their industries. In spite of this, state responsibility cannot altogether be ignored. State should not sit tight fist by handing over such a big mission to the investors who have come here more on the gorund of making profit. The financial commitments made by the Government of Orissa in its annual budgets for environment are the rough indicators of its concern for the environmental management and preservation. Accordingly, the annual budget data on revenue and capital accounts of the Government of Orissa relating to forestry and wild life, ecology and environment, environmental forestry and wild life, ecological research and ecological regeneration, prevention and control of pollution during 2000-01 to 2008-09 has been used in the study.

STATE OF ENVIRONMENT IN ORISSA—AN OVERVIEW

A quantitative increase in state domestic product (SDP) of Orissa in recent time is undoubtedly one of the principal causes of the growing concern for its environment. While agriculture and manufacturing are the main sources of increase in SDP of Orissa, they themselves turn the agents of environmental damage in the absence of rational approach to the issues. Green Revolution had the tremendous contribution to the productivity gains in Indian agriculture. But it has left behind it certain problems of which environmental probems are of greater dimension. Green Revolution emphasises the application of high input intensive technology. This technology available to farmers in a package which contains high yielding variety (HYV) seeds, chemical fertilisers, pesticides, water and such other ingredients relevant for modern system of farming. Among them, chemical fertilisers and pesticides whose uses are unavoidable in the cultivation of HYV are known to have adverse consequences for environment and ecology for which they are recommended to be used under controlled conditions for reaping optimum results from the use of HYV seeds. The relentless drawing of ground water for cultivation throughout the year, indiscriminate uses of chemical fertilisers and pesticides result in numerous environmental problems. These problems include degradation of ground water and surface water quality and fall of water table, degradation in the quality of surface soil of cultivable land and the growth of new pests and plant diseases. Though Orissa agriculture is not that developed as the agriculture of the states of Green Revolution belt, a section of farmers in Orissa whose number is on increase use high input intensive technology. The area under HYV paddy in Orissa as a percentage of net area sown has increased from 22.17 in 1991-92 to 61.21 in 2008-09. During this period, the fertiliser consumption per hectare has increased from 19.96 kilograms to 62.00 kilo grams. Total pesticide consumption has increased from 1 thousand metric tons in 2000-01 to 1.15 thousand metric tons in 2008-09.[2]

Environmental problems caused by large industries and units under informal manufacturing sector in urban centres of the state are quite visible. Vyas and Reddy (1998) in their study have quoted from the work of Council of Professional Social Workers, Bhubaneswar[3] to highlight how urbanisation and industrialisation in Orissa has affected local environment. The Talcher-Angul industrial belt in Orissa presents a grim reality. Air, water and land in the area all have been polluted by the concentration of coal mines, thermal power plants and fertiliser plant (which is closed now) in this industrial belt. A rivulet Nandira which was providing water to the people few years ago is now used as a natural sewerage to drain-off industrial wastes into river Brahmani. The rivulet now is a virtual death trap for the men and animals of the area. Regarding air pollution, the report of the Council says that the norms fixed by the Pollution Control Board in respect of the 'suspended particular matter' (SPM) content in the air have been far exceeded in this industrial belt. The norms prescribe that SPM content in the air should not exceed 200 micrograms per cubic meter in residential areas and 500 micro grams per cubic meter in factory areas. But the SPM content in the air of the residential areas in the proximity of the Talcher Thermal Power Plant is 1430 micro grams per cubic meter. Pollution in the area has assumed a worst dimension after switching over from underground to open cast mining. Due to this, the top soil containing the nutrients which take centuries to form has been removed and has been thrown as huge over burdens. This has resulted in destroying the rich vegetation on the top soil on the one hand and rendering the tract of land on which the over burdens were heaped useless. In the mean time, a good number of mega projects under private initiatives have been made functional from this area. It is quite well-imagined the extent to which it might have affected air, water and land quality in this area.

For the preservation of environment and ecological balance, forests act as catalysts. In Orissa, forests are unevenly distributed. While the coastal districts have comparatively smaller area under forest coverage, the forests are concentrated in Central, Southern and Western parts of the state. The Resolution on National Forest Policy 1988 has fixed a norm of at least 33 per cent of the geographical area should be under forest and/or tree coverage. In Orissa, such norm is slightly over met during 2007-08. Of the total geographical area of the state, 37.34 per cent is recorded to have been under forest coverage and 3.10 per cent is under miscellaneous tree coverage. But during the same year, 1802.58 hectares of the forest area has been diverted into non-forest uses. Afforestation programme has been implemented in the state under State Plan, Central Plan and other Schemes over a total area of 59,127 hectares. Further afforestation over 18,260 hectares with the Twelfth Finance Commission Grant and 4200 hectares under Economic Plantation Scheme has been taken up by the Department of Forest and Environment of the Government of Orissa during 2007-08. Among all the environmental resources, forests are the principal ones. Diversion of forest land into non-forest uses without compensatory afforestation would be having serious implications for the state of environment in Orissa.

APPROACHES TO ENVIRONMENTAL FINANCING— A REVIEW

For quite a long time, it was believed by the developing world that environmental financing was the responsibility of the international donor community. Since environmental services are global public goods, the beneficiaries from such services must pay for in the form of transferring a part of their resources to those who provide and maintain such services. But such view has changed greatly in recent years. These countries have understood the relevance of environmental management and financing for sustaining their own growth process. If they depend and wait for the financial support from the developed world and/or from the international community and the grants and supports (if, at all) reach them later than required, it would be inevitable to encounter the reversal of the growth process that they have already initiated. Accordingly, environmental financing has been conceived by all the countries irrespective of their level of development as an endogenous process to be supported out of their own resources through provisions in their annual budgets. For additional resources beyond their budgetary provisions, the less developed countries (LDCs) may depend on the international community and the developed world.

Business world has a responsibility towards environment. During 1990s, the research on the link between business and environment has confirmed that corporate environmental performance and profitability must go hands in gloves (Rivera and Delmas, 2004). But among the believers in such causality, there is difference in opinion. While one group holds the corporate environmental management efforts to be directly linked with profitability, the other group holds the link to be indirect. The former is based on the belief that pro-environmental voluntary strategy of the businesses would be more profitable by reducing liability and risk that might have been there due to environmental hazards. The latter is supported by the results of the empirical studies that businesses show their concern for environment on the pressure from the regulators particularly public authorities, civil societies and the customers (Delmas, 2002). In spite of the conflicting views on the link between business profitability and corporate environmental responsibility, it is generally held that corporate world should have green responsibility. In order to promote green business responsibility, the Wells Fargo Bank (North-America) has opened a new branch of 'public finance services in the environmental finance area' in addition to the other two conventional ones namely, commercial banking and investment banking. Till January 2010, the Bank has made more than 6 billion dollars loans and investments to 'environmentally beneficial businesses and projects'. Through this arrangement, the Bank has been able to create some sort of spirit among the corporate bodies to grow their businesses while improving the environment.

Environmental financing is a part of development financing. Since development in its most recent version is understood as sustainable development, the financing for environment must be made in that mode and magnitude which would be capable of meeting the end result of sustainable development. Thus, financing for environment must be designed and channelled in a manner that would be most conducive to sustainable development. Though the objectives of sustainable development were iterated in the Earth Summit held at Rio de Janeiro in 1992, they were repeated in the Millennium Declaration and Millennium Development Goals, 2000. Various international institutions like International Monetary Fund, World Bank, United Nations Organisartion and Non-Governmental Organisations functioning from foreign soil particularly those operating from the developed market economies have worked as the channels for financing sustainable development and environment. It has been experienced that such financing has often resulted in some controversy or the other in countries of the donee. Apart from this drawback of the environmental financing through international communities and institutions, other three are marked in respect of scale, accessibility and accountability (Najam, 2002). So far scale of operation is concerned these institutions specialise in funding million (Dollar) plus projects. But many projects particularly in less developed countries do not fall in this million (Dollar) plus category and hence accessibility of these countries to environmental financing by international communities gets restricted. The solution lies in creating a set of intermediate institutions operating from grass root level of sustainable development (local NGOs) to carry on the mission of environmental financing. Another problem is the problem of transparency tagged with accessibility. Once the donor organisations/countries feel that the use of their resources for environmental purposes is not transparent, such funding is turned inaccessible irrespective of the positive intention of both donors and donees. The problems relating to transparency could have been overcome had there been a mechanism through which accountability could have been fixed in the event of any misappropriation or diversion of environmental resources of the donor organisations and counties at the donee level. Since it is too difficult and sometimes impossible to fix the accountability, environmental financing through international communities has been of doubtful durability. This requires the creation of some domestic institutions both at the government and private levels over and above the international communities whose activities would ensure consistent flow of public and private financing to deal with accumulated environmental liabilities. Thus, success of environmental financing for sustainable development requires on the one hand development of certain institutions which should have integrity and mindset to work towards ensuring flow of international resources to those heads which would conserve and sustain management of the natural endowment in the countries of the donee (UNDP, 2000).

Effective sustainability finance for development and environment of the developing countries can come, as is seen above, from the agents like businesses, world communities and organisations and non-governmental organisations over and above out of their own budgetary provisions. Private sector's role and the role of public-private partnership in this context should not be ignored. Hamilton and Dixon (2002) have emphasised three sources of private sector financing of environment and development. These three sources comprise Foreign Direct Investment (FDI), Socially Responsible Investing (SRI) and Microfinance. FDI would serve a genuinely good reason if it is directed for the purpose. But most of the FDI flows are object-oriented from both exporting and importing countries and one of the drawbacks is that FDI flows in many cases are seen to have been the sources of environmental pollution in importing countries. SRI which is defined as that amount of investment which has the capability of effecting positive social change could be a potential source of channelling additional resources available with developed word to developing countries. SRI has gown very fast in Europe and North-America and lately in Japan. It is estimated that in the United States investment market alone, the SRI is more than 13 per cent, that is, out of total investible resources, more than 13 per cent constitute socially responsible (Hamilton and Dixon, 2002). But the difficult aspect of financing environment of the developing countries through SRI is the management of risk of all types of tagged with SRI like social, financial and environmental. Through Microfinance, services like credit, savings and insurance are provided to those who are otherwise partly or wholly deprived of these services due their ill economic conditions. Deployment of capital on micro-size in the developing world could be an asset for growth with social justice. Such type of investment will be undoubtedly environmental-friendly. But one of the impediments on the success of Microfinance lies in several regulations particularly from the side of banks unlike the functioning of private sector. Public and private sectors if work independently towards green development, it is unlikely to reap optimum results. On the contrary, if they work together, green development could not only be advanced but could be made sustainable. Hamilton and Dixon (2002) have taken the example of water resources management in order to show the efficacy of public-private partnership. Thus, green management could be made more effective under public-private partnership mode.

FINANCING ENVIRONMENT IN ORISSA

Financing environment is undoubtedly a broad concept which may include several forms of financing environment and ecology like financing by pubic authorities, by private entities comprising individuals and businesses, and finally, by public-private partnership. In the current context, it is desired to highlight the financing of environment by the Government of Orissa alone during the period 2000-01 to 2008-09. Definition of government environmental expenditure is a complex one in the sense that a part of expenditure by one Department may look like the expenditure of some other Departments. For example, the budgetary provisions for the disposal of hospital wastes though are coming under the expenditure by the Health and Family Welfare Department, it is more of an expenditure for environment and however it is not provided for in the Demand for Grants of the Forest and Environment Department. In spite of this, environmental expenditure on the basis of an item-wise arrangement, includes the expenditure on (1) Air and water pollution control,

(2) Hazardous waste management, (3) Mitigation of greenhouse gas emissions and ozone-depleting substances, (4) Sanitation and solid waste management, (5) Water supply, (6) Watershed management, (7) Water resources management, (8) Soil degradation control, (9) Controlling deforestation, and (10) Protecting bio-diversity and landscapes. For the purpose of an in-depth study, functional classification is more relevant and effective for policy than mere item-wise arrangement. In the budget papers of the Government of Orissa, environmental expenditures both on revenue and capital accounts in its Consolidated Fund are provided under the Major Heads: (1) Forestry and Wild Life, and (2) Ecology and Environment. Under the former, the Minor Head is 'Environmental Forestry and Wild Life' and under the latter, the Minor Heads are: (i) Environmental Research and Ecological Regeneration, and (ii) Prevention and Control of Pollution over and above the others which are unrelated to environment but related to forest. Table-1 presents the disbursements of the Government of Orissa exclusively on the head 'Environment' during the period under study.

The budgetary provisions of the Government of Orissa for 'Environment and Ecology' during the nine-year period 2000-01—2008-09, have been made basically in revenue account. There have been some allocations in capital account (capital outlay) only for 'Forestry and Wild Life'. That again the latter has exhibited a negative growth rate of 0.84 per cent per annum. This disproportionality in allocation of funds (between revenue and capital accounts) is obvious. 'Environment and Ecology' is a social good which requires constant monitoring for its preservation and hence, most part of the budgetary allocations is required to be provided for maintenance. Though the expenditure in revenue account on 'Forestry and Wild Life' and 'Ecology and Environment' taken together has increased by 16.88 per cent per annum, the former has increased by 17.64 per cent against the latter having increased by 6.91 per cent per annum. Thus between 'Forestry and Wild Life' and 'Ecology and Environment', state sector financing has tilted more in favour of the former. Of course, a massive 'Forestry and Wild Life' is required for healthy 'Ecology and Environment'. One discouraging fact is noticed in respect of financing of 'Prevention and Control of Pollution' by the Government of Orissa in revenue account. It is not only remarkably small (Table 1) but also has diminished from Rs. 7 lakh in 2000-01 to Rs. 3.01 lakh in 2008-09. Since prevention is better than cure, before any colossal damage augments due to pollution which is very likely due to rapid industrialisation of the state, necessary precautionary measures need to be taken at a larger scale. Moreover, the functioning of several Iron and Steel Industries from Central Orissa and

TABLE 1

Environmental Financing of the Government of Orissa Head-wise in Revenue and Capital Accounts

(In Rs. Lakh)

Year	2406	2406 (02)	3435	3435 (03)	3435 (04)	2406+3435	4406
(1)	(2)	(3)	(4)	(5)	(6)	(7)	(8)
2000-01	9661.19	2109.97	1003.68	992.84	7.00	10664.87	22214.37
2001-02	15727.26	2497.54	1592.95	1585.95	7.00	17320.21	13052.90
2002-03	16980.38	2126.37	620.92	613.92	7.00	17601.30	13837.95
2003-04	11714.88	2000.65	503.50	496.50	7.00	12218.38	14511.49
2004-05	10946.11	3263.89	892.59	387.73	7.00	11838.70	11589.82
2005-06	14305.14	2919.16	1341.85	1338.84	3.01	15646.99	12282.82
2006-07	22842.54	3437.64	2907.09	2960.08	3.01	25749.63	12244.28
2007-08	28903.69	5157.81	1633.02	1630.01	3.01	30536.71	18589.72
2008-09	35440.18	6361.58	1712.54	1709.53	3.01	37152.72	20763.16
Growth Rates	17.64	14.79	6.91	7.03	- 10.01	16.88	- 0.84

Sources : Demand for Grants 2002-03 to 2010-11, Finance Department, Government of Orissa.
2406: Forestry and Wild Life (Revenue Account).
2406 (02): Environmental Forestry and Wild Life (Revenue Account).
3435: Ecology and Environment (Revenue Account).
3435 (03): Environmental Research and Ecological Regeneration (Revenue Account).
3435 (04): Prevention and Control of Pollution (Revenue Account).
4406: Forestry and Wild Life (Capital Account).
Growth Rates: Annual average compound growth rates.

excavation of minerals (Angul-Talcher Belt) for augmenting growth is undoubtedly a source of environmental pollution. These industries are under the private sector. As a matter of agreement, they are fulfilling their commitments in respect of implementing anti-pollution measures. But it does not seem adequate. For the 'Prevention and Control of Pollution' in the vicinity of these industries, the Government of Orissa should have made some more efforts through some sort of perspective planning out of its own resources for controlling pollution. This requires additional budgetary commitments.

Budgetary provisions for a particular 'Demand' as such do not reflect any substantive meaning as these provisions are made as a matter of routine. If such provisions are evaluated as a percentage of gross state domestic product (GSDP) (in the case of the states in a federation) or as a percentage of gross national product (GNP) (in the case of the country), it could be an effective way of ascertaining the approach and attitude of the government to the particular head of expenditure. As stated earlier in this study, the UNDP (2008) has pointed to the skewed difference that exists between developed and developing countries in respect of their financial commitments as percentages of their respective GNP for environment and ecology. While developed countries spend within 3 to 5 per cent of their GNP on environment, it is hardy 1 per cent in developing countries. Orissa is one of the less developed states of India. The state not only exhibits the highest concentration poverty but it is having poverty in human development and stock of physical infrastructure. During the study period, the budgetary allocations (actuals) both in revenue and capital accounts together to environment and ecology (2406+3435+4406: Tabe 1) as percentages of GSDP have diminished from 0.76 in 2000-01 to 0.43 in 2008-09. This is not a healthy symptom indeed. In recent time, the craze for accelerated growth among the developing countries has been a deepening source of environmental degradation and depletion. This does not mean that developed countries' economic activities are environment-neutral. But they have not only plenty of resources to counter the problem but also a specialised institutional arrangement to deal with. There always exists a trade-off relationship between growth and environment. The fruits of accelerated growth would be more than neutralised if separate institutional arrangements at the cost of the state are not made for the preservation of environment. In view of the increased probabilities of natural disasters due to overuse of environmental resources beyond their carrying capacity, even the market economies advocate direct state intervention towards formulating a rational method of risk management and implementing such method out of its own resources.

CONCLUSION

Public expenditure on environment and ecology presents government resource allocation to this 'Demand'. These allocations are made on prioritised basis which are meant for management and preservation of environmental resources. These allocations are to be effective, efficient and sustainable. But it is very difficult to ascertain whether these criteria have been met. World Bank (1998) has laid down three selective criteria for public expenditure in general. These criteria include fiscal discipline, allocative efficiency and cost-effectiveness. Given the fiscal discipline and allocative efficiency, it is cost-effectiveness that really matters. Orissa finance after reforms particularly after the compliance of the clauses in the Fiscal Responsibility and Budget Management Act, 2003 which was made effective in Orissa from June 14, 2005, has shown fabulous improvements. This is corroborative of fiscal prudence or fiscal discipline in Orissa. Since the 'Demands for Grants' by the Departments of the Government are put into voting process in the State Legislative Assembly in which people's representatives participate, it would be inferred that such 'Demands' comply with allocative efficiency. It is the cost-effectiveness that really matters. The Government of Orissa, for larger welfare interest of the residents, must set-up institutions for evaluating the cost-effectiveness of such expenditure. In view of the scanty allocations, the Government of Orissa should not only increase the budgetary provisions to environment and ecology but also to set-up a well-equipped independent system to evaluate the cost-effectiveness of the allocations made to environment and ecology at the cost of its own resources.

NOTES AND REFERENCES

1. Agenda of reforms in India broadly comprise three items, namely, privatisation, liberalisation and globalisation. The New Economic Policy that was implemented in mid-July 1991 was the beginning of the steps towards the completion of the Agenda. All measures relating to these three broad items were completed in 1994-95 and hence, India entered into full-fledged reforms from this year only.
2. These data have been collected from the Economic Survey, 2009-10, Government of Orissa, Planning and Coordination Department and Directorate of Economics and Statistics, June 2010.
3. Council of Professional Social Workers is a voluntary organisation that operates from the state capital Bhubaneswar.

REFERENCES

Delmas, M. (2002): The Diffusion of Environmental Management Standards in Europe and in the United States: an Institutional Perspective, *Policy Sciences*, Vol. 35, No. 1.

Hamilton, Kirk and John Dixon (2002): Financing Sustainable Development (Financing Environmental Expenditure: Theory and Practice), Course on 'Environmental Economics and Development Policy-V', July 15-26, World Bank Institute, Washington, D.C.

Najam, Adil (2002): Financing Sustainable Development: Crises of Legitimacy, Progress in Development Studies, Vol. 2, No. 2.

Rivera, Jorge and Magali Delmas (2004): Business and Environmental Protection, *Human Ecology Review*, Vol. 11, No. 3.

UNDP (2002): Financing for Sustainable Development in Latin America and the Caribbean: From Monterrey to Johannesburg, Word Summit on Sustainable Development, Johannesburg.

UNDP (2008): Environmental Financing: A UNDP Perspective—Draft, March 16.

Vyas, V.S. and V. Ratna Reddy (1998): Assessment of Environmental Policies and Policy Implementation in India, *Economic and Political Weekly*, Vol. 33, Nos. 1 and 2.

Wells Fargo Bank (2010): Wells Fargo Environmental Report, January, North-America.

World Bank (1998): Public Expenditure Management Handbook, Washington, D.C.

Index